LIST OF THE ELEMENTS WITH THEIR SYMBOLS AND ATOMIC MASSES*

Element	Symbol	Atomic Number	Atomic Mass†	Element	Symbol	Atomic Number	Atomic Mass†
Actinium	Ac	89	(227)	Neon	Ne	10	20.18
Aluminum	Al	13	26.98	Neptumium	Np	93	(237)
Americium	Am	95	(243)	Nickel	Ni	28	58.69
Antimony	Sb	51	121.8	Nielsbohrium	Ns	107	(262)
Argon	Ar	18	39.95	Niobium	Nb	41	92.91
Arsenic	As	33	74.92	Nitrogen	N	7	14.01
Astatine	At	85	(210)	Nobelium	No	102	(253)
Barium	Ba	56	137.3	Osmium	Os	76	190.2
Berkelium	Bk	97	(247)	Oxygen	O	8	16.00
Beryllium	Be	4	9.012	Palladium	Pd	46	106.4
Bismuth	Bi	83	209.0	Phosphorus	P	15	30.97
Boron	B	5	10.81	Platinum	Pt	78	195.1
Bromine	Br	35	79.90	Plutonium	Pu	94	(242)
Cadmium	Cd	48	112.4	Polonium	Po	84	(210)
Calcium	Ca	20	40.08	Potassium	K	19	39.10
Californium	Cf	98	(249)	Praseodymium	Pr	59	140.9
Carbon	C	6	12.01	Promethium	Pm	61	(147)
Cerium	Ce	58	140.1	Protactinium	Pa	91	(231)
Cesium	Cs	55	132.9	Radium	Ra	88	(226)
Chlorine	Cl	17	35.45	Radon	Rn	86	(222)
Chromium	Cr	24	52.00	Rhenium	Re	75	186.2
Cobalt	Co	27	58.93	Rhodium	Rh	45	102.9
Copper	Cu	29	63.55	Rubidium	Rb	37	85.47
Curium	Cm	96	(247)	Ruthenium	Ru	44	101.1
Dysprosium	Dy	66	162.5	Rutherfordium	Rf	104	(257)
Einsteinium	Es	99	(254)	Samarium	Sm	62	150.4
Erbium	Er	68	167.3	Scandium	Sc	21	44.96
Europium	Eu	63	152.0	Seaborgium	Sg	106	(263)
Fermium	Fm	100	(253)	Selenium	Se	34	78.96
Fluorine	F	9	19.00	Silicon	Si	14	28.09
Francium	Fr	87	(223)	Silver	Ag	47	107.9
Gadolinium	Gd	64	157.3	Sodium	Na	11	22.99
Gallium	Ga	31	69.72	Strontium	Sr	38	87.62
Germanium	Ge	32	72.59	Sulfur	S	16	32.07
Gold	Au	79	197.0	Tantalum	Ta	73	180.9
Hafnium	Hf	72	178.5	Technetium	Tc	43	(99)
Hahnium	Ha	105	(260)	Tellurium	Te	52	127.6
Hassium	Hs	108	(265)	Terbium	Tb	65	158.9
Helium	He	2	4.003	Thallium	Tl	81	204.4
Holmium	Ho	67	164.9	Thorium	Th	90	232.0
Hydrogen	H	1	1.008	Thulium	Tm	69	168.9
Indium	In	49	114.8	Tin	Sn	50	118.7
Iodine	I	53	126.9	Titanium	Ti	22	47.88
Iridium	Ir	77	192.2	Tungsten	W	74	183.9
Iron	Fe	26	55.85	Uranium	U	92	238.0
Krypton	Kr	36	83.80	Ununbium	Uub	112	(277)
Lanthanum	La	57	138.9	Ununbexium	Uuh	116	(289)
Lawrencium	Lr	103	(257)	Unannilium	Uun	110	(269)
Lead	Pb	82	207.2	Ununoctium	Uuo	118	(293)
Lithium	Li	3	6.941	Ununquadium	Uuq	114	(285)
Lutetium	Lu	71	175.0	Unununium	Uuu	111	(272)
Magnesium	Mg	12	24.31	Vanadium	V	23	50.94
Manganese	Mn	25	54.94	Xenon	Xe	54	131.3
Meitnerium	Mt	109	(266)	Ytterbium	Yb	70	173.0
Mendelevium	Md	101	(256)	Yttrium	Y	39	88.91
Mercury	Hg	80	200.6	Zinc	Zn	30	65.39
Molybdenum	Mo	42	95.94	Zirconium	Zr	40	91.22
Neodymium	Nd	60	144.2	Ionization constants of monoprotic acids			

*All atomic masses have four significant figures. These values are recommended by the Committee on Teaching of Chemistry, International Union of Pure and Applied Chemistry.

†Approximate values of atomic masses for radioactive elements are given in parentheses.

General, Organic, and Biochemistry

General, Organic, and Biochemistry

Third Edition

Katherine J. Denniston
Towson University

Joseph J. Topping
Towson University

Robert L. Caret
San José State University

Boston   Burr Ridge, IL   Dubuque, IA   Madison, WI   New York   San Francisco   St. Louis
Bangkok   Bogotá   Caracas   Lisbon   London   Madrid
Mexico City   Milan   New Delhi   Seoul   Singapore   Sydney   Taipei   Toronto

McGraw-Hill Higher Education ⚛

A Division of The **McGraw-Hill** Companies

GENERAL, ORGANIC, AND BIOCHEMISTRY
THIRD EDITION

Published by McGraw-Hill, an imprint of The McGraw-Hill Companies, Inc., 1221 Avenue of the Americas, New York, NY 10020. Copyright © 2001, 1997 by The McGraw-Hill Companies, Inc. All rights reserved. No part of this publication may be reproduced or distributed in any form or by any means, or stored in a database or retrieval system, without the prior written consent of The McGraw-Hill Companies, Inc., including, but not limited to, in any network or other electronic storage or transmission, or broadcast for distance learning.

Some ancillaries, including electronic and print components, may not be available to customers outside the United States.

This book is printed on acid-free paper.

1 2 3 4 5 6 7 8 9 0 VNH/VNH 0 9 8 7 6 5 4 3 2 1 0

ISBN 0–07–231784–1
ISBN 0–07–118073–7 (ISE)

Vice president and editor-in-chief: *Kevin T. Kane*
Publisher: *James M. Smith*
Sponsoring editor: *Kent A. Peterson*
Developmental editor: *Shirley R. Oberbroeckling*
Editorial assistant: *Jennifer L. Bensink*
Senior marketing manager: *Martin J. Lange*
Senior marketing assistant: *Tami Petsche*
Senior project manager: *Marilyn M. Sulzer*
Production supervisor: *Sandy Ludovissy*
Design manager: *Stuart D. Paterson*
Cover/interior designer: *Elise Lansdon*
Cover image: *FPG International*
Senior photo research coordinator: *Lori Hancock*
Photo research: *Feldman and Associates*
Senior supplement coordinator: *Brenda A. Ernzen*
Compositor: *GAC—Indianapolis*
Typeface: *10/12 Palatino*
Printer: *Von Hoffman Press, Inc.*

The credits section for this book begins on page 823 and is considered an extension of the copyright page.

Library of Congress Cataloging-in-Publication Data

Denniston, K. J. (Katherine J.)
    General, organic, and biochemistry / Katherine J. Denniston, Joseph J. Topping,
    Robert L. Caret. — 3rd ed.
        p. ; cm.
    Rev. ed. of : Principles & applications of organic & biological chemistry / Robert L.
    Caret, Katherine J. Denniston, Joseph J. Topping. 2nd ed. © 1997.
    Includes index.
    ISBN 0–07–231784–1
    1. Chemistry, Organic. 2. Biochemistry. I. Topping, Joseph J. II. Caret, Robert L.,
    1947– . III. Caret, Robert L., 1947– Principles & applications of organic & biological
    chemistry. IV. Title.
    [DNLM: 1. Chemistry. 2. Biochemistry. 3. Chemistry, Organic. QD 33 D411g 2001]
    QD253 . C27   2001
    547—dc21   99–088202
                                                                                    CIP

INTERNATIONAL EDITION ISBN 0–07–118073–7
Copyright © 2001. Exclusive rights by The McGraw-Hill Companies, Inc., for manufacture and export. This book cannot be re-exported from the country to which it is sold by McGraw-Hill. The International Edition is not available in North America.

www.mhhe.com

Brief Contents

v

Contents

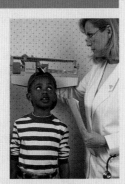

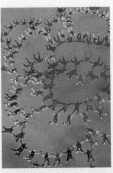

8 Chemical and Physical Change: Energy, Rate, and Equilibrium 205

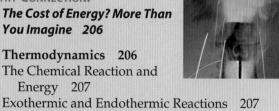

CHEMISTRY CONNECTION:
The Cost of Energy? More Than You Imagine 206

9 Charge-Transfer Reactions: Acids and Bases and Oxidation–Reduction 237

CHEMISTRY CONNECTION:
Drug Delivery 238

10 The Nucleus, Radioactivity, and Nuclear Medicine 269

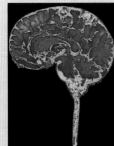

Organic Chemistry

11 An Introduction to Organic Chemistry: The Saturated Hydrocarbons 295

12 The Unsaturated Hydrocarbons: Alkenes, Alkynes, and Aromatics   323

13 Alcohols, Phenols, Thiols, and Ethers   355

14 Aldehydes and Ketones   383

Biochemistry

18 Lipids and Their Functions in Biochemical Systems   495

19 Protein Structure and Function   533

22 Aerobic Respiration and Energy Production 635

CHEMISTRY CONNECTION:

***Mitochondria from Mom* 636**

A HUMAN PERSPECTIVE:

***Exercise and Energy Metabolism* 638**

A HUMAN PERSPECTIVE:

***Brown Fat: The Fat That Makes You Thin?* 650**

23 Fatty Acid Metabolism 665

CHEMISTRY CONNECTION:

***Obesity: A Genetic Disorder?* 666**

A HUMAN PERSPECTIVE:

***Losing Those Unwanted Pounds of Adipose Tissue* 672**

A CLINICAL PERSPECTIVE:

***Diabetes Mellitus and Ketone Bodies* 680**

24 Introduction to Molecular Genetics 691

CHEMISTRY CONNECTION:

***Molecular Genetics and Detection of Human Genetic Disease* 692**

A CLINICAL PERSPECTIVE:

***Fooling the AIDS Virus with "Look-Alike" Nucleotides* 698**

Chemistry Connections and Perspectives

Chemistry Connection

A Human Perspective

A Clinical Perspective

A Medical Perspective

An Environmental Perspective

Preface

Ours is an age when an understanding of chemistry has become an increasingly important aspect of medicine. The third edition of *General, Organic, and Biochemistry* has been designed to help undergraduate health-related majors and students of all majors understand key concepts and appreciate the significant connections between chemistry, health, disease, and the treatment of disease. This text strikes a balance between theoretical and practical chemistry, while emphasizing material that is unique to health-related studies. It is written at a level intended for students whose professional goals do not include a mastery of chemistry, but for whom an understanding of the principles of chemistry and their practical ramifications is a necessity.

Key Changes to the Third Edition

In the preparation of the third edition, we have been guided by the collective wisdom of reviewers who represent the diversity of higher education experiences, including two-year and four-year colleges and universities. Over fifty different reviewers participated in the review process. We also received very valuable comments from a focus group of faculty who regularly teach this material.

New Organization

Recognizing that courses based on this textbook are organized in a variety of formats within both quarter and semester systems, we have reorganized the chapter sequence into three sections: inorganic chemistry (Chapters 1–10), organic chemistry (Chapters 11–16), and biochemistry (Chapters 17–24). The new organization will allow tremendous latitude in usage. The course may be taught in a traditional sense, following the new chapter order, within either the semester or quarter system. If you prefer the organization in the second edition, you may continue to teach using the different teaching order without affecting your students' understanding of the material. Many users of the second edition choose to integrate traditional organic and biochemistry and we have constructed the chapters to allow for alternate order of coverage. Frequent use of cross-referencing and reviewing of material discussed earlier in the book support this needed flexibility.

Clear Presentation

Today's students have numerous demands on their time. Many students are nontraditional students, working full time, who have families. Students need to be able to identify important concepts quickly and easily. Each section of the book was reviewed with the goal of becoming more concise while retaining the intellectual rigor of a college textbook. Some of the ways we have accomplished this goal include:

- **New design** facilitates access to information and engages student interest.
- **Key terms** are bolded and are immediately defined.
- **In-chapter examples** provide stepwise guidance to problem-solving strategies.
- **New tables** were created allowing easier access to information.
- **More headings** allow students to find important material faster.

Key Features of the Third Edition

Engaging Applications

We believe that there are a variety of factors in a text that can promote student learning and facilitate teaching. It is important to engage the interest of the student, especially when the subject may appear difficult and does not seem directly related to the student's career goals. We have included a diverse array of applications to accomplish our goals.

- **Perspectives:** We have added eleven new Clinical, Medical, and Human Perspectives throughout the book. A list of Chemistry Connections and Perspectives is provided on page xviii. These provide new and updated applications of chemistry to engage the students' interest and help them understand chemistry in the context of their daily lives.

- **Learning goal icons** 1 : These icons help to alert the student to the important concepts covered in the text. An icon is placed next to the textual material that supports the learning goal.

- **Integration of chemistry in all disciplines:** Students must understand the inter-relatedness of the three subdisciplines: general, organic, and biochemistry. Students need to visualize chemistry as one interconnected discipline. *Margin notes* highlight the relationships between the areas.

Clear Approach to Problem Solving

Students often have trouble with quantitative problem solving. Consequently, we have increased the number of in-chapter examples dramatically and have made them more user-friendly by a careful step-by-step explanation of the logic needed to answer the question. We recognize that students need support in the development of problem-solving skills.

- **In-Chapter Examples, Solutions, and Problems:** Each chapter includes examples that show the student, step by step, precisely how to develop problem-solving strategies. Whenever possible, the solved examples are followed by in-text problems that allow students to test their mastery of information and to build self-confidence.
- **In-Chapter and End-of-Chapter Problems:** We have created a wide variety of paired concept problems. The answers to the odd-numbered questions will be found in the book, providing reinforcement for students as they develop problem-solving skills. The students will be challenged to apply the same principles to the related even-numbered problem.
- **Critical Thinking Problems:** Each chapter includes a set of critical thinking problems. These problems provide an opportunity for the students to integrate concepts to solve more complex problems. They make a perfect complement to the classroom lecture, because they foster in-class discussion of complex problems dealing with daily life and the health care sciences.

Dynamic Visual Program

Today's students are much more visually oriented than any previous generation. Television and the computer represent an alternate mode of learning. Consequently, we have attempted to further develop these skills through the expanded use of color and illustrations. Chemical structures rendered by WaveFunction Inc.'s Spartan molecular modeling software are clear, engaging, and instructive.

- **New Design:** The entire book has a new design that allows the professor and the student to easily see the different sections by color tab—general, organic, and biochemistry. The design helps the student to quickly recognize the in-chapter examples and questions.
- **Illustrations:** Each chapter is amply illustrated using figures, tables, and chemical structures and formulas.

All of these illustrations are carefully annotated for clarity.
- **Color-Coding Scheme:** Because it may be difficult for students to understand the chemical changes that occur in complex reactions, we have color-coded the reactions so that chemical groups being added or removed in a reaction can be quickly recognized. In the organic chemistry section of the text, each major reaction type is highlighted on a green background. The color-coding scheme is illustrated in the "Guided Tour" section of this book.
- **Spartan models:** The students' ability to understand the geometry and three-dimensional structure of molecules is essential to the understanding of organic and biochemical reactions. We have used WaveFunction, Inc.'s, cutting edge molecular modeling software, Spartan, to render many of the molecules in the text.

Content Changes in the Third Edition

- **Chapter Three**—Improved coverage of electron configuration and atomic structure; *new* Clinical Perspective on dietary calcium
- **Chapter Four**—Improved coverage of covalent bonding; coverage of electronegativity moved to earlier in the chapter; improved coverage of Lewis structures and their role in portraying the bonding process
- **Chapter Five**—*New* Clinical Perspective on carbon monoxide poisoning; *new* Medical Perspective on pharmaceutical chemistry
- **Chapter Six**—Improved coverage of physical properties and ideal gases; *new* Clinical Perspective on autoclaves and the gas laws
- **Chapter Eight**—Improved coverage of physical equilibrium
- **Chapter Nine**—Improved coverage of Brønsted-Lowry acid–base chemistry; expanded coverage of oxidation–reduction reactions including a discussion of voltaic cells and their application; *new* Medical Perspective on turning the human body into a battery; *new* Clinical Perspective on electrochemical reactions on the Statue of Liberty and in dental fillings
- **Chapter Eleven**—*New* comparison of the major properties of typical organic and inorganic compounds; *new* coverage of alkyl groups; improved coverage of geometric isomers
- **Chapter Twelve**—Expanded coverage of aromatic hydrocarbons; deleted coverage of hydration reaction mechanism; *new* Human Perspective on life without polymers
- **Chapter Thirteen**—Improved and doubled the number of examples; deleted coverage of dehydration reaction mechanism

- **Chapter Fourteen**—Expanded coverage of addition reactions and their biological significance
- **Chapter Fifteen**—Expanded coverage of acid anhydrides
- **Chapter Sixteen**—Improved coverage of nomenclature
- **Chapter Seventeen**—Deleted discussion of Fischer projections; *new* Medical Perspective on monosaccharide derivatives
- **Chapter Eighteen**—*New* Clinical Perspective on disorders of sphingolipid metabolism
- **Chapter Nineteen**—*New* Human Perspective on collagen

- **Chapter Twenty**—*New* Clinical Perspective on enzymes, nerve transmission, and nerve agents
- **Chapter Twenty-one**—*New* coverage on the regulation of glycolysis
- **Chapter Twenty-two**—Coverage of the conversion of pyruvate to acetyl CoA (formerly in Chapter 21)
- **Chapter Twenty-four**—*New* and expanded coverage of DNA and RNA; *new* and expanded coverage of the polymerase chain reaction; *new* Human Perspective on DNA fingerprinting

This text has a complete support package for instructors and students. Several print and media supplements have been prepared to accompany the text and make learning as meaningful and up-to-date as possible.

For the Instructor:

1. **Instructor's Manual:** The Instructor's Manual contains the printed test item file and solutions to the even-numbered problems. Written by the authors, this ancillary also contains suggestions for organizing lectures, additional "Perspectives," and a list of each chapter's key problems and concepts.

2. **Transparencies:** A set of 100 transparencies is available to help the instructor coordinate the lecture with key illustrations from the text.

3. **Computerized Test Bank:** This computerized classroom management system/service includes a database of test questions, reproducible student self-quizzes, and a grade-recording program. Disks are available for IBM and Macintosh computers, and require no programming experience.

4. **Laboratory Resource Guide:** This helpful prep guide contains the hints that the authors have learned over the years to ensure students' success.

5. **Book-Specific Website:** A book-specific website is available to students and instructors using this text. The website will offer quizzes, key definitions, and interesting links for the students. The instructor will find a downloadable version of the Test Bank, the transparencies in a PowerPoint Presentation, the Instructor's Manual, and Solutions Manual. Also available for the instructor is PageOut, which allows the instructor to create his or her own personal course website. The address for the book-specific website is http://www.mhhe.com/physsci/chemistry/denniston.

For the Students:

1. **Student Study Guide/Solutions Manual:** A separate Student Study Guide/Solutions Manual is available. It contains the answers and complete solutions for the odd-numbered problems. It also offers students a variety of exercises and keys for testing their comprehension of basic, as well as difficult, concepts.

2. **Laboratory Manual:** Written by Charles H. Henrickson, Larry C. Byrd, and Norman W. Hunter, all of Western Kentucky University, *Experiments in General, Organic, and Biochemistry,* carefully and safely guides students through the process of scientific inquiry. The manual features self-contained experiments that can easily be reorganized to suit individual course needs.

3. **Is Your Math Ready for Chemistry?** Developed by Walter Gleason of Bridgewater State College, this unique booklet provides a diagnostic test that measures the student's math ability. Part II of the booklet provides helpful hints in the math skills needed to successfully complete a chemistry course.

4. **Problem Solving Guide to General Chemistry:** Written by Ronald DeLorenzo of Middle Georgia College, this exceptional supplement provides the student with over 2500 problems and questions. The guide holds the student's interest by integrating the solution of chemistry problems with real-life applications, analogies, and anecdotes.

5. **Schaum's Outline of General, Organic, and Biological Chemistry:** Written by George Odian and Ira Blei, this supplement provides students with over 1400 solved problems with complete solutions. It also teaches effective problem-solving techniques.

6. **How to Study Science:** Written by Fred Drewes of Suffolk County Community College, this excellent workbook offers students helpful suggestions for meeting the considerable challenges of a science course. It offers tips on how to take notes and how to overcome science anxiety. The book's unique design helps to stir critical thinking skills, while facilitating careful note taking on the part of the student.

7. **Book-Specific Website:** A book-specific website is available to students and instructors using this text. The website will offer quizzes, key definitions, and interesting links for the students. The address for the book-specific website is http://www.mhhe.com/physsci/chemistry/denniston.

As a full-service publisher of quality educational products, McGraw-Hill does much more than just sell textbooks to your students. We create and publish an extensive array of print, video, and digital supplements to support instruction on your campus. Orders of new (versus used) textbooks help us to defray the cost of developing such supplements, which is substantial. Please consult your local McGraw-Hill representative to learn about the availability of the supplements that accompany **General, Organic, and Biochemistry.**

Acknowledgements

We are grateful to our families, whose patience and support made it possible for us to undertake this project. We are grateful to our colleagues at McGraw-Hill, especially Jim Smith, publisher, and Kent Peterson, sponsoring editor, for their support of our book. We would like to thank Shirley Oberbroeckling, developmental editor, for her guidance during the reviewing and writing process. We also would like to express our appreciation to Marilyn Sulzer, project manager, for her skilled assistance throughout production.

A revision cannot move forward without the feedback of professors teaching the course. The reviewers have our gratitude and assurance that their comments received serious consideration.

The following professors provided reviews, participated in a focus group, or gave valuable advice for the preparation of the third edition:

Hugh Akers, *Lamar University*
Catherine A. Anderson, *San Antonio College*
A. G. Andrewes, *Saginaw Valley State University*
Mark A. Benvenuto, *University of Detroit-Mercy*
Warren L. Bosch, *Elgin Community College*
James R. Braun, *Clayton College and State University*
Philip A. Brown, *Barton College*
Teresa L. Brown, *Rochester Community College*
Scott Carr, *Trinity Christian College*
Bernadette Corbett, *Metropolitan Community College*
Wayne B. Counts, *Georgia Southwestern State University*
Robert P. Dixon, *Southern Illinois University*
Wes Fritz, *College of DuPage*
Edwin J. Geels, *Dordt College*
Deepa Godambe, *William Rainey Harper College*
Judith M. Iriarte-Gross, *Middle Tennessee State University*
T. G. Jackson, *University of South Alabama*
Paul G. Johnson, *Duquesne University*
Warren Johnson, *University of Wisconsin-Green Bay*
James F. Kirby, *Quinnipiac College*
Roscoe E. Lancaster, *Golden West College*
Richard H. Langley, *Stephen F. Austin State University*
Julie E. Larson, *Bemidji State University*
K. W. Loach, *Plattsburgh State University*
Ralph Martinez, *Humboldt State University*
John Mazzella, *William Paterson University*

Lawrence McGahey, *College of St. Scholastica*
Cleon McKnight, *Hinds Community College*
Melvin Merken, *Worcester State College*
Robert Midden, *Bowling Green State University*
David Millsap, *South Plains College*
Ellen M. Mitchell, *Bridgewater College*
Jay Mueller, *Green River Community College*
Lynda P. Nelson, *Westark College*
Richard E. Parent, *Housatonic Community Technical College*
Chetna Patel, *Aurora University*
Jeffrey A. Rahn, *Eastern Washington University*
B. R. Ramachandran, *Indiana State University*
John W. Reasoner, *Western Kentucky University*
Rill Ann Reuter, *Winona State University*
Terry Salerno, *Minnesota State University-Mankato*
Karen Sanchez, *Florida Community College at Jacksonville*
Sarah Selfe, *University of Washington*
Kevin R. Siebenlist, *Marquette University*
Steven M. Socal, *Southern Utah University*
Gordon Sproul, *University of South Carolina-Beaufort*
Pratibha Varma-Nelson, *Saint Xavier University*
Robert T. Wang, *Salem State University*
Steven Weitstock, *Indiana University*
Catherine Woytowicz, *Loyola University*
Jesse Yeh, *South Plains College*
Edward P. Zovinka, *St. Francis College*

The following professors provided reviews and other valuable advice for the previous editions:

Raymond D. Baechler, *Russell Sage College*
Satinder Bains, *Arkansas State University-Beebe*
Sister Marjorie Baird, O.P., *West Virginia Northern Community College*
Ronald Bost, *North Central Texas College*
Fred Brohn, *Oakland Community College*
Sister Helen Burke, *Chestnut Hill College*
Sharmaine S. Cady, *East Stroudsburg University*
Robert C. Costello, *University of South Carolina-Sumter*
Marianne Crocker, *Ozarks Technical Community College*
Peter DiMaria, *Delaware State University*
Ronald Dunsdon, *Iowa Central Community College*
Donald R. Evers, *Iowa Central Community College*
Patrick Flash, *Kent State University-Ashtabula*

Shelley Gaudia, *Lane Community College*
W. M. Hemmerlin, *Pacific Union College*
Hildegard Hof, *College of Misericordia*
Rosalind Humerick, *St. Johns River Community College*
Devin Iimoto, *Whittier College*
Michael A. Janusa, *Nicholls State University*
Donald R. Jones, *Ozarks Technical Community College*
Lidija Kampa, *Kean College of New Jersey*
Judith Kasperek, *Pitt Community College*
Kennan Kellaris, *Georgetown University*
Barid W. Lloyd, *Miami University*
William Moeglein, *Northland Community College*
Donal P. O'Mathuna, *Mount Carmel College of Nursing*
John A. Paparelli, *San Antonio College*

Richard E. Parent, *Housatonic Community Technical College*
Mona Y. Rampy, *Columbia Basin College*
George A. Schwarzmann, Jr., *Southwest Technical College*
Mary Selman, *Charminade University*
Courtney J. Smith, *Tuskegee University*
Gordon Sproul, *University of South Carolina-Beaufort*
Ronald H. Swisher, *Oregon Institute of Technology*
C. G. Vlassis, *Keystone College*
Janet R. Waldeck, *The Ellis School*
Larry Williams, *Golden West College*
Les Wynston, *California State University-Long Beach*
Gordon T. Yee, *University of Colorado-Boulder*
Carolyn S. Yoder, *Millersville University*

The *General, Organic, and Biochemistry* Learning System

The *General, Organic, and Biochemistry* Learning System is easy to follow, and will allow the student to excel in this course. The materials are presented in such a way that the student will effectively learn and retain the important information.

Clear Approach to Solving Problems

Because problem solving is most efficiently learned by a combination of studying examples and practicing, problems with step-by-step solutions are provided wherever appropriate. Examples are followed by a question requiring the student to integrate the newly learned material.

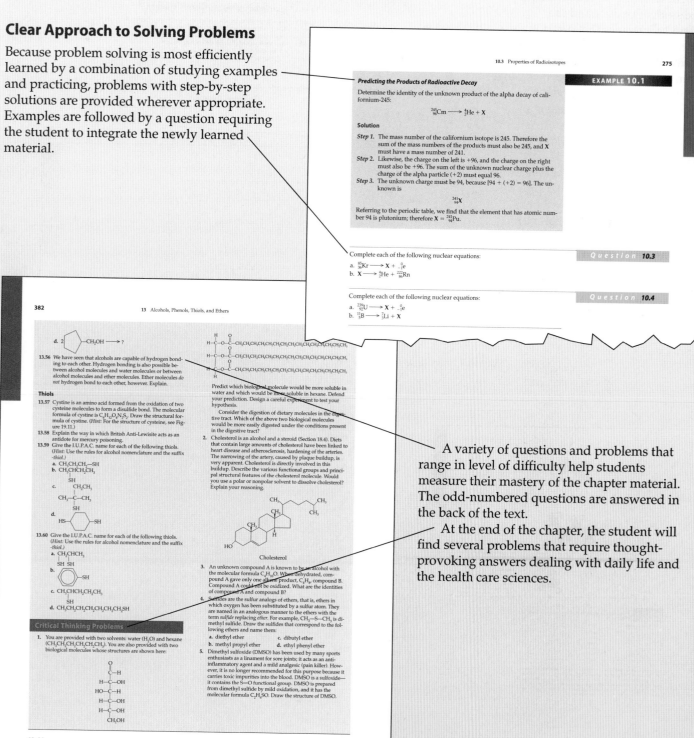

A variety of questions and problems that range in level of difficulty help students measure their mastery of the chapter material. The odd-numbered questions are answered in the back of the text.

At the end of the chapter, the student will find several problems that require thought-provoking answers dealing with daily life and the health care sciences.

Dynamic Visuals

Many of the equations and reactions are color coded to help the student understand the chemical changes that occur in complex reactions. The student can easily recognize the chemical groups being added or removed in a reaction by the color coding. Green background illustrates an important equation or key reaction; yellow background illustrates energy in the general and biochemistry sections and reveals the parent chain of a compound in the organic section; red and blue lettering distinguish two or more compounds that appear similar.

The art program has been significantly updated with the use of molecular art and drawings. The students will gain a better perspective and understanding of a molecule with a Spartan computer-generated model.

1.6 Experimental Quantities

Learning Goal
9

Thus far we have discussed the scientific method and its role in acquiring data and converting the data to obtain the results of the experiment. We have seen that such data must be reported in the proper units with the appropriate number of significant figures. The quantities that are most often determined include mass, length, volume, time, temperature, and energy. Now let's look at each of these quantities in more detail.

Mass

Mass describes the quantity of matter in an object. The terms *weight* and *mass*, in common usage, are often considered synonymous. They are not, in fact. **Weight** is the force of gravity on an object:

$$\text{Weight} = \text{mass} \times \text{acceleration due to gravity}$$

When gravity is constant, mass and weight are directly proportional. But gravity is not constant; it varies as a function of the distance from the center of the earth. Therefore weight cannot be used for scientific measurement because the weight of an object may vary from one place on the earth to the next.

Mass, on the other hand, is independent of gravity; it is a result of a comparison of an unknown mass with a known mass

---

The Chemical Reaction

Consider the exothermic reaction that we discussed in Section 8.1:

$$CH_4(g) + 2O_2(g) \longrightarrow CO_2(g) + 2H_2O(l) + 211 \text{ kcal}$$

For the reaction to proceed, C—H and O—O bonds must be broken, and C—O and H—O bonds must be formed. Sufficient energy must be available to cause

---

Now add the substituents. In this example a bromine atom is bonded to carbon-1, and a methyl group is bonded to carbon-4:

Finally, add the correct number of hydrogen atoms so that each carbon has four covalent bonds:

As a final check of your accuracy, use the I.U.P.A.C. system to name the compound that you have drawn

---

EXAMPLE 9.10

Calculating the pH of a Buffer Solution

Calculate the pH of a buffer solution similar to that described in Example 9.9 except that the acid concentration is doubled, while the salt concentration remains the same.

Solution

Acetic acid is the acid; [acid] = $2.00 \times 10^{-1} M$ (remember, the acid concentration is twice that of Example 9.9; $2 \times [1.00 \times 10^{-1}] = 2.00 \times 10^{-1} M$
Sodium acetate is the salt; [salt] = $1.00 \times 10^{-1} M$
The equilibrium is

$$CH_3COOH(aq) + H_2O(l) \rightleftharpoons H_3O^+(aq) + CH_3COO^-(aq)$$
$$\text{acid} \qquad\qquad\qquad\qquad\qquad\qquad \text{salt}$$

and the hydronium ion concentration,

$$[H_3O^+] = \frac{[\text{acid}]K_a}{[\text{salt}]}$$

Substituting the values given in the problem

$$[H_3O^+] = \frac{[2.00 \times 10^{-1}]1.75 \times 10^{-5}}{[1.00 \times 10^{-1}]}$$

$$[H_3O^+] = 3.50 \times 10^{-5}$$

and because

$$pH = -\log[H_3O^+]$$
$$pH = -\log 3.50 \times 10^{-5}$$
$$= 4.456$$

The pH of the buffer solution is 4.456.

Figure 11.2
(a) Drawing and (b) ball-and-stick model of ethane. All the carbon atoms have a tetrahedral arrangement, and all bond angles are approximately 109.5°. (c) Drawing and (d) ball-and-stick model of a more complex alkane, butane.

Figure 11.3
The tetrahedral carbon atom: (a) a tetrahedron; (b) the tetrahedral carbon drawn with dashes and wedges; (c) the stick drawing of the tetrahedral carbon atom; (d) drawing of a ball-and-stick model of methane.

Table 11.4	Names and Formulas of the First Five Continuous-Chain Alkyl Groups
Alkyl Group Structure	**Name**
$CH_3—$	Methyl
$CH_3CH_2—$	Ethyl
$CH_3CH_2CH_2—$	Propyl
$CH_3CH_2CH_2CH_2—$	Butyl
$CH_3CH_2CH_2CH_2CH_2—$	Pentyl

Carbon atoms are classified according to the number of other carbon atoms to which they are attached. A **primary carbon (1°)** is directly bonded to one other carbon. A **secondary carbon (2°)** is bonded to two other carbon atoms; a **tertiary carbon (3°)** is bonded to three other carbon atoms, and a **quaternary carbon** to four.
Alkyl groups are classified according to the number of carbons attached to the carbon atom that joins the alkyl group to a molecule.

Primary alkyl group   Secondary alkyl group   Tertiary alkyl group

Health/Life Related Applications

There are four different Perspective boxes in the text. Chemistry Connections provides an introductory scenario for the chapter, Medical Perspective and Clinical Perspective demonstrate use of the chapter material in an allied health field, Environmental Perspective demonstrates chapter concepts in ecological problems, and Human Perspective demonstrates how important chemistry is in our day to day lives.

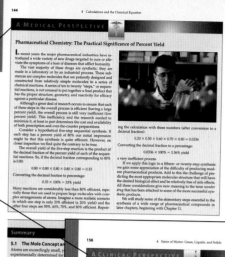

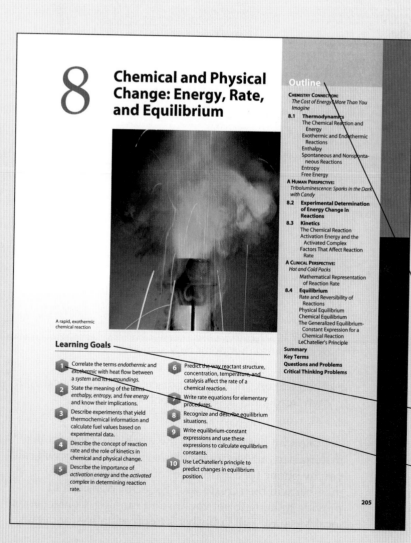

A rapid, exothermic chemical reaction

8

Chemical and Physical Change: Energy, Rate, and Equilibrium

Learning Goals

1. Correlate the terms *endothermic* and *exothermic* with heat flow between a system and its surroundings.

2. State the meaning of the terms *enthalpy*, *entropy*, and *free energy* and know their implications.

3. Describe experiments that yield thermochemical information and calculate fuel values based on experimental data.

4. Describe the concept of reaction rate and the role of kinetics in chemical and physical change.

5. Describe the importance of *activation energy* and the *activated complex* in determining reaction rate.

6. Predict the way reactant structure, concentration, temperature, and catalysis affect the rate of a chemical reaction.

7. Write rate equations for elementary processes.

8. Recognize and describe equilibrium situations.

9. Write equilibrium-constant expressions and use these expressions to calculate equilibrium constants.

10. Use LeChatelier's principle to predict changes in equilibrium position.

Outline

CHEMISTRY CONNECTION:
The Cost of Energy? More Than You Imagine

8.1 Thermodynamics
  The Chemical Reaction and Energy
  Exothermic and Endothermic Reactions
  Enthalpy
  Spontaneous and Nonspontaneous Reactions
  Entropy
  Free Energy

A HUMAN PERSPECTIVE:
Triboluminescence: Sparks in the Dark with Candy

8.2 Experimental Determination of Energy Change in Reactions

8.3 Kinetics
  The Chemical Reaction
  Activation Energy and the Activated Complex
  Factors That Affect Reaction Rate

A CLINICAL PERSPECTIVE:
Hot and Cold Packs
  Mathematical Representation of Reaction Rate

8.4 Equilibrium
  Rate and Reversibility of Reactions
  Physical Equilibrium
  Chemical Equilibrium
  The Generalized Equilibrium-Constant Expression for a Chemical Reaction
  LeChatelier's Principle

Summary
Key Terms
Questions and Problems
Critical Thinking Problems

205

Clear Presentation

Each chapter begins with an outline that introduces students to the topics to be presented. This outline also provides the instructor with a quick topic summary to organize lecture material.

A list of learning goals, based on the major concepts covered in the chapter, enables students to preview the material and become aware of the topics they are expected to master.

This icon is found within the chapters wherever the associated learning goal is first presented, allowing the student to focus attention on the major concepts.

Margin notes direct the student to a reference in the book for further material or review.

All bold-faced terms in the chapter are listed at the end of each chapter and defined in the Glossary at the end of the text. The student can easily find important terms when reading and studying.

At the end of each chapter is a summary designed to help students more easily identify important concepts and help them review for quizzes and tests.

Because k_2 is a constant, we may equate them, resulting in

$$\frac{V_i}{T_i} = \frac{V_f}{T_f}$$

and use this expression to solve some practical problems.

Consider a gas occupying a volume of 10.0 L at 273 K. The ratio V/T is a constant, k_2. Doubling the temperature, to 546 K, increases the volume to 20.0 L as shown here:

Appendix A contains a review of the mathematics used here.

$$\frac{10.0\,L}{273\,K} = \frac{V_f}{546\,K}$$

$$V_f = 20.0\,L$$

Tripling the temperature, to 819 K, increases the volume by a factor of 3:

$$\frac{10.0\,L}{273\,K} = \frac{V_f}{819\,K}$$

Questions and Problems   **379**

Summary

13.1 Alcohols: Structure and Physical Properties

Alcohols are characterized by the *hydroxyl group (—OH)* and have the general formula R—OH. They are very polar, owing to the polar hydroxyl group, and are able to form intermolecular hydrogen bonds. Because of hydrogen bonding between alcohol molecules, they have higher boiling points than hydrocarbons of comparable molecular weight. The smaller alcohols are very water soluble.

13.2 Alcohols: Nomenclature

In the I.U.P.A.C. system, alcohols are named by determining the parent compound and replacing the *-e* ending with *-ol*. The chain is numbered to give the hydroxyl group the lowest possible number. Common names are derived from the alkyl group corresponding to the parent compound.

13.3 Medically Important Alcohols

Methanol is a toxic alcohol that is used as a solvent. Ethanol is the alcohol consumed in beer, wine, and distilled liquors. Isopropanol is used as a disinfectant. Ethylene glycol (1,2-ethanediol) is used as antifreeze, and glycerol (1,2,3-propanetriol) is used in cosmetics and pharmaceuticals.

13.4 Classification of Alcohols

Alcohols may be classified as *primary*, *secondary*, or *tertiary*, depending on the number of alkyl groups attached to the *carbinol* carbon, the carbon bearing the hydroxyl group. A primary alcohol has a single alkyl group bonded to the *carbinol carbon*. Secondary and tertiary alcohols have two and three alkyl groups, respectively.

13.5 Reactions Involving Alcohols

Alcohols can be prepared by the *hydration* of alkenes. Alcohols can undergo *dehydration* to yield alkenes. Primary and secondary alcohols undergo oxidation reactions to yield aldehydes and ketones, respectively. Tertiary alcohols do not undergo oxidation.

13.6 Oxidation and Reduction in Living Systems

In organic and biological systems *oxidation* involves the gain of oxygen or loss of hydrogen. *Reduction* involves the loss of oxygen or gain of hydrogen. Nicotinamide adenine dinucleotide, NAD+, is a coenzyme involved in many biological oxidation and reduction reactions.

13.7 Phenols

Phenols are compounds in which the hydroxyl group is attached to a benzene ring; they have the general formula Ar—OH. Many phenols are important as antiseptics and disinfectants.

13.8 Ethers

Ethers are characterized by the R—O—R functional group. Ethers are generally nonreactive but are extremely flammable. Diethyl ether was the first general anesthetic used in medical practice. It has since been replaced by penthrane and enthrane, which are less flammable.

13.9 Thiols

Thiols are characterized by the sulfhydryl group (—SH). The amino acid cysteine is a thiol that is extremely important for maintaining the correct shapes of proteins. Coenzyme A is a thiol that serves as a "carrier" of acetyl groups in biochemical reactions.

Key Terms

alcohol (13.1)
carbinol carbon (13.4)
dehydration (13.5)
disulfide (13.9)
elimination reaction (13.5)
ether (13.8)
fermentation (13.3)
hydration (13.5)
hydroxyl group (13.1)

oxidation (13.6)
phenol (13.7)
primary (1°) alcohol (13.4)
reduction (13.6)
secondary (2°) alcohol (13.4)
tertiary (3°) alcohol (13.4)
thiol (13.9)
Zaitsev's rule (13.5)

Questions and Problems

Alcohols: Structure and Physical Properties

13.11 Arrange the following compounds in order of increasing boiling point, beginning with the lowest:
a. $CH_3CH_2CH_2CH_2CH_3$ b. $CH_3CHCH_2CHCH_3$
c. $CH_3CHCH_2CH_2CH_3$ d. $CH_3CH_2CH_2—O—CH_2CH_3$

13.12 Why do alcohols have higher boiling points than alkanes? Why are small alcohols readily soluble in water whereas large alcohols are much less soluble?

13.13 Which member of each of the following pairs is more soluble in water?
a. CH_3CH_2OH or $CH_3CH_2CH_2CH_2OH$
b. $CH_3CH_2CH_2CH_2CH_3$ or $CH_3CH_2CH_2CH_2—OH$
c.

13–25

Media

The website can be found at
http://www.mhhe.com/physsci/chemistry/denniston.
The student will find quizzes and math help a benefit in understanding and studying chemistry. There are also interesting links to chemistry areas and also links to various health related careers to help the student make career decisions.

The instructor will have available the Instructor's Manual and Solution Manual, the Test Bank, a PowerPoint demonstration of the transparency set and PageOut.

Over 6,000 professors have chosen PageOut to create course Websites. And for good reason: PageOut offers powerful features, yet is incredibly easy to use.

New Features based on customer feedback:

- Specific question selection for quizzes
- Ability to copy your course and share it with colleagues or as a foundation for a new semester
- Enhanced grade book with reporting features
- Ability to use the PageOut discussion area, or add your own third party discussion tool
- Password protected courses

Short on time? Let us do the work. Send your course materials to our McGraw-Hill service team. They will call you by phone for a 30 minute consultation. A team member will then create your PageOut Website and provide training to get you up and running. Contact your local McGraw-Hill Publisher's Representative for details.

1 Chemistry: Methods and Measurement

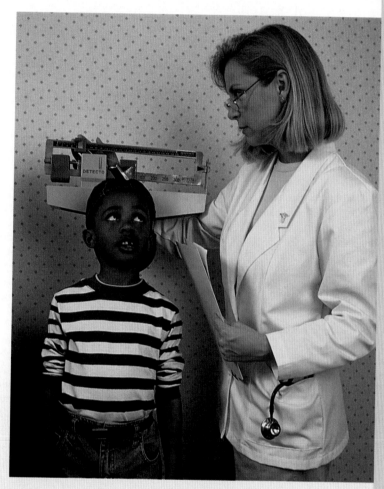

Accurate measurement is an essential part of good health care.

Learning Goals

1 State the definition of chemistry and discuss its interrelationship with other fields of science and medicine.

2 Describe the approach to science, the scientific method.

3 Distinguish among the terms *hypothesis, theory,* and *scientific law.*

4 State both the differences and relationships between science and technology.

5 Distinguish between data and results.

6 Learn the major units of measure in the English and metric systems, and be able to convert from one system to another.

7 Compare and contrast the terms *error, accuracy, precision,* and *uncertainty.*

8 Report data and results using scientific notation and the proper number of significant figures.

9 Use appropriate experimental quantities in problem solving.

10 Calculate the density of an object from mass and volume data and calculate the specific gravity of an object from its density.

CHEMISTRY CONNECTION

Chance Favors the Prepared Mind

Most of you have chosen a career in medicine because you want to help others. In medicine, helping others means easing pain and suffering by treating or curing diseases. One important part of the practice of medicine involves observation. The physician must carefully observe the patient and listen to his or her description of symptoms to arrive at a preliminary diagnosis. Then appropriate tests must be done to determine whether the diagnosis is correct. During recovery the patient must be carefully observed for changes in behavior or symptoms. These changes are clues that the treatment or medication needs to be modified.

These practices are also important in science. The scientist makes an observation and develops a preliminary hypothesis or explanation for the observed phenomenon. Experiments are then carried out to determine whether the hypothesis is correct. When performing the experiment and analyzing the data, the scientist must look for any unexpected results that indicate that the original hypothesis must be modified.

Several important discoveries in medicine and the sciences have arisen from accidental observations. A health care worker or scientist may see something quite unexpected. Whether this results in an important discovery or is ignored depends on the training and preparedness of the observer.

It was Louis Pasteur, a chemist and microbiologist, who said, "Chance favors the prepared mind." In the history of science and medicine there are many examples of individuals who have made important discoveries because they recognized the value of an unexpected observation.

One such example is the use of ultraviolet (UV) light to treat infant jaundice. Infant jaundice is a condition in which the skin and the whites of the eyes appear yellow because of high levels of the bile pigment bilirubin in the blood. Bilirubin is a breakdown product of the oxygen-carrying blood protein hemoglobin. If bilirubin accumulates in the body, it can cause brain damage and death. The immature liver of the baby cannot remove the bilirubin.

An observant nurse in England noticed that when jaundiced babies were exposed to sunlight, the jaundice faded. Research based on her observation showed that the UV light changes the bilirubin into another substance that can be excreted. To this day, jaundiced newborns are treated with UV light.

The Pap smear test for the early detection of cervical and uterine cancer was also developed because of an accidental observation. Dr. George Papanicolaou, affectionately called Dr. Pap, was studying changes in the cells of the vagina during the stages of the menstrual cycle. In one sample he recognized cells that looked like cancer cells. Within five years, Dr. Pap had perfected a technique for staining cells from vaginal fluid and observing them microscopically for the presence of any abnormal cells. The lives of countless women have been saved because a routine Pap smear showed early stages of cancer.

In this first chapter of your study of chemistry you will learn more about the importance of observation and accurate, precise measurement in medical practice and scientific study. You will also study the scientific method, the process of developing hypotheses to explain observations, and the design of experiments to test those hypotheses.

Introduction

When you awoke this morning, a flood of chemicals called neurotransmitters was sent from cell to cell in your nervous system. As these chemical signals accumulated, you gradually became aware of your surroundings. Chemical signals from your nerves to your muscles propelled you out of your warm bed to prepare for your day.

For breakfast you had a glass of milk, two eggs, and buttered toast, thus providing your body with needed molecules in the form of carbohydrates, proteins, lipids, vitamins, and minerals. As you ran out the door, enzymes of your digestive tract were dismantling the macromolecules of your breakfast. Other enzymes in your cells were busy converting the chemical energy of food molecules into adenosine triphosphate (ATP), the universal energy currency of all cells.

As you continue through your day, thousands of biochemical reactions will keep your cells functioning optimally. Hormones and other chemical signals will regulate the conditions within your body. They will let you know if you are hungry or thirsty. If you injure yourself or come into contact with a disease-causing microorganism, chemicals in your body will signal cells to begin the necessary repair or defense processes.

Life is an organized array of large, carbon-based molecules maintained by biochemical reactions. To understand and appreciate the nature of a living being, we must understand the principles of science and chemistry as they apply to biological molecules.

1.1 The Discovery Process

Chemistry

Chemistry is the study of matter, its chemical and physical properties, the chemical and physical changes it undergoes, and the energy changes that accompany those processes. **Matter** is anything that has mass and occupies space. The changes that matter undergoes always involve either gain or loss of energy. **Energy** is the ability to do work to accomplish some change. The study of chemistry involves matter, energy, and their interrelationship. Matter and energy are at the heart of chemistry.

Learning Goal

1

Major Areas of Chemistry

Chemistry is a broad area of study covering everything from the basic parts of an atom to interactions between huge biological molecules. Because of this, chemistry encompasses the following specialties.

Biochemistry is the study of life at the molecular level and the processes associated with life, such as reproduction, growth, and respiration. *Organic chemistry* is the study of matter that is composed principally of carbon and hydrogen. Organic chemists study methods of preparing such diverse substances as plastics, drugs, solvents, and a host of industrial chemicals. *Inorganic chemistry* is the study of matter that consists of all of the elements other than carbon and hydrogen and their combinations. Inorganic chemists have been responsible for the development of unique substances such as semiconductors and high-temperature ceramics for industrial use. *Analytical chemistry* involves the analysis of matter to determine its composition and the quantity of each kind of matter that is present. Analytical chemists detect traces of toxic chemicals in water and air. They also develop methods to analyze human body fluids for drugs, poisons, and levels of medication. *Physical chemistry* is a discipline that attempts to explain the way in which matter behaves. Physical chemists develop theoretical concepts and try to prove them experimentally. This helps us understand how chemical systems behave.

Over the last thirty years, the boundaries between the traditional sciences of chemistry and biology, mathematics, physics, and computer science have gradually faded. Medical practitioners, physicians, nurses, and medical technologists use therapies that contain elements of all these disciplines. The rapid expansion of the pharmaceutical industry is based on a recognition of the relationship between the function of an organism and its basic chemical make-up. Function is a consequence of changes that chemical substances undergo.

For these reasons, an understanding of basic chemical principles is essential for anyone considering a medically related career; indeed, a worker in any science-related field will benefit from an understanding of the principles and applications of chemistry.

The Scientific Method

The **scientific method** is a systematic approach to the discovery of new information. How do we learn about the properties of matter, the way it behaves in nature, and how it can be modified to make useful products? Chemists do this by using the scientific method to study the way in which matter changes under carefully controlled conditions.

Learning Goal

2

A HUMAN PERSPECTIVE

The Scientific Method

The discovery of penicillin by Alexander Fleming is an example of the scientific method at work. Fleming was studying the growth of bacteria. One day, his experiment was ruined because colonies of mold were growing on his plates. From this failed experiment, Fleming made an observation that would change the practice of medicine: Bacterial colonies could not grow in the area around the mold colonies. Fleming hypothesized that the mold was making a chemical compound that inhibited the growth of the bacteria. He performed a series of experiments designed to test this hypothesis.

The key to the scientific method is the design of carefully controlled experiments that will either support or disprove the hypothesis. This is exactly what Fleming did.

In one experiment he used two sets of tubes containing sterile nutrient broth. To one set he added mold cells. The second set (the control tubes) remained sterile. The mold was allowed to grow for several days. Then the broth from each of the tubes (experimental and control) was passed through a filter to remove any mold cells. Next, bacteria were placed in each tube. If Fleming's hypothesis was correct, the tubes in which the mold had grown would contain the chemical that inhibits growth, and the bacteria would not grow. On the other hand, the control tubes (which were never used to grow mold) would allow bacterial growth. This is exactly what Fleming observed.

Within a few years this *antibiotic*, penicillin, was being used to treat bacterial infections in patients.

The scientific method is not a "cookbook recipe" that, if followed faithfully, will yield new discoveries; rather, it is an organized approach to solving scientific problems. Every scientist brings his or her own curiosity, creativity, and imagination to scientific study. But scientific inquiry still involves some of the "cookbook approach." For example, in the laboratory you may use a procedure that was developed by others to measure certain physical properties of matter, such as measuring the density of urine to determine the amount of sugar present. By doing this, you are applying part of the scientific method.

Characteristics of the scientific process include the following:

1. *Observation.* The description of, for example, the color, taste, or odor of a substance is a result of observation. The measurement of the temperature of a liquid or the size or mass of a solid results from observation.
2. *Formulation of a question.* Humankind's fundamental curiosity motivates questions of why and how things work.
3. *Pattern recognition.* If a scientist finds a cause-and-effect relationship, it may be the basis of a generalized explanation of substances and their behavior. The observation that certain bacteria thrive in an oxygen environment whereas others cease to multiply and, in fact, die under the same conditions leads to the conclusion that a cause-and-effect relationship certainly exists between bacterial metabolism and oxygen levels.

Learning Goal

4. *Developing theories.* When scientists observe a phenomenon, they want to explain it. The process of explaining observed behavior begins with a hypothesis. A **hypothesis** is simply an attempt to explain an observation, or series of observations, in a common-sense way. If many experiments support a hypothesis, it may attain the status of a theory. A **theory** is a hypothesis supported by extensive testing (experimentation) that explains scientific facts and can predict new facts.
5. *Experimentation.* Demonstrating the correctness of hypotheses and theories is at the heart of the scientific method. This is done by carrying out carefully designed experiments that will either support or disprove the theory or hypothesis.
6. *Summarizing information.* A scientific **law** is nothing more than the summary of a large quantity of information. For example, the law of conservation of

matter states that matter cannot be created or destroyed, only converted from one form to another. This statement represents a massive quantity of chemical information gathered from experiments.

The scientific process often begins with an observation that gives rise to questions such as, "Why does this happen?" or "Why is this phenomenon only observed under certain conditions?" As in everyday life, a question demands an answer, the hypothesis. Real life also teaches us that such an approach generally meets with some skepticism: "Show me!" The scientific response is an experiment (or series of experiments) *designed* to support or disprove the hypothesis. Depending on the outcome of the experiments the hypothesis is supported, leading to theory and subsequent investigation, or rejected, calling for a new hypothesis and more experimentation.

A properly designed experiment looks at only one variable at a time. To ensure this, a second experiment, the control, is also performed. This experiment is identical to the first in every way but one: The variable being studied is not present. So if the results of the two experiments differ, the difference must be due to the variable. In Fleming's experiment, described in A Human Perspective: The Scientific Method, the variable was the culture of mold cells.

We have noted that a law summarizes the similar behavior of all matter. Isaac Newton's theory of gravity illustrates the development of a scientific law. According to the story, Newton saw an apple fall from a tree while having tea in the garden. From this observation he developed a cause-and-effect relationship: There is a universal attractive force between bodies. He called this force *gravity*. His experiments and calculations supported his hypothesis. It became the *theory* of gravitational attraction, and centuries of experiments have supported this theory. We now call Newton's observed relationship *the law of gravity*.

The scientific method involves the interactive use of hypotheses, development of theories, and thorough testing of theories using well-designed experiments and is summarized in Figure 1.1.

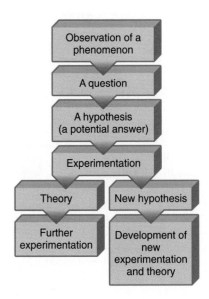

Figure 1.1
The scientific method, an organized way of doing science. A degree of trial and error is apparent here. If experimentation does not support the hypothesis, one must begin the cycle again.

Models in Chemistry

Hypotheses, theories, and laws are frequently expressed using mathematical equations. These equations may confuse all but the best of mathematicians. For this reason a *model* of a chemical unit or system is often used to make ideas more clear. A good model based on everyday experience, although imperfect, gives a great deal of information in a simple fashion. Consider the fundamental unit of methane, the major component of natural gas, which is composed of one carbon atom (symbolized by C) and four hydrogen atoms (symbolized by H).

A geometrically correct model of methane can be constructed from balls and sticks. The balls represent the individual units (atoms) of hydrogen and carbon, and the sticks correspond to the attractive forces that hold the hydrogen and carbon together. The model consists of four balls representing hydrogen symmetrically arranged around a center ball representing carbon. The "carbon" ball is attached to each "hydrogen" ball by sticks, as shown:

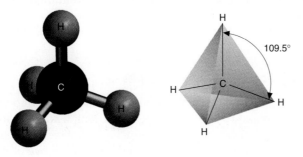

Color-coding the balls distinguishes one type of matter from another; the geometrical form of the model, all of the angles and dimensions of a tetrahedron, are the same for each methane unit found in nature. Methane is certainly not a collection of balls and sticks; but such models are valuable because they help us understand the chemical behavior of methane and other, more complex substances.

Science and Technology

Learning Goal

Science can be broadly defined as the systematic study of humankind and our surrounding environment. We constantly ask how and why we and our surroundings behave as they do. The development of the scientific method as a way to do science has played a major role in the rapid growth of civilization during the past two centuries. In fact, more profound changes have occurred in the short time since the Industrial Revolution in eighteenth century Europe than in all of previous history. It is true that the human race had achieved a certain level of sophistication in its earlier history. Fire, the wheel, manufacturing of tools and weapons from metals, and tanning of hides for protective clothing are prominent examples. In later times the process of photography was developed without any real understanding of the chemical process. However, this technological progress was slow and sporadic without science to aid in the understanding of the underlying principles.

Technology is the conversion of a material or information from its current form to a more useful form. Many would describe technology as *applied science,* the use of scientific principles to meet human needs. For example, a few drops of petroleum may be converted into a life-saving drug. Some do not view all technological conversions as "more useful." Is the paper used for advertising "more useful" than the tree from which it was made? All of these processes do, however, fulfill some human need.

Clearly, science breeds technology with both its benefits and its potential dangers. For example, Enrico Fermi's discovery that the atom could be "split," producing enormous amounts of energy, has given rise to clearly beneficial applications (nuclear radiation used as a treatment for cancer) as well as developments that pose serious threats to society (nuclear weapons). Many spinoffs of the space program are in routine use today; flame-resistant materials, voice-controlled wheelchairs, reading machines for the blind, solar energy devices, and portable X-ray units are only a few of the technological rewards of this scientific venture. Figure 1.2 depicts some of these technologies.

1.2 Data, Results, and Units

A scientific experiment produces **data.** Each piece of data is the individual result of a single measurement or observation. Examples include the *mass* of a sample and the *time* required for a chemical reaction to occur. Mass, length, volume, time, temperature, and energy are common types of data obtained from chemical experiments.

Learning Goal

Results are the outcome of an experiment. Data and results may be identical, but more often several related pieces of data are combined, and logic is used to produce a result.

EXAMPLE 1.1

Distinguishing between Data and Results

In many cases, a drug is less stable if moisture is present, and excess moisture can hasten the breakdown of the active ingredient, leading to loss of potency. Therefore we may wish to know how much water a certain quantity of a drug gains when exposed to air. To do this experiment, we must first weigh the drug sample, then expose it to the air for a period and reweigh it. The change in weight,

$$[weight_{final} - weight_{initial}] = weight\ difference$$

indicates the weight of water taken up by the drug formulation. The initial and final weights are individual bits of *data;* by themselves they do not answer the question, but they do provide the information necessary to calculate the answer: the results. The difference in weight and the conclusions based on the observed change in weight are the *results* of the experiment.

The experiment described in Example 1.1 was really not a very good experiment because many other environmental conditions were not measured. Measurement of the temperature and humidity of the atmosphere and the length of time that the drug was exposed to the air (the creation of a more complete set of data) would make the results less ambiguous.

Any measurement made in the experiment must also specify the units of that measurement. An initial weight of three *ounces* is clearly quite different than three *pounds.* A **unit** defines the basic quantity of mass, volume, time, or whatever quantity is being measured. A number that is not followed by the correct unit usually conveys no useful information.

Proper use of units is central to all aspects of science. Sections 1.3, 1.4, and 1.5 are designed to develop a fundamental understanding of this vital topic.

1.3 Measurement in Chemistry

English and Metric Units

The **English system** is a collection of functionally unrelated units. In the *English system of measurement* the standard *pound* (lb) is the basic unit of weight. The

Figure 1.2
Examples of technology originating from scientific inquiry. (a) Synthesis of a new drug. (b) Solar energy cells. (c) Preparation of solid-state electronics. (d) Use of a gypsy moth sex attractant for insect control.

fundamental unit of *length* is the standard *yard* (yd), and the basic unit of *volume* is the standard *gallon* (gal). The English system is used in the United States in business and industry. However, it is not used in scientific work, primarily because it is difficult to convert from one unit to another. For example,

$$1 \text{ foot} = 12 \text{ inches} = 0.33 \text{ yard} = \frac{1}{5280} \text{ mile} = \frac{1}{6} \text{ fathom}$$

Clearly, operations such as the conversion of 1.62 yards to units of miles are not straightforward. In fact, the English "system" is not really a system at all. It is simply a collection of measures accumulated throughout English history. Because they have no functional relationship, it is not surprising that conversion from one unit to another is not straightforward.

Learning Goal

6

The United States, the last major industrial country to retain the English system, has begun efforts to convert to the metric system. The **metric system** is truly "systematic." It is composed of a set of units that are related to each other decimally, in other words, as powers of ten. Because the *metric system* is a decimal-based system, it is inherently simpler to use and less ambiguous. For example, the length of an object may be represented as

$$1 \text{ meter} = 10 \text{ decimeters} = 100 \text{ centimeters} = 1000 \text{ millimeters}$$

Only the decimal point moves in the conversion from one unit to another, simplifying many calculations.

The metric system was originally developed in France just before the French Revolution in 1789. The modern version of this system is the *Système International,* or *S.I. system.* Although the S.I. system has been in existence for over forty years, it has yet to gain widespread acceptance. To make the S.I. system truly systematic, it utilizes certain units, especially those for pressure, that many people find difficult to use.

In this text we will use the metric system, not the S.I. system, and we will use the English system only to the extent of converting *from* it to the more scientifically useful metric system.

Other metric units, for time, temperature, and energy, will be treated in Section 1.6.

In the metric system there are three basic units. Mass is represented as the *gram,* length as the *meter,* and volume as the *liter.* Any subunit or multiple unit contains one of these units preceded by a prefix indicating the power of ten by which the base unit is to be multiplied to form the subunit or multiple unit. The most common metric prefixes are shown in Table 1.1.

The same prefix may be used for volume, mass, length, time, and so forth. Consider the following examples:

See Appendix A for a review of the mathematics involved.

$$1 \text{ milliliter (mL)} = \frac{1}{1000} \text{ liter} = 0.001 \text{ liter} = 10^{-3} \text{ liter}$$

or

$$1 \text{ microliter } (\mu\text{L}) = \frac{1}{1,000,000} \text{ liter} = 0.000001 \text{ liter} = 10^{-6} \text{ liter}$$

A volume unit is indicated by the base unit, liter, and the prefix *milli-,* which indicates that the unit is one thousandth of the base unit, or *micro-,* which indicates that the unit is one millionth of the base unit. In the same way,

$$1 \text{ milligram (mg)} = \frac{1}{1000} \text{ gram} = 0.001 \text{ gram} = 10^{-3} \text{ gram}$$

or

$$1 \text{ microgram } (\mu\text{g}) = \frac{1}{1,000,000} \text{ gram} = 0.000001 \text{ gram} = 10^{-6} \text{ gram}$$

Table 1.1	Some Common Prefixes Used in the Metric System	
Prefix	**Multiple**	**Decimal Equivalent**
mega (M)	10^6	1,000,000.
kilo (k)	10^3	1,000.
deka (da)	10^1	10.
deci (d)	10^{-1}	0.1
centi (c)	10^{-2}	0.01
milli (m)	10^{-3}	0.001
micro (μ)	10^{-6}	0.000001
nano (n)	10^{-9}	0.000000001

and

$$1 \text{ millimeter (mm)} = \frac{1}{1000} \text{ meter} = 0.001 \text{ meter} = 10^{-3} \text{ meter}$$

or

$$1 \text{ micrometer (}\mu\text{m)} = \frac{1}{1,000,000} \text{ meter} = 0.000001 \text{ meter} = 10^{-6} \text{ meter}$$

Additionally,

$$1 \text{ kilogram (kg)} = 1000 \text{ grams} = 10^3 \text{ grams}$$

$$1 \text{ decigram (dg)} = \frac{1}{10} \text{ gram} = 0.1 \text{ gram} = 10^{-1} \text{ gram}$$

The representation of numbers as powers of ten may be unfamiliar to you. This useful notation is discussed in Section 1.5.

Other measurements can be treated in the same way; for example, a millisecond is $\frac{1}{1000}$ of a second, and so forth.

Unit Conversion: English and Metric Systems

To convert from one unit to another, we must have a **conversion factor** or series of conversion factors that relate two units. The proper use of these conversion factors is called the *factor-label method*. This method is also termed *dimensional analysis*.

This method is used for two kinds of conversions: to convert from one unit to another within the *same system* or to convert units from *one system to another.*

Conversion of Units within the Same System

We know, for example, that in the English system,

$$1 \text{ gallon} = 4 \text{ quarts}$$

Because dividing both sides of the equation by the same term does not change its identity,

$$\frac{1 \text{ gallon}}{1 \text{ gallon}} = \frac{4 \text{ quarts}}{1 \text{ gallon}}$$

The expression on the left is equal to unity (1); therefore

$$1 = \frac{4 \text{ quarts}}{1 \text{ gallon}} \quad \text{or} \quad 1 = \frac{1 \text{ gallon}}{4 \text{ quarts}}$$

Learning Goal

6

Now, multiplying any other expression by the ratio 4 quarts/1 gallon will not change the value of the term, because multiplication of any number by 1 produces the original value. However, there is one important difference: The units will have changed.

EXAMPLE 1.2

Using Conversion Factors

Convert 12 gallons to units of quarts.

Solution

$$12 \text{ gal} \times \frac{4 \text{ qt}}{1 \text{ gal}} = 48 \text{ qt}$$

The conversion factor, 4 qt/1 gal, serves as a bridge, or linkage, between the unit that was given (gallons) and the unit that was sought (quarts).

The conversion factor in Example 1.2 may be written as 4 qt/1 gal or 1 gal/4 qt, because both are equal to 1. However, only the first factor, 4 qt/1 gal, will give us the units we need to solve the problem. If we had set up the problem incorrectly, we would get

$$12 \text{ gal} \times \frac{1 \text{ gal}}{4 \text{ qt}} = 3 \frac{\text{gal}^2}{\text{qt}}$$

Incorrect units

Clearly, units of gal²/qt are not those asked for in the problem, nor are they reasonable units. The factor-label method is therefore a self-indicating system; the correct units (those required by the problem) will result only if the factor is set up properly.

Table 1.2 lists a variety of commonly used English system relationships that may serve as the basis for useful conversion factors.

Conversion of units within the metric system may be accomplished by using the factor-label method as well. Unit prefixes that dictate the conversion factor facilitate unit conversion (refer to Table 1.1).

EXAMPLE 1.3

Using Conversion Factors

Convert 10.0 centimeters to meters.

Solution

First, recognize that the prefix *centi-* means ¹⁄₁₀₀ of the base unit, the meter. Thus our conversion factor is either

$$\frac{1 \text{ meter}}{100 \text{ cm}} \quad \text{or} \quad \frac{100 \text{ cm}}{1 \text{ meter}}$$

each being equal to 1. Only one, however, will result in proper cancellation of units, producing the correct answer to the problem. If we proceed as follows:

$$10.0 \text{ cm} \times \frac{1 \text{ meter}}{100 \text{ cm}} = 0.100 \text{ meter}$$

| Data given | Conversion factor | Desired result |

we obtain the desired units, meters (m). If we had used the conversion factor 100 cm/1 m, the resulting units would be meaningless and the answer would have been incorrect:

$$10.0 \text{ cm} = \frac{100 \text{ cm}}{1 \text{ m}} = 1000 \frac{\text{cm}^2}{\text{m}}$$

Incorrect units

Convert 1.0 liter to each of the following units, using the factor-label method:

a. milliliters   d. centiliters
b. microliters   e. dekaliters
c. kiloliters

*Question **1.1***

Convert 1.0 gram to each of the following units:

a. micrograms   d. centigrams
b. milligrams   e. decigrams
c. kilograms

*Question **1.2***

Conversion of Units from One System to Another

The conversion of a quantity expressed in units of one system to an equivalent quantity in the other system (English to metric or metric to English) requires a *bridging* conversion unit. Examples are shown in Table 1.3.

English and metric conversions are shown in Tables 1.1 and 1.2.

Table 1.2	Some Common Relationships Used in the English System
A. Weight	1 pound = 16 ounces
	1 ton = 2000 pounds
B. Length	1 foot = 12 inches
	1 yard = 3 feet
	1 mile = 5280 feet
C. Volume	1 gallon = 4 quarts
	1 quart = 2 pints
	1 quart = 32 fluid ounces

Table 1.3	Commonly Used "Bridging" Units for Intersystem Conversions		
Quantity	English		Metric
Mass	1 pound	=	454 grams
	2.2 pounds	=	1 kilogram
Length	1 inch	=	2.54 centimeters
	1 yard	=	0.91 meter
Volume	1 quart	=	0.946 liter
	1 gallon	=	3.78 liters

Learning Goal

The conversion may be represented as a three-step process:

1. Conversion from the units given in the problem to a bridging unit.
2. Conversion to the other system using the bridge.
3. Conversion within the desired system to units required by the problem.

EXAMPLE 1.4

Using Conversion Factors between Systems

Convert 4.00 ounces to kilograms.

Solution

Step 1. A convenient bridging unit for mass is 1 lb = 454 grams. To use this conversion factor, we relate ounces (given in the problem) to pounds:

$$4.00 \text{ ounces} \times \frac{1 \text{ pound}}{16 \text{ ounces}} = 0.250 \text{ pound}$$

Step 2. Using the bridging unit conversion, we get

$$0.250 \text{ pound} \times \frac{454 \text{ grams}}{1 \text{ pound}} = 114 \text{ grams}$$

Refer to the discussion of rounding off numbers on p. 19.

Step 3. Grams may then be directly converted to kilograms, the desired unit:

$$114 \text{ grams} \times \frac{1 \text{ kilogram}}{1000 \text{ grams}} = 0.114 \text{ kilogram}$$

The calculation may also be done in a single step by arranging the factors in a chain:

$$4.00 \text{ oz} \times \frac{1 \text{ lb}}{16 \text{ oz}} \times \frac{454 \text{ g}}{1 \text{ lb}} \times \frac{1 \text{ kg}}{1000 \text{ g}} = 0.114 \text{ kg}$$

EXAMPLE 1.5

Using Conversion Factors

Convert 1.5 meters^2 to centimeters^2.

Solution

The problem is similar to the conversion performed in Example 1.3. However, we must remember to include the exponent in the units. Thus

$$1.5 \text{ m}^2 \times \left(\frac{10^2 \text{ cm}}{1 \text{ m}}\right)^2 = 1.5 \text{ m}^2 \times \frac{10^4 \text{ cm}^2}{1 \text{ m}^2} = 1.5 \times 10^4 \text{ cm}^2$$

Note: The exponent affects both the number *and* unit within the parentheses.

Question 1.3

a. Convert 0.50 inch to meters.
b. Convert 0.75 quart to liters.
c. Convert 56.8 grams to ounces.
d. Convert 1.5 cm^2 to m^2.

Question 1.4

a. Convert 0.50 inch to centimeters.
b. Convert 0.75 quart to milliliters.
c. Convert 56.8 milligrams to ounces.
d. Convert 3.6 m^2 to cm^2.

1.4 Error, Accuracy, Precision, and Uncertainty

Error is the difference between the true value and our estimation, or measurement, of the value. All measurements are associated with some degree of error. Two types of error exist: random error and systematic error. *Random error* causes data from multiple measurements of the same quantity to be scattered in a more or less uniform way around some average value. *Systematic error* causes data to be either smaller or larger than the accepted value. Random error is inherent in the experimental approach to the study of matter and its behavior; systematic error can be found and, in many cases, removed or corrected.

Learning Goal

7

Examples of systematic error include such situations as:

- Dust on the balance pan, causing all objects weighed to appear heavier than they really are.
- Impurities in chemicals used for the analysis of materials, which may interfere with (or block) the desired process.

Accuracy is the degree of agreement between the true value and the measured value. **Uncertainty** is the degree of doubt in a single measurement.

Only discrete quantities, such as the number of pages in this book or the number of quarters in your pocket, can be measured with certainty; they are exact numbers. For example, there are 50 pages in your notebook, *exactly* 50, not 50½ or 49½. In measuring quantities that show continuous variation, for example, the weight of this page or the volume of one of your quarters, some doubt or uncertainty is present because the answer cannot be expressed with an infinite number of meaningful digits. The number of meaningful digits is determined by the measuring device. The presence of some error is a natural consequence of any measurement.

The simple process of converting the fraction ⅔ to its decimal equivalent can produce a variety of answers that depend on the device used to perform the calculation: pencil and paper, calculator, computer. The answer might be

0.67

0.667

0.6667

and so forth. All are correct, but each value has a different level of uncertainty. The first number listed, 0.67, has the greatest uncertainty.

It is always best to measure a quantity several times. Modern scientific instruments are designed to perform measurements rapidly; this allows many more measurements to be completed in a reasonable period. Replicate measurements of the same quantity minimize the uncertainty of the result. **Precision** is a measure of the agreement of replicate measurements.

It is important to recognize that accuracy and precision are not the same thing. It is possible to have one without the other. However, when scientific measurements are carefully made, the two most often go hand in hand; high-quality data are characterized by high levels of precision and accuracy.

In Figure 1.3, bullseye (a) shows the goal of all experimentation: accuracy *and* precision. Bullseye (b) shows the results to be repeatable (good precision); however, some error in the experimental procedure has caused the results to center on an incorrect value. This error is systematic, occurring in each replicate measurement. Occasionally, an experiment may show "accidental" accuracy. The precision is poor, but the average of these replicate measurements leads to a correct value. We don't want to rely on accidental success; the experiment should be repeated until the precision inspires faith in the accuracy of the method. Modern measuring devices in chemistry, equipped with powerful computers with immense storage capacity, are capable of making literally thousands of individual replicate measurements to

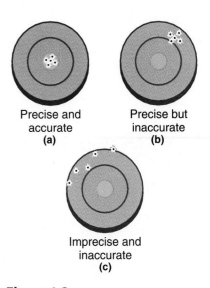

Precise and accurate
(a)

Precise but inaccurate
(b)

Imprecise and inaccurate
(c)

Figure 1.3

An illustration of precision and accuracy in replicate experiments.

enhance the quality of the result. Bullseye (c) describes the most common situation. A low level of precision is all too often associated with poor accuracy.

1.5 Significant Figures and Scientific Notation

Information-bearing figures in a number are termed *significant figures*. Data and results arising from a scientific experiment convey information about the way in which the experiment was conducted. The degree of uncertainty or doubt associated with a measurement or series of measurements is indicated by the number of figures used to represent the information.

Significant Figures

Learning Goal

8

Consider the following situation: A student was asked to obtain the length of a section of wire. In the chemistry laboratory, several different types of measuring devices are usually available. Not knowing which was most appropriate, the student decided to measure the object using each device that was available in the laboratory. The following data were obtained:

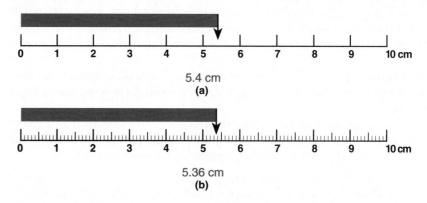

5.4 cm
(a)

5.36 cm
(b)

Two questions should immediately come to mind:

>   Are the two answers equivalent?

>   If not, which answer is correct?

In fact, the two answers are *not* equivalent, but *both* are correct. How do we explain this apparent contradiction?

The data are not equivalent because each is known to a different degree of certainty. The answer 5.36 cm, containing three significant figures, specifies the length of the object more exactly than 5.4 cm, which contains only two significant figures. The term **significant figures** is defined to be all digits in a number representing data or results that are known with certainty *plus one uncertain digit.*

In case (a) we are certain that the object is at least 5 cm long and equally certain that it is *not* 6 cm long because the end of the object falls between the calibration lines 5 and 6. We can only estimate between 5 and 6, because there are no calibration indicators between 5 and 6. The end of the wire appears to be approximately four-tenths of the way between 5 and 6, hence 5.4 cm. The 5 is known with certainty, and 4 is estimated; there are two significant figures.

In case (b) the ruler is calibrated in tenths of centimeters. The end of the wire is at least 5.3 cm and not 5.4 cm. Estimation of the second decimal place between the two closest calibration marks leads to 5.36 cm. In this case, 5.3 is certain, and the 6 is estimated (or uncertain), leading to three significant digits.

Both answers are correct because each is consistent with the measuring device used to generate the data. An answer of 5.36 cm obtained from a measurement using ruler (a) would be *incorrect* because the measuring device is not capable of that exact specification. On the other hand, a value of 5.4 cm obtained from ruler (b) would be erroneous as well; in that case the measuring device is capable of generating a higher level of certainty (more significant digits) than are actually reported.

In summary, the number of significant figures associated with a measurement is determined by the measuring device. Conversely, the number of significant figures reported is an indication of the sophistication of the measurement itself.

Recognition of Significant Figures

Only *significant* digits should be reported as data or results. However, are all digits, as written, significant digits? Let's look at a few examples illustrating the rules that are used to represent data and results with the proper number of significant digits.

EXAMPLE 1.6

RULE: All nonzero digits are significant. ■

7.314 has *four* significant digits.

EXAMPLE 1.7

RULE: The number of significant digits is independent of the position of the decimal point. ■

73.14 has *four* significant digits.

EXAMPLE 1.8

RULE: Zeros located between nonzero digits are significant. ■

60.052 has *five* significant figures.

EXAMPLE 1.9

RULE: Zeros at the end of a number (often referred to as trailing zeros) are significant if the number contains a decimal point. ■

4.70 has *three* significant figures.

Trailing zeros are ambiguous; the next section offers a solution for this ambiguity.

EXAMPLE 1.10

RULE: Trailing zeros are insignificant if the number does not contain a decimal point and are significant if a decimal point is indicated. ■

100 has *one* significant figure; 100. has three significant figures.

EXAMPLE 1.11

RULE: Zeros to the left of the first nonzero integer are not significant; they serve only to locate the position of the decimal point. ■

0.0032 has *two* significant figures.

How many significant figures are contained in each of the following numbers?
a. 7.26
b. 726
c. 700.2
d. 7.0
e. 0.0720

How many significant figures are contained in each of the following numbers?
a. 0.042
b. 4.20
c. 24.0
d. 240
e. 204

Scientific Notation

Learning Goal

It is often difficult to express very large numbers to the proper number of significant figures using conventional notation. Consider the number fifteen thousand, which might have resulted from a measurement capable of generating three significant figures. Writing the number in a conventional fashion as 15,000. (note the presence of a decimal point) shows five digits that, if written, must be assumed to be significant. Alternatively, writing 15,000 (no decimal point) indicates only two significant figures. How can we express such a number using only the desired number (in this case, three) of significant figures? The solution lies in the use of **scientific notation,** also referred to as *exponential notation,* which involves the representation of a number as a power of ten.

See Appendix A for a review of the mathematics involved.

If we consider the number 15,000, it is equivalent to 15×1000, which is also 15×10^3. In fact 15,000 could be represented as

$$15{,}000 \times 10^0$$

$$1500 \quad \times 10^1 \qquad \text{(decimal moved } \textit{one} \text{ place to the left)}$$

$$150 \quad \times 10^2 \qquad \text{(decimal moved } \textit{two} \text{ places to the left)}$$

$$15 \quad \times 10^3 \qquad \text{(decimal moved } \textit{three} \text{ places to the left)}$$

$$1.5 \quad \times 10^4 \qquad \text{(decimal moved } \textit{four} \text{ places to the left)}$$

All appear to be correct; however, when we superimpose our initial requirement that the data point, fifteen thousand, has three significant figures, 15.0×10^3 or 1.50×10^4 is the logical choice. In this case 1.50×10^4 would be preferred; by convention scientific notation shows the decimal point to the right of the leading digit.

RULE: To convert a number greater than 1 to scientific notation, the original decimal point is moved x places to the left, and the resulting number is multiplied by 10^x. The exponent (x) is a *positive* number equal to the number of places the original decimal point was moved. ∎

Scientific notation is also useful in representing numbers less than 1. For example, the mass of a single helium atom is

$$0.0000000000000000000000006692 \text{ gram}$$

a rather cumbersome number as written. Scientific notation would represent the mass of a single helium atom as 6.692×10^{-24} gram. The conversion is illustrated by using a simpler number:

$$0.0062 = 6.2 \times \frac{1}{1000} = 6.2 \times \frac{1}{10^3} = 6.2 \times 10^{-3}$$

or

$$0.0534 = 5.34 \times \frac{1}{100} = 5.34 \times \frac{1}{10^2} = 5.34 \times 10^{-2}$$

RULE: To convert a number less than 1 to scientific notation, the original decimal point is moved x places to the right, and the resulting number is multiplied by 10^{-x}. The exponent $(-x)$ is a *negative* number equal to the number of places the original decimal point was moved. ■

Q u e s t i o n **1.7**

Represent each of the following numbers in scientific notation, showing only significant digits:

a. 0.0024
b. 0.0180
c. 224

Q u e s t i o n **1.8**

Represent each of the following numbers in scientific notation, showing only significant digits:

a. 48.20
b. 480.0
c. 0.126

Significant Figures in Calculation of Results

Addition and Subtraction

If we combine the following numbers:

37.68	liters
108.428	liters
6.71862	liters

our calculator will show a final result of

152.82662	liters

Clearly, the answer, with eight digits, defines the volume of total material much more accurately than *any* of the individual quantities being combined. This cannot be correct; *the answer cannot have greater significance than any of the quantities that produced the answer.* We rewrite the problem:

37.68xxx	liters
108.428xx	liters
+ 6.71862	liters
152.82662	(should be 152.83) liters

where x = no information; x may be any integer from 0 to 9. Adding 2 to two unknown numbers (in the right column) produces no information. Similar logic prevails for the next two columns. Thus five digits remain, all of which are significant. Conventional rules for rounding off would dictate a final answer of 152.83.

> Remember the distinction between the words *zero* and *nothing*. *Zero* is one of the ten digits and conveys as much information as 1, 2, and so forth. *Nothing* implies no information; the digits in the positions indicated by x's could be 0, 1, 2, or any other.

> *See rules for rounding off discussed on page 19.*

Q u e s t i o n **1.9**

Report the result of each of the following to the proper number of significant figures:

a. 4.26 = 3.831 =
b. 8.321 − 2.4 =
c. 16.262 + 4.33 − 0.40 =

Question 1.10

Report the result of each of the following to the proper number of significant figures:
a. $7.939 + 6.26 =$
b. $2.4 - 8.321 =$
c. $2.333 + 1.56 - 0.29 =$

Multiplication and Division

In the preceding discussion of addition and subtraction the position of the decimal point in the quantities being combined has a bearing on the number of significant figures in the answer. In multiplication and division this is not the case. The decimal point position is irrelevant when determining the number of significant figures in the answer. It is the number of significant figures in the data that is important. Consider

$$\frac{4.237 \times 1.21 \times 10^{-3} \times 0.00273}{11.125} = 1.26 \times 10^{-6}$$

The answer is limited to three significant figures; the answer can have *only* three significant figures because two numbers in the calculation, 1.21×10^{-3} and 0.00273, have three significant figures and "limit" the answer. Remember, *the answer can be no more precise than the least precise number from which the answer is derived.* The *least precise number* is the number with the fewest significant figures.

Question 1.11

Report the results of each of the following operations using the proper number of significant figures:
a. $63.8 \times 0.80 =$
b. $\dfrac{63.8}{0.80} =$
c. $\dfrac{53.8 \times 0.90}{0.3025} =$

Question 1.12

Report the results of each of the following operations using the proper number of significant figures:
a. $\dfrac{27.2 \times 15.63}{1.84} =$
b. $\dfrac{13.6}{18.02 \times 1.6} =$
c. $\dfrac{12.24 \times 6.2}{18.02 \times 1.6} =$

See Appendix A for a review of the mathematics involved.

Exponents

Now consider the determination of the proper number of significant digits in the results when a value is multiplied by any power of ten. In each case the number of significant figures in the answer is identical to the number contained in the original term. Therefore

$$(8.314 \times 10^2)^3 = 574.7 \times 10^6 = 5.747 \times 10^8$$

and

$$(8.314 \times 10^2)^{1/2} = 2.883 \times 10^1$$

Each answer contains four significant figures.

It is important to note, in operating with significant figures, that defined or counted numbers do *not* determine the number of significant figures.

For example,

How many grams are contained in 0.240 kg?

$$0.240 \text{ kg} \times \frac{1000 \text{ g}}{1 \text{ kg}} = 240. \text{ g}$$

The "1" in the conversion factor is defined, or exact, and does not limit the number of significant digits.

Exact numbers are counting numbers or defined numbers. They have infinitely many significant figures. Consequently they do not limit the number of significant figures in the result of the calculation. You should *recognize* and *ignore* exact numbers when assigning significant figures.

A good rule of thumb to follow is: In the metric system the quantity being converted, not the conversion factor, generally determines the number of significant figures.

Rounding Off Numbers

The use of an electronic calculator generally produces more digits for a result than are justified by the rules of significant figures on the basis of the data input. For example, on your calculator,

$$3.84 \times 6.72 = 25.8048$$

The most correct answer would be 25.8, dropping 048.

RULE: When the number to be dropped is less than 5 the preceding number is not changed. When the number to be dropped is 5 or larger, the preceding number is increased by one unit.   ∎

EXAMPLE 1.12

Rounding Numbers

Round off each of the following to three significant figures.

Solution

a. 63.669 becomes 63.7. *Rationale:* 6 > 5.
b. 8.7715 becomes 8.77. *Rationale:* 1 < 5.
c. 2.2245 becomes 2.22. *Rationale:* 4 < 5.
d. 0.0004109 becomes 0.000411. *Rationale:* 9 > 5.

Symbol $x > y$ implies "x greater than y."
Symbol $x < y$ implies "x less than y."

Question 1.13

Round off each of the following numbers to three significant figures.

a. 61.40
b. 6.171
c. 0.066494

Question 1.14

Round off each of the following numbers to three significant figures.

a. 6.2262
b. 3895
c. 6.885

1.6 Experimental Quantities

Learning Goal

9

Thus far we have discussed the scientific method and its role in acquiring data and converting the data to obtain the results of the experiment. We have seen that such data must be reported in the proper units with the appropriate number of significant figures. The quantities that are most often determined include mass, length, volume, time, temperature, and energy. Now let's look at each of these quantities in more detail.

Mass

Mass describes the quantity of matter in an object. The terms *weight* and *mass*, in common usage, are often considered synonymous. They are not, in fact. **Weight** is the force of gravity on an object:

> Weight = mass × acceleration due to gravity

When gravity is constant, mass and weight are directly proportional. But gravity is not constant; it varies as a function of the distance from the center of the earth. Therefore weight cannot be used for scientific measurement because the weight of an object may vary from one place on the earth to the next.

Mass, on the other hand, is independent of gravity; it is a result of a comparison of an unknown mass with a known mass called a *standard mass*. Balances are instruments used to measure the mass of materials.

Examples of common balances used for the determination of mass are shown in Figure 1.4.

The common conversion units for mass are as follows:

$$1 \text{ gram (g)} = 10^{-3} \text{ kilogram (kg)} = \frac{1}{454} \text{ pound (lb)}$$

In chemistry, when we talk about incredibly small bits of matter such as individual atoms or molecules, units such as grams and even micrograms are much too large. We don't say that a 100-pound individual weighs 0.0500 ton; the unit does not fit the quantity being described. Similarly, an atom of a substance such as hydrogen is very tiny. Its mass is only 1.661×10^{-24} gram.

Figure 1.4

Illustration of three common balances that are useful for the measurement of mass. (a) A two-pan comparison balance for approximate mass measurement suitable for routine work requiring accuracy to 0.1 g (or perhaps 0.01 g). (b) A top-loading single-pan electronic balance that is similar in accuracy to (a) but has the advantages of speed and ease of operation. The revolution in electronics over the past twenty years has resulted in electronic balances largely supplanting the two-pan comparison balance in routine laboratory usage. (c) An analytical balance that is capable of precise mass measurement (three to five significant figures beyond the decimal point). A balance of this type is used when the highest level of precision and accuracy is required.

(a)

(c)

(b)

One *atomic mass unit* (amu) is a more convenient way to represent the mass of one hydrogen atom, rather than 1.661×10^{-24} gram:

$$1 \text{ amu} = 1.661 \times 10^{-24} \text{ g}$$

Units should be chosen to suit the quantity being described. This can easily be done by choosing a unit that gives an exponential term closest to 10^0.

Length

The standard metric unit of *length*, the distance between two points, is the meter. Large distances are measured in kilometers; smaller distances are measured in millimeters or centimeters. Very small distances such as the distances between atoms on a surface are measured in *nanometers* (nm):

$$1 \text{ nm} = 10^{-7} \text{ cm} = 10^{-9} \text{ m}$$

Common conversions for length are as follows:

$$1 \text{ meter (m)} = 10^2 \text{ centimeters (cm)} = 3.94 \times 10^1 \text{ inch (in)}$$

Volume

The standard metric unit of *volume*, the space occupied by an object, is the liter. A liter is the volume occupied by 1000 grams of water at 4 degrees Celsius (°C). The volume, 1 liter, also corresponds to:

$$1 \text{ liter (L)} = 10^3 \text{ milliliters (mL)} = 1.06 \text{ quarts (qt)}$$

The relationship between the liter and the milliliter is shown in Figure 1.5.

Typical laboratory glassware used for volume measurement is shown in Figure 1.6. The volumetric flask is designed to *contain* a specified volume, and the graduated cylinder, pipet, and buret *dispense* a desired volume of liquid.

Volume: 1000 cm³;
1000 mL;
1 dm³;
1 L

← 1 cm
← 10 cm = 1 dm →

Volume: 1 cm³;
1 mL

← 1 cm

Figure 1.5
The relationship among various volume units.

The *milliliter* and the *cubic centimeter* are equivalent.

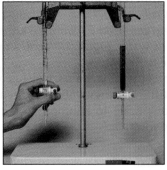
(a)

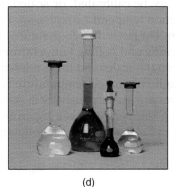
(b)

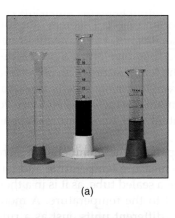
(c)

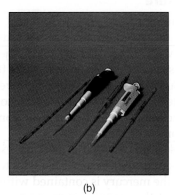
(d)

Figure 1.6
Common laboratory equipment used for the measurement of volume. Graduated cylinders (a), pipets (b), and burets (c) are used for the delivery of liquids; volumetric flasks (d) are used to contain a specific volume. A graduated cylinder is usually used for measurement of approximate volumes; it is less accurate and precise than either pipets or burets.

Questions and Problems

Fundamental Concepts

1.19 Define each of the following terms:
  a. chemistry
  b. matter
  c. energy
1.20 Define each of the following terms:
  a. hypothesis
  b. theory
  c. law
1.21 Define each of the following terms:
  a. precision
  b. accuracy
  c. data
1.22 Define each of the following terms:
  a. results
  b. mass
  c. weight
1.23 Give the base unit for each of the following in the metric system:
  a. mass
  b. volume
  c. length
1.24 Give the base unit for each of the following in the metric system:
  a. time
  b. temperature
  c. energy
1.25 Discuss the difference between the terms *mass* and *weight*.
1.26 Discuss the difference between the terms *data* and *results*.
1.27 Distinguish between specific gravity and density.
1.28 Distinguish between kinetic energy and potential energy.
1.29 Discuss the meaning of the term *scientific method*.
1.30 Describe an application of reasoning involving the scientific method that has occurred in your day-to-day life.

Conversion Factors

1.31 Convert 2.0 pounds to:
  a. ounces   d. milligrams
  b. tons   e. dekagrams
  c. grams
1.32 Convert 5.0 quarts to:
  a. gallons   d. milliliters
  b. pints   e. microliters
  c. liters
1.33 Convert 3.0 grams to:
  a. pounds   d. centigrams
  b. ounces   e. milligrams
  c. kilograms
1.34 Convert 3.0 meters to:
  a. yards   d. centimeters
  b. inches   e. millimeters
  c. feet
1.35 Convert 50.°F to:
  a. °C
  b. K
1.36 Convert −10.°F to:
  a. °C
  b. K

1.37 Convert 20.°C to:
  a. K
  b. °F
1.38 Convert 300. K to:
  a. °C
  b. °F
1.39 A 150-lb adult has approximately 9 pints of blood. How many liters of blood does the individual have?
1.40 If a drop of blood has a volume of 0.05 mL, how many drops of blood are in the adult described in Problem 1.39?
1.41 A patient's temperature is found to be 38.5°C. To what Fahrenheit temperature does this correspond?
1.42 A newborn is 21 inches in length and weighs 6 lb 9 oz. Describe the baby in metric units.

Significant Figures

1.43 How many significant figures are contained in each of the following numbers?
  a. 10.0   d. 2.062
  b. 0.214   e. 10.50
  c. 0.120   f. 1050
1.44 How many significant figures are contained in each of the following numbers?
  a. 3.8×10^{-3} d. 24
  b. 5.20×10^2 e. 240
  c. 0.00261   f. 2.40
1.45 Round the following numbers to three significant figures:
  a. 3.873×10^{-3} d. 24.3387
  b. 5.202×10^{-2} e. 240.1
  c. 0.002616   f. 2.407
1.46 Round the following numbers to three significant figures:
  a. 123700   d. 53.2995
  b. 0.00285792   e. 16.96
  c. 1.421×10^{-3} f. 507.5
1.47 Perform each of the following arithmetic operations, reporting the answer with the proper number of significant figures:
  a. (23)(657)   d. $1157.23 - 17.812$
  b. $0.00521 + 0.236$ e. $\dfrac{(1.987)(298)}{0.0821}$
  c. $\dfrac{18.3}{3.0576}$
1.48 Perform each of the following arithmetic operations, reporting the answer with the proper number of significant figures:
  a. $\dfrac{(16.0)(0.1879)}{45.3}$ d. $18 + 52.1$
  b. $\dfrac{(76.32)(1.53)}{0.052}$ e. $58.17 - 57.79$
  c. $(0.0063)(57.8)$

The Factor-Label Method and Scientific (Exponential) Notation

1.49 Express the following numbers in scientific notation (use the proper number of significant figures):
  a. 12.3   e. 92,000,000
  b. 0.0569   f. 0.005280
  c. −1527   g. 1.279
  d. 0.000000789   h. −531.77

1.50 Using scientific notation, express the number two thousand in terms of:
- **a.** one significant figure
- **b.** two significant figures
- **c.** three significant figures
- **d.** four significant figures
- **e.** five significant figures

1.51 Express each of the following numbers in decimal notation:
- **a.** 3.24×10^3
- **b.** 1.50×10^{-4}
- **c.** 4.579×10^{-1}
- **d.** -6.83×10^5
- **e.** -8.21×10^{-2}
- **f.** 2.9979×10^8
- **g.** 1.50×10^0
- **h.** 6.02×10^{23}

1.52 Which of the following numbers have two significant figures? Three significant figures? Four significant figures?
- **a.** 327
- **b.** 1.049×10^4
- **c.** 1.70
- **d.** 0.000570
- **e.** 7.8×10^3
- **f.** 1507
- **g.** 4.8×10^2
- **h.** 7.389×10^{15}

Experimental Quantities

1.53 Calculate the density of a 3.00×10^2-g object that has a volume of 50.0 mL.

1.54 What volume, in liters, will 8.00×10^2 g of air occupy if the density of air is 1.29 g/L?

1.55 What is the mass, in grams, of a piece of iron that has a volume of 1.50×10^2 mL and a density of 7.20 g/mL?

1.56 What is the mass of a femur (leg bone) having a volume of 118 cm^3? The density of bone is 1.8 g/cm^3.

1.57 You are given a piece of wood that is maple, teak, or oak. The piece of wood has a volume of 1.00×10^2 cm^3 and a mass of 98 g. The densities of maple, teak, and oak are as follows:

Wood	Density (g/cm^3)
Maple	0.70
Teak	0.98
Oak	0.85

What is the identity of the piece of wood?

1.58 The specific gravity of a patient's urine sample was measured to be 1.008. Given that the density of water is 1.000 g/mL at 4°C, what is the density of the urine sample?

1.59 The density of grain alcohol is 0.789 g/mL. Given that the density of water at 4°C is 1.00 g/mL, what is the specific gravity of grain alcohol?

1.60 The density of mercury is 13.6 g/mL. If a sample of mercury weighs 272 g, what is the volume of the sample in milliliters?

1.61 You are given three bars of metal. Each is labeled with its identity (lead, uranium, platinum). The lead bar has a mass of 5.0×10^1 g and a volume of 6.36 cm^3. The uranium bar has a mass of 75 g and a volume of 3.97 cm^3. The platinum bar has a mass of 2140 g and a volume of 1.00×10^2 cm^3. Which of these metals has the lowest density? Which has the greatest density?

1.62 Refer to Problem 1.61. Suppose that each of the bars had the same mass. How could you determine which bar had the lowest density or highest density?

Critical Thinking Problems

1. An instrument used to detect metals in drinking water can detect as little as one microgram of mercury in one liter of water. Mercury is a toxic metal; it accumulates in the body and is responsible for the deterioration of brain cells. Calculate the number of mercury atoms you would consume if you drank one liter of water that contained only one microgram of mercury. [The mass of one mercury atom is 3.3×10^{-22} grams.]

2. Yesterday's temperature was 40°F. Today it is 80°F. Bill tells Sue that it is twice as hot today. Sue disagrees. Do you think Sue is correct or incorrect? Why or why not?

3. Aspirin has been recommended to minimize the chance of heart attacks in persons who have already had one or more occurrences. If a patient takes one aspirin tablet per day for ten years, how many pounds of aspirin will the patient consume? [Assume that each tablet is approximately 325 mg.]

4. Design an experiment that will allow you to measure the density of your favorite piece of jewelry.

5. The diameter of an aluminum atom is 250 picometers (1 picometer = 10^{-12} meters). How many aluminum atoms must be placed end to end to make a "chain" of aluminum atoms one foot long?

2 The Composition and Structure of the Atom

The spectrum of visible light displayed by a prism.

Learning Goals

1. Describe the properties of the solid, liquid, and gaseous states.

2. Classify properties as chemical or physical.

3. Classify observed changes in matter as chemical or physical.

4. Provide specific examples of physical and chemical properties.

5. Distinguish between intensive and extensive properties.

6. Classify matter as element, compound, or mixture.

7. Recognize the interrelationship of the structure of matter and its physical and chemical properties.

8. Describe the important properties of protons, neutrons, and electrons.

9. Calculate the number of protons, neutrons, and electrons in any atom.

10. Distinguish among atoms, ions, and isotopes.

11. Trace the history of the development of atomic theory, beginning with Dalton.

12. Summarize the experimental basis for the discovery of charged particles and the nucleus.

13. Explain the critical role of spectroscopy in the development of atomic theory *and* in our everyday lives.

14. State the basic postulates of Bohr's theory.

15. Compare and contrast Bohr's theory and the more sophisticated "wave-mechanical" approach.

CHEMISTRY CONNECTION

Curiosity, Science, and Medicine

Curiosity is one of the most important human traits. Small children constantly ask "why?". As we get older, our questions become more complex, but the curiosity remains.

Curiosity is also the basis of the scientific method. A scientist observes an event, wonders why it happens, and sets out to answer the question. Dr. Michael Zasloff's curiosity may lead to the development of an entirely new class of antibiotics. When he was a geneticist at the National Institutes of Health, his experiments involved the surgical removal of the ovaries of African clawed frogs. After surgery he sutured (sewed up) the incision and put the frogs back in their tanks. These water-filled tanks were teeming with bacteria, but the frogs healed quickly, and the incisions did not become infected!

Of all the scientists to observe this remarkable healing, only Zasloff was curious enough to ask whether there were chemicals in the frogs' skin that defended the frogs against bacterial infections—a new type of antibiotic. All currently used antibiotics are produced by fungi or are synthesized in the laboratory. One big problem in medicine today is more and more pathogenic (disease-causing) bacteria are becoming resistant to these antibiotics. Zasloff hoped to find an antibiotic that

worked in an entirely new way so the current problems with antibiotic resistance might be overcome.

Dr. Zasloff found two molecules in frog skin that can kill bacteria. Both are small proteins. Zasloff named them *magainins*, from the Hebrew word for shield. Most of the antibiotics that we now use enter bacteria and kill them by stopping some biochemical process inside the cell. Magainins are more direct; they simply punch holes in the bacterial membrane, and the bacteria explode.

One of the magainins, now chemically synthesized in the laboratory so that no frogs are harmed, may be available to the public in the near future. This magainin can kill a wide variety of bacteria (broad-spectrum antibiotic), and it has passed the Phase I human trials. If this compound passes all the remaining tests, it will be used in treating deep infected wounds and ulcers, providing an alternative to traditional therapy.

The curiosity that enabled Zasloff to advance the field of medicine also catalyzed the development of atomic structure. We will see the product of this fundamental human characteristic in the work of Crookes, Bohr, and others throughout this chapter.

Introduction

In Chapter 1 we described chemistry as the study of matter and the changes that matter undergoes. In this chapter we will expand and enhance our understanding of matter. We can deal with visible quantities of matter such as an ounce of silver or a pint of blood. However, we can also describe matter at the level of individual particles that make it up. For instance, one atom of silver is the smallest amount of silver that retains the properties of the bulk material. This is important because the description of matter at the atomic level can be used to explain the behavior of the larger, visible quantities of the same material.

Why does ice float on water? Why don't oil and water mix? Why does blood transport oxygen to our cells, whereas carbon monoxide inhibits this process? Questions such as these are best explained by understanding the behavior of substances at the atomic level.

2.1 Matter and Properties

Properties are characteristics of matter and are classified as either physical or chemical. In this section we will learn the meaning of physical and chemical properties and how they are used to characterize matter.

Matter and Physical Properties

There are three **states of matter:** the **gaseous state,** the **liquid state,** and the **solid state.** A gas is made up of particles that are widely separated. In fact, a gas will

Learning Goal
1

Learning Goal
2

Learning Goal
3

Learning Goal
4

(a)

(b)

(c)

Figure 2.1
The three states of matter exhibited by water: (a) solid, as ice; (b) liquid, as ocean water; (c) gas, as humidity in the air.

expand to fill any container; it has no definite shape or volume. In contrast, particles of a liquid are closer together; a liquid has a definite volume but no definite shape; it takes on the shape of its container. A solid consists of particles that are close together and that often have a regular and predictable pattern of particle arrangement (crystalline). A solid has both fixed volume and fixed shape. Attractive forces, which exist between all particles, are very pronounced in solids and much less so in gases.

Water is the most common example of a substance that can exist in all three states over a reasonable temperature range (Figure 2.1). Conversion of water from one state to another constitutes a *physical change*. A **physical change** produces a recognizable difference in the appearance of a substance without causing any change in its composition or identity. For example, we can warm an ice cube and it will melt, forming liquid water. Clearly its appearance has changed; it has been transformed from the solid to the liquid state. It is, however, still water; its composition and identity remain unchanged. A physical change has occurred. We could in fact demonstrate the constancy of composition and identity by refreezing the liquid water, re-forming the ice cube. This melting and freezing cycle could be repeated over and over. This very process is a hallmark of our global weather changes. The continual interconversion of the three states of water in the environment (snow, rain, and humidity) clearly demonstrates the retention of the identity of water particles or *molecules*.

A **physical property** can be observed or measured without changing the composition or identity of a substance. As we have seen, melting ice is a physical change. We can measure the temperature when melting occurs; this is the *melting point* of water. We can also measure the *boiling point* of water, when liquid water becomes a gas. Both the melting and boiling points of water, and of any other substance, are physical properties.

A practical application of separation of materials based upon their differences in physical properties is shown in Figure 2.2.

Matter and Chemical Properties

We have noted that physical properties can be exhibited, measured, or observed without any change in identity or composition. In contrast, **chemical properties** do result in a change in composition and can be observed only through chemical reactions. A **chemical reaction** is a process of rearranging, replacing, or adding atoms to produce new substances. For example, the process of photosynthesis can be shown as

$$\text{carbon dioxide} + \text{water} \xrightarrow[\text{Chlorophyll}]{\text{Light}} \text{sugar} + \text{oxygen}$$

Learning Goal Learning Goal Learning Goal

Light is the energy needed to make the reaction happen. Chlorophyll is the energy absorber, converting light energy to chemical energy.

Figure 2.2
An example of separation based on differences in physical properties. Magnetic iron is separated from other non-magnetic substances. A large-scale version of this process is important in the recycling industry.

Chapter 8 discusses the role of energy in chemical reactions.

This chemical reaction involves the conversion of carbon dioxide and water (the **reactants**) to a sugar and oxygen (the **products**). The products and reactants are clearly different. We know that carbon dioxide and oxygen are gases at room temperature and water is a liquid at this temperature; the sugar is a solid white powder. A chemical property of carbon dioxide is its ability to form sugar under certain conditions. The process of formation of this sugar is the *chemical change*.

EXAMPLE 2.1

Identifying Properties

Can the process that takes place when an egg is fried be described as a physical or chemical change?

Solution

Examine the characteristics of the egg before and after frying. Clearly, some significant change has occurred. Furthermore the change appears irreversible. More than a simple physical change has taken place. A chemical reaction (actually, several) must be responsible; hence chemical change.

Question 2.1

Classify each of the following as either a chemical property or a physical property:

a. color   d. odor
b. flammability   e. taste
c. hardness

Question 2.2

Classify each of the following as either a chemical change or a physical change:

a. water boiling to become steam
b. butter becoming rancid
c. combustion of wood
d. melting of ice in spring
e. decay of leaves in winter

Intensive and Extensive Properties

It is important to recognize that properties can also be classified according to whether they depend on the size of the sample. Consequently, there is a fundamental difference between properties such as density and specific gravity and properties such as mass and volume.

An **intensive property** is a property of matter that is *independent* of the *quantity* of the substance. Density and specific gravity are intensive properties. For example, the density of one single drop of water is exactly the same as the density of a liter of water.

An **extensive property** *depends* on the *quantity* of a substance. Mass and volume are extensive properties. There is an obvious difference between 1 g of silver and 1 kg of silver; the quantities and, incidentally, the value, differ substantially.

Learning Goal

5

Differentiating between Intensive and Extensive Properties

EXAMPLE 2.2

Is temperature an extensive or intensive property?

Solution

Imagine two glasses each containing 100 g of water, and each at 25°C. Now pour the contents of the two glasses into a larger glass. You would predict that the mass of the water in the larger glass would be 200 g (100 g + 100 g) because mass is an extensive property, dependent on quantity. However, we would expect the temperature of the water to remain the same (not 25°C + 25°C); hence temperature is *an intensive property* . . . independent of quantity.

Classification of Matter

Chemists look for similarities in properties among various types of materials. Recognizing these likenesses simplifies learning the subject and allows us to predict the behavior of new substances on the basis of their relationship to substances already known and characterized.

Learning Goal

6

Many classification systems exist. The most useful system, based on composition, is described in the following paragraphs (see also Figure 2.3).

All matter is either *a pure substance or a mixture*. A **pure substance** is a substance that has only one component. Pure water is a pure substance. It is made up only of particles containing two hydrogen atoms and one oxygen atom, that is, water molecules (H_2O).

There are different types of pure substances. Elements and compounds are both pure substances. An **element** is a pure substance that cannot be changed into a simpler form of matter by any chemical reaction. Hydrogen and oxygen, for example, are elements. Alternatively, a **compound** is a substance resulting from the combination of two or more elements in a definite, reproducible way. The elements hydrogen and oxygen, as noted earlier, may combine to form the compound water, H_2O.

At present, more than 100 elements have been characterized. A complete listing of the elements and their symbols is found on the inside front cover of this textbook.

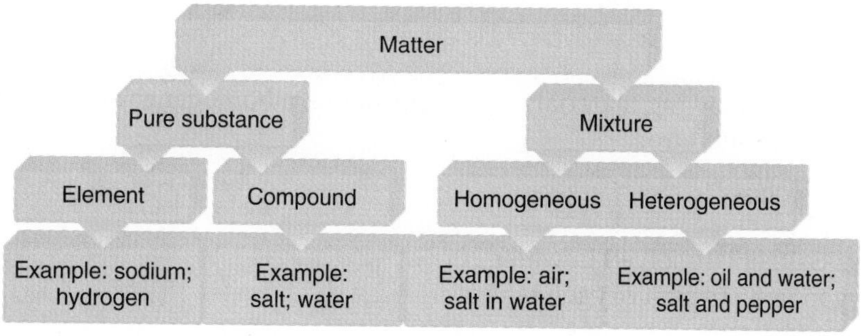

Figure 2.3

Classification of matter. All matter is either a pure substance or a mixture of pure substances. Pure substances are either elements or compounds, and mixtures may be either homogeneous (uniform composition) or heterogeneous (nonuniform composition).

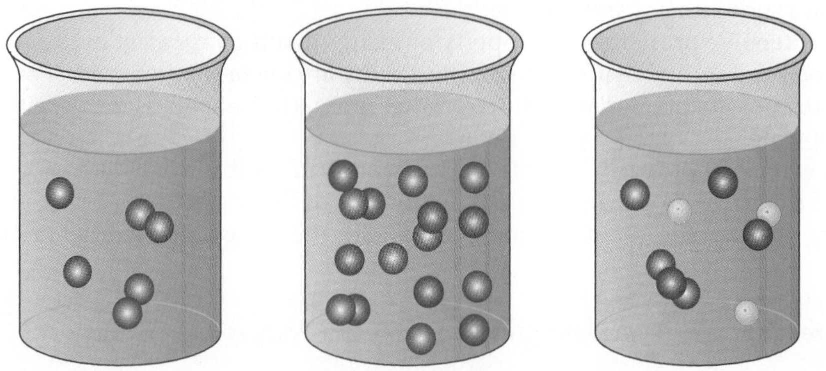

(a) Pure substance   (b) Homogeneous mixture   (c) Heterogeneous mixture

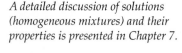

Water   Sugar and water   Salt, water, sand, toxic waste, etc.
                                                          Seawater

Figure 2.4
Schematic representation of some classes of matter. A pure substance (a) consists of a single component. A homogeneous mixture (b) has a uniform distribution of components. A heterogeneous mixture (c) has a nonuniform distribution of components.

A **mixture** is a combination of two or more pure substances in which each substance retains its own identity. Alcohol and water can be combined in a mixture. They coexist as pure substances because they do not undergo a chemical reaction; they exist as thoroughly mixed discrete molecules. This collection of dissimilar particles is the mixture. A mixture has variable composition; there are an infinite number of combinations of quantities of alcohol and water that can be mixed. For example, the mixture may contain a small amount of alcohol and a large amount of water or vice versa. Each is, however, an alcohol–water mixture.

A mixture may be either *homogeneous or heterogeneous* (Figure 2.4). A **homogeneous mixture** has uniform composition. Its particles are well mixed, or thoroughly intermingled. A homogeneous mixture, such as alcohol and water, is described as a *solution*. Air, a mixture of gases, is an example of a gaseous solution. A **heterogeneous mixture** has a nonuniform composition. A mixture of salt and pepper is a good example of a heterogeneous mixture. Concrete is also composed of a heterogeneous mixture of materials (various types and sizes of stone and sand present with cement in a nonuniform mixture).

A detailed discussion of solutions (homogeneous mixtures) and their properties is presented in Chapter 7.

EXAMPLE 2.3

Categorizing Matter

Is seawater a pure substance, a homogeneous mixture, or a heterogeneous mixture?

Solution

Imagine yourself at the beach, filling a container with a sample of water from the ocean. Examine it. You would see a variety of solid particles suspended in the water: sand, green vegetation, perhaps even a small fish! Clearly, it is a mixture, and one in which the particles are not uniformly distributed throughout the water; hence a heterogeneous mixture.

Question **2.3**

Is each of the following materials a pure substance, a homogeneous mixture, or a heterogeneous mixture?

a. ethyl alcohol
b. blood
c. Alka-Seltzer® dissolved in water
d. oxygen in a hospital oxygen tank

Is each of the following materials a pure substance, a homogeneous mixture, or a heterogeneous mixture?

a. air
b. paint
c. perfume
d. carbon monoxide

2.2 Matter and Structure

Chemists and physicists have used the observed properties of matter to develop models of the individual units of matter. These models collectively make up what we now know as the atomic theory of matter.

These models have developed from experimental observations over the past 200 years. Thus theory and experiment reinforce each other. We must gain some insight into atomic structure to appreciate the behavior of the atoms themselves as well as larger aggregates of atoms: compounds.

The structure–properties concept has advanced so far that compounds are designed and synthesized in the laboratory with the hope that they will perform very specific functions, such as curing diseases that have been resistant to other forms of treatment.

As we saw in Section 1.1 models allow us to make ideas more clear and enable us to predict behavior; this is their main value.

Learning Goal

Atomic Structure

The theory of atomic structure has progressed rapidly, from a very primitive level to its present point of sophistication, in a relatively short time. We next briefly summarize our present knowledge of the composition of the atom. Then in Section 2.3 we present an outline of the significant scientific discoveries that spurred development of atomic theory. Before we proceed, let us insert a note of caution. We must not think of the present picture of the atom as final. Scientific inquiry continues, and we should view the present theory as a step in an evolutionary process. *Theories are subject to constant refinement,* as was noted in our discussion of the scientific method.

Section 1.1 discusses the scientific method.

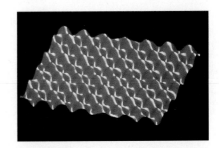

Figure 2.5
Sophisticated techniques, such as scanning tunneling electron microscopy, provide visual evidence for the structure of atoms and molecules. The planar nature of graphite, a commonly used lubricant, is shown here; the peaks are images of carbon atoms.

Electrons, Protons, and Neutrons

The basic structural unit of an element is the **atom,** which is the smallest unit of an element that retains the chemical properties of that element. A tiny sample of the element copper, too small to be seen by the naked eye, is composed of billions of copper atoms arranged in some orderly fashion. Each atom is incredibly small. Only recently have we been able to "see" atoms using modern instruments such as the scanning tunneling microscope (Figure 2.5).

We know from experience that certain kinds of atoms can "split" into smaller particles and release large amounts of energy; this process is *radioactive decay.* We also know that the atom is composed of three primary particles: the *electron,* the *proton,* and the *neutron.* Although other subatomic fragments with unusual names (neutrinos, gluons, quarks, and so forth) have also been discovered, we shall concern ourselves only with the primary particles: the protons, neutrons, and electrons.

We can consider the atom to be composed of two distinct regions:

1. The **nucleus** is a small, dense, positively charged region in the center of the atom. The nucleus is composed of positively charged **protons** and uncharged **neutrons.**

Radioactivity and radioactive decay are discussed in Chapter 10.

Learning Goal

Table 2.1	Selected Properties of the Three Basic Subatomic Particles		
Name	Charge	Mass (amu)	Mass (grams)
Electron (e)	-1	5.4×10^{-4}	9.1095×10^{-28}
Proton (p)	$+1$	1.00	1.6725×10^{-24}
Neutron (n)	0	1.00	1.6750×10^{-24}

2. Surrounding the nucleus is a diffuse region of negative charge populated by **electrons,** the source of the negative charge. Electrons are very low in mass in contrast to the protons and neutrons.

The properties of these particles are summarized in Table 2.1.

Atoms of various types differ in their number of protons, neutrons, and electrons. The number of protons determines the identity of the atom. As such, the number of protons is *characteristic* of the element. When the number of protons is equal to the number of electrons, the atom is neutral because the charges are balanced and effectively cancel one another.

We may represent an element symbolically as follows:

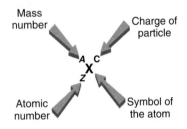

Learning Goal

The **atomic number** (Z) is equal to the number of protons in the atom, and the **mass number** (*A*) is equal to the *sum* of the number of protons and neutrons (the mass of the electrons is so small as to be insignificant in comparison to that of the nucleus).

If

number of protons + number of neutrons = mass number

then, if the number of protons is subtracted from each side,

number of neutrons = mass number − number of protons

or, because the number of protons equals the atomic number,

number of neutrons = mass number − atomic number

For an atom, in which positive and negative charges cancel, the number of protons and electrons must be equal and identical to the atomic number.

EXAMPLE 2.4

Calculating the Composition of an Atom

Calculate the numbers of protons, neutrons, and electrons in an atom of fluorine.

Solution

The atomic symbol for the fluorine atom is

$$^{19}_{9}\text{F}$$

The mass number 19 tells us that the total number of protons + neutrons is 19. The atomic number, 9, represents the number of protons. The difference, 19 − 9, or 10, is the number of neutrons. The number of electrons must be the same as the number of protons, hence 9, for a neutral fluorine atom.

Calculate the number of protons, neutrons, and electrons in each of the following atoms:

Question 2.5

a. $^{32}_{16}S$

b. $^{23}_{11}Na$

Calculate the number of protons, neutrons, and electrons in each of the following atoms:

Question 2.6

a. $^{1}_{1}H$

b. $^{244}_{94}Pu$

Isotopes

Isotopes are atoms of the same element having different masses *because they contain different numbers of neutrons*. In other words, isotopes have different mass numbers. For example, all of the following are isotopes of hydrogen:

Learning Goal

$$^{1}_{1}H \qquad\qquad ^{2}_{1}H \qquad\qquad ^{3}_{1}H$$

Hydrogen   Deuterium   Tritium

(Hydrogen-1)   (Hydrogen-2)   (Hydrogen-3)

Isotopes are often written with the name of the element followed by the mass number. For example, the isotopes $^{12}_{6}C$ and $^{14}_{6}C$ may be written as carbon-12 (or C-12) and carbon-14 (or C-14), respectively.

Certain isotopes (radioactive isotopes) of elements emit particles and energy that can be used to trace the behavior of biochemical systems. These isotopes otherwise behave identically to any other isotope of the same element. Their chemical behavior is identical; it is their nuclear behavior that is unique. As a result, a radioactive isotope can be substituted for the "nonradioactive" isotope, and its biochemical activity can be followed by monitoring the particles or energy emitted by the isotope as it passes through the body.

A detailed discussion of the use of radioactive isotopes in the diagnosis and treatment of disease is found in Chapter 10.

The existence of isotopes explains why the average masses, measured in atomic mass units (amu), of the various elements are not whole numbers. This is contrary to what we would expect from proton and neutron masses, which are whole numbers to three significant figures.

See Section 1.6 for the definition of the atomic mass unit.

Consider, for example, the mass of one chlorine atom, containing 17 protons (atomic number) and 18 neutrons:

$$17 \ \cancel{\text{protons}} \times \frac{1.00 \ \text{amu}}{\cancel{\text{proton}}} = 17.00 \ \text{amu}$$

$$18 \ \cancel{\text{neutrons}} \times \frac{1.00 \ \text{amu}}{\cancel{\text{neutron}}} = 18.00 \ \text{amu}$$

$$17.00 \ \text{amu} + 18.00 \ \text{amu} = 35.00 \ \text{amu} \ (\text{mass of chlorine atom})$$

Inspection of the periodic table reveals that the mass number of chlorine is actually 35.45 amu, *not* 35.00 amu. The existence of isotopes accounts for this difference. A natural sample of chlorine is composed principally of two isotopes, chlorine-35 and chlorine-37, in approximately a 3:1 ratio, and the tabulated mass is the *weighted average* of the two isotopes. In our calculation the chlorine atom referred to was the isotope that has a mass number of 35 amu.

The weighted average is not a true average but is corrected by the relative amounts (the weighting factor) of each isotope present in nature.

The weighted average of the masses of all of the isotopes of an element is the **atomic mass** and should be distinguished from the mass number, which is the sum of the number of protons and neutrons in a single isotope of the element.

Example 2.5 demonstrates the calculation of the atomic mass of chlorine.

EXAMPLE 2.5

Determining Atomic Mass

Calculate the atomic mass of naturally occurring chlorine if 75.77% of chlorine atoms are ${}^{35}_{17}Cl$ (chlorine-35) and 24.23% of chlorine atoms are ${}^{37}_{17}Cl$ (chlorine-37).

Solution

Step 1. Convert each percentage to a decimal fraction.

$$75.77\% \text{ chlorine-35} \times \frac{1}{100\%} = 0.7577 \text{ chlorine-35}$$

$$24.23\% \text{ chlorine-37} \times \frac{1}{100\%} = 0.2423 \text{ chlorine-37}$$

Step 2. Multiply the decimal fraction of each isotope by the mass of that isotope to determine the isotopic contribution to the average atomic mass.

$$\begin{array}{l} \text{contribution to} \\ \text{atomic mass} \\ \text{by chlorine-35} \end{array} = \begin{array}{l} \text{fraction of all} \\ \text{Cl atoms that} \\ \text{are chlorine-35} \end{array} \times \begin{array}{l} \text{mass of a} \\ \text{chlorine-35} \\ \text{atom} \end{array}$$

$$= 0.7577 \times 35.00 \text{ amu}$$

$$= 26.52 \text{ amu}$$

$$\begin{array}{l} \text{contribution to} \\ \text{atomic mass} \\ \text{by chlorine-37} \end{array} = \begin{array}{l} \text{fraction of all} \\ \text{Cl atoms that} \\ \text{are chlorine-37} \end{array} \times \begin{array}{l} \text{mass of a} \\ \text{chlorine-37} \\ \text{atom} \end{array}$$

$$= 0.2423 \times 37.00 \text{ amu}$$

$$= 8.965 \text{ amu}$$

Step 3. The weighted average is:

$$\begin{array}{l} \text{atomic mass} \\ \text{of naturally} \\ \text{occurring Cl} \end{array} = \begin{array}{l} \text{contribution} \\ \text{of} \\ \text{chlorine-35} \end{array} + \begin{array}{l} \text{contribution} \\ \text{of} \\ \text{chlorine-37} \end{array}$$

$$= 26.52 \text{ amu} + 8.965 \text{ amu}$$

$$= 35.49 \text{ amu}$$

which is very close to the tabulated value of 35.45 amu. An even more exact value would be obtained by using a more exact value of the mass of the proton and neutron (experimentally known to a greater number of significant figures).

Whenever you do calculations such as those in Example 2.5, before even beginning the calculation you should look for an approximation of the value sought. Then do the calculation and see whether you obtain a reasonable number (similar to your anticipated value). In the preceding problem, if the two isotopes have masses of 35 and 37, the atomic mass must lie somewhere between the two extremes. Furthermore, because the majority of a naturally occurring sample is chlorine-35 (about 75%), the value should be closer to 35 than to 37. An analysis of the results often avoids problems stemming from untimely events such as pushing the wrong button on a calculator.

> A hint for numerical problem solving: Estimate (at least to an order of magnitude) your answer before beginning the calculation using your calculator.

EXAMPLE 2.6

Determining Atomic Mass

Calculate the atomic mass of naturally occurring carbon if 98.90% of carbon atoms are $^{12}_{6}C$ (carbon-12) with a mass of 12.00 amu and 1.11% are $^{13}_{6}C$ (carbon-13) with a mass of 13.00 amu. (Note that a small amount of $^{14}_{6}C$ is also present but is small enough to ignore in a calculation involving three or four significant figures.)

Solution

Step 1. Convert each percentage to a decimal fraction.

$$98.90\% \text{ carbon-12} \times \frac{1}{100\%} = 0.9890 \text{ carbon-12}$$

$$1.11\% \text{ carbon-13} \times \frac{1}{100\%} = 0.0111 \text{ carbon-13}$$

Step 2.

$$
\begin{aligned}
&\text{contribution to} && \text{(fraction of all} && \text{(mass of a} \\
&\text{atomic mass} &=& \text{ C atoms that} && \times \text{ carbon-12} \\
&\text{by carbon-12} && \text{are carbon-12)} && \text{atom)} \\
&&=& \ 0.9890 && \times \ 12.00 \text{ amu} \\
&&=& \ 11.87 \text{ amu}
\end{aligned}
$$

$$
\begin{aligned}
&\text{contribution to} && \text{(fraction of all} && \text{(mass of a} \\
&\text{atomic mass} &=& \text{ C atoms that} && \times \text{ carbon-13} \\
&\text{by carbon-13} && \text{are carbon-13)} && \text{atom)} \\
&&=& \ 0.0111 && \times \ 13.00 \text{ amu} \\
&&=& \ 0.144 \text{ amu}
\end{aligned}
$$

Step 3. The weighted average is:

$$
\begin{aligned}
&\text{atomic mass} && \text{(contribution} && \text{(contribution} \\
&\text{of naturally} &=& \text{ of} && + \text{ of} \\
&\text{occurring carbon} && \text{carbon-12)} && \text{carbon-13)} \\
&&=& \ 11.87 \text{ amu} && + \ 0.144 \text{ amu} \\
&&=& \ 12.01 \text{ amu}
\end{aligned}
$$

Because most of the carbon is carbon-12, with very little carbon-13 present, the atomic mass should be very close to that of carbon-12. Approximations, before performing the calculation, provide another check on the accuracy of the final result.

Question 2.7

The element neon has three naturally occurring isotopes. One of these has a mass of 19.99 amu and a natural abundance of 90.48%. A second isotope has a mass of 20.99 amu and a natural abundance of 0.27%. A third has a mass of 21.99 amu and a natural abundance of 9.25%. Calculate the atomic mass of neon.

Question 2.8

The element nitrogen has two naturally occurring isotopes. One of these has a mass of 14.003 amu and a natural abundance of 99.63%; the other isotope has a mass of 15.000 amu and a natural abundance of 0.37%. Calculate the atomic mass of nitrogen.

Ions

Learning Goal

Ions are electrically charged particles that result from a gain of one or more electrons by the parent atom (forming negative ions, or **anions**) or a loss of one or more electrons from the parent atom (forming positive ions, or **cations**).

Formation of an anion may occur as follows:

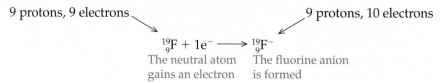

9 protons, 9 electrons   9 protons, 10 electrons

$$^{19}_{9}\text{F} + 1e^- \longrightarrow {}^{19}_{9}\text{F}^-$$

The neutral atom   The fluorine anion
gains an electron   is formed

Alternatively, formation of a cation of sodium may proceed as follows:

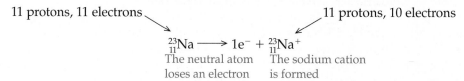

11 protons, 11 electrons   11 protons, 10 electrons

$$^{23}_{11}\text{Na} \longrightarrow 1e^- + {}^{23}_{11}\text{Na}^+$$

The neutral atom   The sodium cation
loses an electron   is formed

Note that the electrons gained are written to the left of the reaction arrow (they are reactants), whereas the electrons lost are written as products to the right of the reaction arrow. For simplification the atomic and mass numbers are often omitted, because they do not change during ion formation. For example, the sodium cation would be written as Na^+ and the anion of fluorine as F^-.

2.3 Development of the Atomic Theory

Learning Goal

With this overview of our current understanding of the structure of the atom, we now look at a few of the most important scientific discoveries that led to modern atomic theory.

Dalton's Theory

The first experimentally based theory of atomic structure was proposed by John Dalton, an English schoolteacher, in the early 1800s. Dalton proposed the following description of atoms:

1. All matter consists of tiny particles called atoms.
2. An atom cannot be created, divided, destroyed, or converted to any other type of atom.
3. Atoms of a particular element have identical properties.
4. Atoms of different elements have different properties.
5. Atoms of different elements combine in simple whole-number ratios to produce compounds (stable aggregates of atoms).
6. Chemical change involves joining, separating, or rearranging atoms.

Although Dalton's theory was founded on meager and primitive experimental information, we regard much of it as correct today. Postulates 1, 4, 5, and 6 are

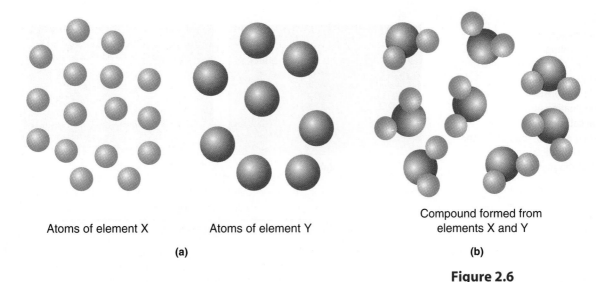

Atoms of element X   Atoms of element Y

Compound formed from elements X and Y

(a)

(b)

Figure 2.6

An illustration of John Dalton's atomic theory. (a) Atoms of the same element are identical, but different from atoms of any other element. (b) Atoms combine in whole-number ratios to form compounds.

currently regarded as true. The discovery of the processes of nuclear fusion, fission ("splitting" of atoms), and radioactivity has disproved the postulate that atoms cannot be created or destroyed. Postulate 3, that all the atoms of a particular element are identical, was disproved by the discovery of isotopes.

Fusion, fission, radioactivity, and isotopes are discussed in some detail in Chapter 10. Figure 2.6 uses a simple model to illustrate Dalton's theory.

Subatomic Particles: Electrons, Protons, and Neutrons

The next major discoveries occurred almost a century later (1879–1897). Although Dalton pictured atoms as indivisible, various experiments, particularly those of William Crookes and Eugene Goldstein, indicated that the atom is composed of charged (+ and −) particles.

Crookes connected two metal electrodes (metal discs connected to a source of electricity) at opposite ends of a sealed glass vacuum tube. When the electricity was turned on, rays of light were observed to travel between the two electrodes. They were called **cathode rays,** because they traveled from the **cathode** (the negative electrode) to the **anode** (the positive electrode). A diagram of the apparatus is shown in Figure 2.7.

Later experiments by J. J. Thomson, an English scientist, demonstrated the electrical and magnetic properties of cathode rays (Figure 2.8). The rays were deflected toward the positive electrode of an external electric field. Because opposite charges attract, this indicates the negative character of the rays. Similar experiments with an external magnetic field showed a deflection as well; hence these cathode rays also have magnetic properties.

A change in the material used to fabricate the electrode discs brought about no change in the experimental results. This suggested that the ability to produce cathode rays is a characteristic of all materials.

In 1897, Thomson announced that cathode rays are streams of negative particles of energy. These particles are *electrons*. Similar experiments, conducted by Goldstein, led to the discovery of particles that are equal in charge to the electron but opposite in sign. These particles, much heavier than electrons (actually 1837 times as heavy), are called *protons*.

As we have seen, the third fundamental atomic particle is the *neutron*. It has a mass virtually identical (it is less than 1% heavier) to that of the proton and has

Crookes's cathode ray tube was the forerunner of the computer screen (often called CRT) and the television.

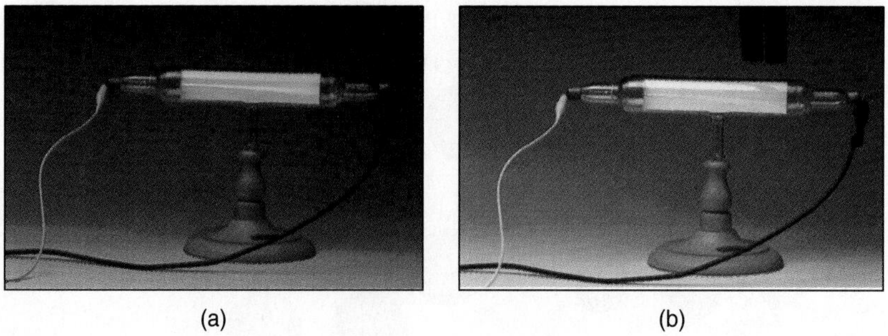

Figure 2.7

Illustration of Crookes's experiment. (a) When a high voltage was applied between two electrodes in a sealed, evacuated tube, a cathode ray (an electron track) was observed between the electrodes, originating at the negative electrode, the cathode; (b) When a magnetic field is applied, the path of the cathode ray shifts, indicating that the cathode ray has magnetic properties.

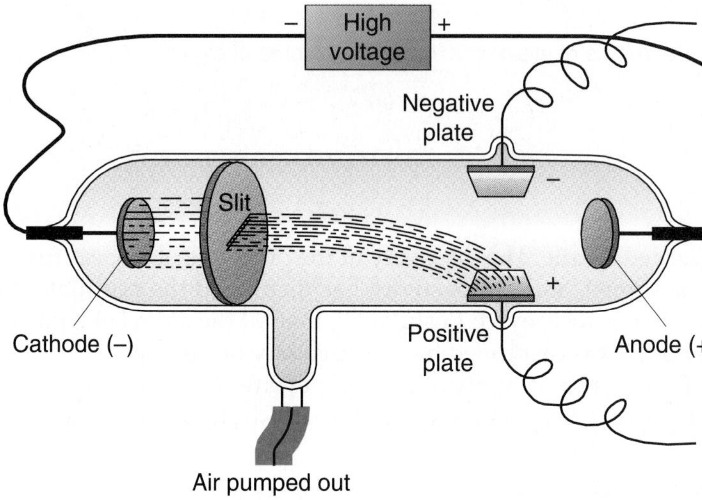

(a)   (b)

Figure 2.8

Illustration of an experiment demonstrating the charge of cathode rays. The application of an external electric field causes the electron beam to deflect toward a positive charge, implying that the cathode ray is negative.

High voltage − +

Negative plate

Slit

Cathode (−)

Positive plate

Anode (+)

Air pumped out

zero charge. The neutron was first postulated in the early 1920s, but it was not until 1932 that James Chadwick demonstrated its existence with a series of experiments involving the use of small particle bombardment of nuclei.

The Nucleus

Learning Goal

In the early 1900s it was believed that protons and electrons were uniformly distributed throughout the atom. However, an experiment by Hans Geiger led Ernest Rutherford (in 1911) to propose that the majority of the mass and positive charge of the atom was actually located in a small, dense region, the *nucleus*, with small, negatively charged electrons occupying a much larger volume outside of the nucleus.

To understand how Rutherford's theory resulted from the experimental observations of Geiger, let us examine this experiment in greater detail. Rutherford and others had earlier demonstrated that some atoms spontaneously "decay" to produce three types of radiation: alpha (α), beta (β), and gamma (γ) radiation. This process is known as **natural radioactivity** (Figure 2.9). Geiger used radioactive materials, such as *radium,* as projectile sources, "firing" alpha particles at a metal foil target. He then observed the interaction of the metal and alpha particles with a detection screen (Figure 2.10) and found that:

a. Most alpha particles pass through the foil without being deflected.
b. A small fraction of the particles were deflected, some even *directly back to the source.*

Rutherford interpreted this to mean that most of the atom is empty space, because most alpha particles were not deflected. Further, most of the mass and positive charge must be located in a small, dense region; collision of the heavy and posi-

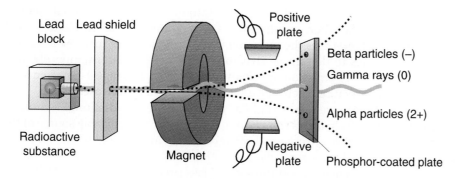

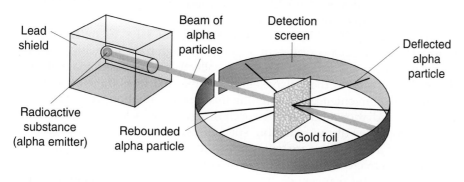

Figure 2.9

Types and characteristics of radioactive emissions. The direction taken by the radioactive emissions indicates the presence of three types of emissions: positive, negative, and neutral components.

Figure 2.10

The alpha particle scattering experiment.

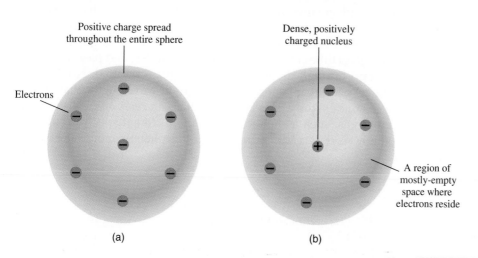

Figure 2.11

(a) A model of the atom (credited to Thomson) prior to the work of Geiger and Rutherford. This was termed the "plum pudding" model. (b) A model of the atom supported by the alpha-particle scattering experiments of Geiger and Rutherford.

tively charged alpha particle with this small dense and positive region (the nucleus) caused the great deflections. Rutherford summarized his astonishment at observing the deflected particles: "It was almost as incredible as if you fired a 15-inch shell at a piece of tissue and it came back and hit you."

The significance of Rutherford's contribution cannot be overstated. It caused a revolutionary change in the way that scientists pictured the atom (Figure 2.11). His discovery of the nucleus is fundamental to our understanding of chemistry. Chapter 10 will provide much more information on a special branch of chemistry: nuclear chemistry.

Light and Atomic Structure

The Rutherford atom leaves us with a picture of a tiny, dense, positively charged nucleus containing protons and surrounded by electrons. The electron arrangement, or configuration, is not clearly detailed. More information is needed regarding the relationship of the electrons to each other and to the nucleus. In dealing

Learning Goal

13

Figure 2.12
The visible spectrum of light. Light passes through a prism, producing a continuous spectrum.

with dimensions on the order of 10^{-8} cm (the atomic level), conventional methods for measurement of location and distance of separation become impossible. An alternative approach involves the measurement of *energy* rather than the *position* of the atomic particles to determine structure. For example, information obtained from the absorption or emission of *light* by atoms (energy changes) can yield valuable insight into structure. Such studies are referred to as **spectroscopy.**

In a general sense we refer to light as **electromagnetic radiation.** Electromagnetic radiation travels in *waves* from a source. The most recognizable source of this radiation is the sun. We are aware of a rainbow, in which visible white light from the sun is broken up into several characteristic bands of different colors. Similarly, visible white light, when passed through a glass prism, is separated into its various component colors (Figure 2.12). These various colors are simply light (electromagnetic radiation) of differing *wavelengths*.

All electromagnetic radiation travels at a speed of 3.0×10^8 m/sec, the **speed of light.** However, each wavelength of light, although traveling with identical velocity, has its own characteristic energy. A collection of all electromagnetic radiation, including each of these wavelengths, is referred to as the **electromagnetic spectrum.** For convenience in discussing this type of radiation we subdivide electromagnetic radiation into various spectral regions, which are characterized by physical properties of the radiation, such as its *wavelength* or its *energy* (Figure 2.13). Some of these regions are quite familiar to us from our everyday experiences; the visible and microwave regions are two common examples.

Light of shorter wavelength has higher energy; in fact, the magnitude of the energy and wavelength are inversely proportional. The wavelength of a particular type of light can be measured, and from this the energy may be calculated.

If we take a sample of some element, such as hydrogen, in the gas phase, place it in an evacuated glass tube containing a pair of electrodes, and pass an electrical charge (cathode ray) through the hydrogen gas, light is emitted. Not all wavelengths (or energies) of light are emitted—only certain wavelengths that are characteristic of the gas under study. This is referred to as an *emission spectrum* (Figure 2.14). If a different gas, such as helium, is used, a different spectrum (different wavelengths of light) is observed. The reason for this phenomenon was explained by Niels Bohr.

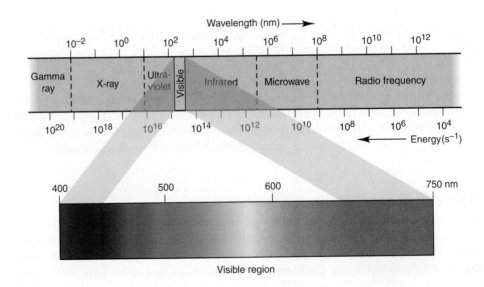

Figure 2.13
The electromagnetic spectrum. Note that the visible spectrum is only a small part of the total electromagnetic spectrum.

The Bohr Atom

Niels Bohr hypothesized that surrounding each atomic nucleus were certain fixed **energy levels** that could be occupied by electrons. He also believed that each level was defined by a circular **orbit** around the nucleus, located at a specific distance from the nucleus. The concept of certain fixed energy levels is referred to as the **quantization** of energy. The implication is that only these **quantum levels,** or orbits, are allowed locations for electrons. If an atom *absorbs* energy, an electron undergoes **promotion** from an orbit closer to the nucleus (lower energy) to one farther from the nucleus (higher energy), creating an **excited state.** Similarly, the release of energy by an atom, or **relaxation,** results from an electron falling into an orbit closer to the nucleus (lower energy level).

Promotion and relaxation processes are referred to as **electronic transitions.** The amount of energy absorbed in jumping from one energy level to a higher energy level is a precise quantity (hence, quantum), and that energy corresponds exactly to the energy differences between the orbits involved. Electron promotion resulting from absorption of energy results in an *excited state atom;* the process of relaxation allows the atom to return to the *ground state* (Figure 2.15) with the simultaneous release of light energy. The **ground state** is the lowest possible

Learning Goal

14

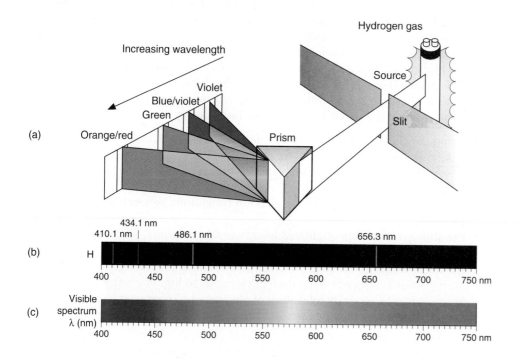

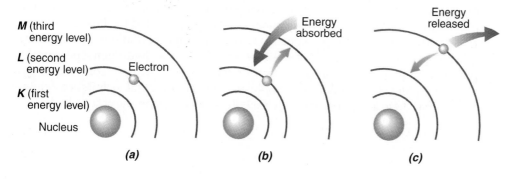

Figure 2.14

(a) The emission spectrum of hydrogen. Certain wavelengths of light, characteristic of the atom, are emitted upon electrical excitation. (b) The line spectrum of hydrogen is compared with the spectrum of visible light, (c).

Figure 2.15

(a) The Bohr atom. (b) Excitation involves promotion of an electron to a higher energy level by absorption of energy. (c) Relaxation is the reverse process, whereby an atom returns to the ground state as the electron moves to a lower energy level and energy is released. [Note that the orbits are not drawn to scale.]

Table 2.2	Electronic Transitions Responsible for the Hydrogen Spectrum			
Line Color	Wavelength Emitted (nm)	Electronic Transition		
		$n =$	to	$n =$
Red	656.4	3		2
Green	486.3	4		2
Blue	434.2	5		2
Violet	410.3	6		2

energy state. This emission process, such as the release of energy after excitation of hydrogen atoms by an electric arc, produces the series of emission lines (emission spectrum). Measurement of the wavelengths of these lines enables the calculation of energy levels in the atom. These energy levels represent the location of the atom's electrons.

We may picture the Bohr atom as a series of concentric orbits surrounding the nucleus. The orbits are identified by two different systems, one using numbers ($n = 1, 2, 3, \ldots$, etc.) and the other using letters (K, L, M, $\ldots$, etc.). The number n is referred to as a **quantum number.** The quantum number $n = 1$ corresponds to a K shell, $n = 2$ is L, and so forth.

The hydrogen spectrum consists of four lines in the visible region of the spectrum. Electronic transitions, calculated from the Bohr theory, account for each of these lines. Table 2.2 gives a summary of the hydrogen spectrum.

A summary of the major features of the Bohr theory is as follows:

- Atoms can absorb and emit energy via *promotion* of electrons to higher energy levels and *relaxation* to lower levels.
- Energy that is emitted upon relaxation is observed as a single wavelength of light.
- These *spectral lines* are a result of electron transitions between *allowed levels* in the atom.
- The allowed levels are quantized energy levels, or orbits.
- Electrons are found only in these energy levels.
- The highest-energy orbits are located farthest from the nucleus.
- Atoms absorb energy by excitation of electrons to higher energy levels.
- Atoms release energy by relaxation of electrons to lower energy levels.
- Energy differences may be calculated from the wavelengths of light emitted.

2.4 Modern Atomic Theory

Learning Goal

The Bohr model was an immensely important contribution to the understanding of atomic structure. The idea that electrons exist in specific energy states and that transitions between states involve quanta of energy provided the linkage between atomic structure and atomic spectra. However, some limitations of this model quickly became apparent. Although it explained the hydrogen spectrum, it provided only a crude approximation of the spectra for more complex atoms. Subsequent development of more sophisticated experimental techniques demonstrated that there are problems with the Bohr theory even in the case of hydrogen.

Atomic Spectra and the Fourth of July

At one time or another we have all marveled at the bright, multicolored display of light and sound that is a fireworks display. These sights and sounds are produced by a chemical reaction that generates the energy necessary to excite a variety of elements to their higher-energy electronic states. Light emission results from relaxation of the excited atoms to the ground state. Each atom releases light of specific wavelengths. The visible wavelengths are seen as colored light.

Fireworks need a chemical reaction to produce energy. We know from common experience that oxygen and a fuel will release energy. The fuel in most fireworks preparations is sulfur or aluminum. Each reacts slowly with oxygen; a more potent solid-state source of oxygen is potassium perchlorate ($KClO_4$). The potassium perchlorate reacts with the fuel (an oxidation–reduction reaction), producing a bright white flash of light. The heat produced excites the various elements packaged with the fuel and oxidant.

Sodium salts, such as sodium chloride, furnish sodium ions, which, when excited, produce yellow light (a wavelength of 589 nm). Red colors arise from salts of strontium, which emit several shades of red corresponding to wavelengths in the 600- to 700-nm region of the visible spectrum. Copper salts produce blue radiation, because copper emits in the 400- to 500-nm spectral region.

A fireworks display is a dramatic illustration of light emission by excited atoms.

The beauty of fireworks is a direct result of the skill of the manufacturer. Selection of the proper oxidant, fuel, and color-producing elements is critical to the production of a spectacular display. Packaging these chemicals in proper quantities so that they can be stored and used safely is an equally important consideration.

A French physicist, Louis deBroglie, noted that in certain situations, energy possessed particlelike properties. He hypothesized that the reverse could be true as well: Electrons could, at times, behave as waves rather than particles. This is known today as deBroglie's *wave–particle duality.*

Werner Heisenberg, a German physicist, building on deBroglie's hypothesis, argued that it would be impossible to exactly specify the location of a particle (such as the electron) because of its wavelike character (remember, a wave travels indefinitely in space). This hypothesis in turn led to the *Heisenberg uncertainty principle* (1927), which states that it is impossible to specify both the location and the momentum (momentum is the product of mass and velocity) of an electron in an atom.

The work of deBroglie and Heisenberg represents a departure from the Bohr theory and the development of modern atomic theory. Although Bohr's concept of principal energy levels is still valid, restriction of electrons to fixed orbits is too rigorous in light of Heisenberg's principle. All current evidence shows that electrons do *not,* in fact, orbit the nucleus. We now speak of the *probability* of finding an electron in a *region* of space within the principal energy level, referred to as an **atomic orbital.** The rapid movement of the electron spreads the charge into a *cloud* of

An Environmental Perspective

Electromagnetic Radiation and Its Effects on Our Everyday Lives

From the preceding discussion of the interaction of electromagnetic radiation with matter—spectroscopy—you might be left with the impression that the utility of such radiation is limited to theoretical studies of atomic structure. Although this is a useful application that has enabled us to learn a great deal about the structure and properties of matter, it is by no means the only application. Useful, everyday applications of the theories of light energy and transmission are all around us. Let's look at just a few examples.

Transmission of sound and pictures is conducted at radio frequencies or radio wavelengths. We are immersed in radio waves from the day we are born. A radio or television is our "detector" of these waves. Radio waves are believed to cause no physical harm because of their very low energy, although some concern for people who live very close to transmission towers has resulted from recent research.

X-rays are electromagnetic radiation, and they travel at the speed of light just like radio waves. However, because of their higher energy, they can pass through the human body and leave an image of the body's interior on a photographic film. X-ray photographs are invaluable for medical diagnosis. However, caution is advised in exposing oneself to X-rays, because the high energy can remove electrons from biological molecules, causing subtle and potentially harmful changes in their chemistry.

The sunlight that passes through our atmosphere provides the basis for a potentially useful technology for providing heat and electricity: *solar energy*. Light is captured by absorbers, referred to as *solar collectors*, which convert the light energy into heat energy. This heat can be transferred to water circulating beneath the collectors to provide heat and hot water for homes or industry. Wafers of a silicon-based material can convert light energy to electrical energy; many believe that if the efficiency of these processes can be improved, such approaches may provide at least a partial solution to the problems of rising energy costs and pollution associated with our fossil fuel–based energy economy.

The intensity of infrared radiation from a solid or liquid is an indicator of relative temperature. This has been used to advantage in the design of infrared cameras, which can obtain images without the benefit of the visible light that is necessary for conventional cameras. The infrared photograph shows the coastline surrounding the city of San Francisco.

charge. This cloud is more dense in certain regions, the **electron density** being proportional to the probability of finding the electron at any point in time. Insofar as these atomic orbitals are part of the principal energy levels, they are referred to as sublevels. In Chapter 3 we will see that the orbital model of the atom can be used to predict how atoms can bond together to form compounds. Furthermore, electron arrangement in orbitals enables us to predict various chemical and physical properties of these compounds.

Microwave radiation for cooking, *infrared* lamps for heating and remote sensing, *ultraviolet* lamps used to kill microorganisms on environmental surfaces, *gamma radiation* from nuclear waste, the *visible* light from the lamp you are using to read this chapter—all are forms of the same type of energy that, for better or worse, plays such a large part in our twentieth century technological society.

Electromagnetic radiation and spectroscopy also play a vital role in the field of diagnostic medicine. They are routinely used as diagnostic and therapeutic tools in the detection and treatment of disease.

The radiation therapy used in the treatment of many types of cancer has been responsible for saving many lives and extending the span of many others. When radiation is used as a treatment, it destroys cancer cells. This topic will be discussed in detail in Chapter 10.

As a diagnostic tool, spectroscopy has the benefit of providing data quickly and reliably; it can also provide information

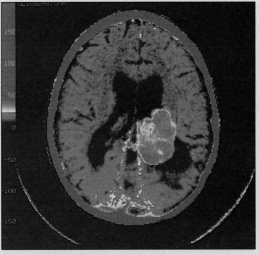

An image of a tumor detected by a CT scan.

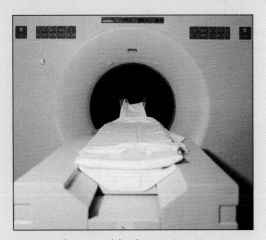

The CT scanner is a device used for diagnostic purposes.

that might not be available through any other means. Additionally, spectroscopic procedures are often nonsurgical, outpatient procedures. Such procedures are safer, can be more routinely performed, and are more acceptable to the general public than surgical procedures. The potential cost savings because of the elimination of many unnecessary surgical procedures is an added benefit.

The most commonly practiced technique uses the CT scanner, an acronym for *computer-accentuated tomography.* In this technique, X-rays are directed at the tissue of interest. As the X-rays pass through the tissue, detectors surrounding the tissue gather the signal, compare it to the original X-ray beam, and, using the computer, produce a three-dimensional image of the tissue.

What was deBroglie's new way of considering matter?

Question **2.9**

Why is it not possible to know both the exact energy and the location of an electron?

Question **2.10**

Summary

2.1   Matter and Properties

Properties (characteristics) of matter may be classified as either physical or chemical. Physical properties can be observed without changing the chemical composition of the sample. *Chemical properties* result in a change in composition and can be observed only through *chemical reactions*. *Intensive properties* are independent of the quantity of the substance. *Extensive properties* depend on the quantity of a substance.

Three states of matter exist (solid, liquid, and gas); these states of matter are distinguishable by differences in physical properties.

All matter is classified as either a *pure substance* or a *mixture*. A pure substance is a substance that has only one component. A mixture is a combination of two or more pure substances in which the combined substances retain their identity.

A *homogeneous mixture* has uniform composition. Its particles are well mixed. A *heterogeneous mixture* has a nonuniform composition.

An *element* is a pure substance that cannot be converted into a simpler form of matter by any chemical reaction. A *compound* is a substance produced from the combination of two or more elements in a definite, reproducible fashion.

2.2   Matter and Structure

The basic structural unit of an element is the *atom*, which is the smallest unit of an element that retains the chemical properties of that element. The atom is composed of three primary particles: the electron, the proton, and the neutron.

The atom has two distinct regions. The *nucleus* is a small, dense, positively charged region in the center of the atom composed of positively charged *protons* and uncharged *neutrons*. Surrounding the nucleus is a diffuse region of negative charge occupied by *electrons*, the source of the negative charge. Electrons are very low in mass in comparison to protons and neutrons.

The *atomic number* (Z) is equal to the number of protons in the atom. The *mass number* (A) is equal to the sum of the protons and neutrons (the mass of the electrons is insignificant).

Isotopes are atoms of the same element that have different masses because they have different numbers of neutrons (different mass numbers). Isotopes have chemical behavior identical to that of any other isotope of the same element.

Ions are electrically charged particles that result from a gain or loss of one or more electrons by the parent atom. *Anions*, negative ions, are formed by a gain of one or more electrons by the parent atom. *Cations*, positive ions, are formed by a loss of one or more electrons from the parent atom.

2.3   Development of the Atomic Theory

The first experimentally based theory of atomic structure was proposed by John Dalton. Although Dalton pictured atoms as indivisible, the experiments of William Crookes, Eugene Goldstein, and J. J. Thomson indicated that the atom is composed of charged particles: protons and electrons. The third fundamental atomic particle is the neutron. An experiment conducted by Hans Geiger led Ernest Rutherford to propose that the majority of the mass and positive charge of the atom is located in a small, dense region, the *nucleus*, with small, negatively charged electrons occupying a much larger, diffuse space outside of the nucleus.

Niels Bohr proposed an atomic model that described the atom as a nucleus surrounded by fixed *energy levels* (or *quantum levels*) that can be occupied by electrons. He believed that each level was defined by a circular *orbit* located at a specific distance from the nucleus.

Promotion and relaxation processes are referred to as *electronic transitions*. Electron promotion resulting from absorption of energy results in an *excited state* atom; the process of relaxation allows the atom to return to the *ground state*.

2.4   Modern Atomic Theory

The modern view of the atom describes the probability of finding an electron in a region of space within the principal energy level, referred to as an *atomic orbital*. The rapid movement of the electrons spreads them into a cloud of charge. This cloud is more dense in certain regions, the density being proportional to the probability of finding the electron at any point in time. The orbital is strikingly different from Bohr's orbit. The electron does not orbit the nucleus; rather, its behavior is best described as that of a wave.

Key Terms

anion (2.2)	atomic orbital (2.4)
anode (2.3)	cathode (2.3)
atom (2.2)	cathode rays (2.3)
atomic mass (2.2)	cation (2.2)
atomic number (2.2)	chemical property (2.1)

chemical reaction (2.1)
compound (2.1)
electromagnetic radiation (2.3)
electromagnetic spectrum (2.3)
electron (2.2)
electron density (2.4)
electronic transitions (2.3)
element (2.1)
energy level (2.3)
excited state (2.3)
extensive property (2.1)
gaseous state (2.1)
ground state (2.3)
heterogeneous mixture (2.1)
homogeneous mixture (2.1)
intensive property (2.1)
ion (2.2)
isotope (2.2)

liquid state (2.1)
mass number (2.2)
mixture (2.1)
natural radioactivity (2.3)
neutron (2.2)
nucleus (2.2)
orbit (2.3)
physical change (2.1)
physical property (2.1)
product (2.1)
promotion (2.3)
properties (2.1)
proton (2.2)
pure substance (2.1)
quantization (2.3)
quantum levels (2.3)
quantum number (2.3)
reactant (2.1)
relaxation (2.3)
solid state (2.1)
spectroscopy (2.3)
speed of light (2.3)
states of matter (2.1)

Questions and Problems

Matter and Properties

2.11   Describe what is meant by a physical property.
2.12   Describe what is meant by a physical change.
2.13   Label each of the following as either a physical change or a chemical reaction:
   a. An iron nail rusts.
   b. An ice cube melts.
   c. A limb falls from a tree.
2.14   Label each of the following as either a physical change or a chemical reaction:
   a. A puddle of water evaporates.
   b. Food is digested.
   c. Wood is burned.
2.15   Label each of the following properties of sodium as either a physical property or a chemical property:
   a. Sodium is a soft metal (can be cut with a knife).
   b. Sodium reacts violently with water to produce hydrogen gas and sodium hydroxide.
2.16   Label each of the following properties of sodium as either a physical property or a chemical property:
   a. When exposed to air, sodium forms a white oxide.
   b. Sodium melts at 98°C.
   c. The density of sodium metal at 25°C is 0.97 g/cm^3.
2.17   Describe several chemical properties of matter.
2.18   Describe what is meant by a chemical reaction.
2.19   Distinguish between a pure substance and a mixture.
2.20   Label each of the following as either a pure substance or a mixture:

   a. water
   b. table salt (sodium chloride)
   c. blood
   d. sucrose (table sugar)
   e. orange juice
2.21   Distinguish between a homogeneous mixture and a heterogeneous mixture.
2.22   Label each of the following as either a homogeneous mixture or a heterogeneous mixture:
   a. a soft drink
   b. a saline solution
   c. gelatin
   d. gasoline
   e. vegetable soup
2.23   Describe the general properties of the gaseous state.
2.24   Contrast the physical properties of the gaseous and solid states.
2.25   Distinguish between an intensive property and an extensive property.
2.26   Label each of the following as either an intensive property or an extensive property.
   a. mass
   b. volume
   c. density
   d. specific gravity
2.27   Describe the difference between the terms *atom* and *element*.
2.28   Give at least one example of each of the following:
   a. an element
   b. a pure substance
   c. a homogeneous mixture
   d. a heterogeneous mixture

Matter and Structure

2.29   Calculate the number of protons, neutrons, and electrons in:
   a. $^{16}_{8}O$
   b. $^{31}_{15}P$
2.30   Calculate the number of protons, neutrons, and electrons in:
   a. $^{136}_{56}Ba$
   b. $^{209}_{84}Po$
2.31   Why are isotopes useful in medicine?
2.32   Describe the similarities and differences among carbon-12, carbon-13, and carbon-14.
2.33   State the mass and charge of the:
   a. electron
   b. proton
   c. neutron
2.34   Calculate the number of protons, neutrons, and electrons in:
   a. $^{37}_{17}Cl$
   b. $^{23}_{11}Na$
   c. $^{84}_{36}Kr$
2.35   **a.** What is an ion?
   b. What process results in the formation of a cation?
   c. What process results in the formation of an anion?
2.36   **a.** What are isotopes?
   b. What is the major difference among isotopes of an element?

c. What is the major similarity among isotopes of an element?

2.37 Fill in the blanks:

Symbol	No. of Protons	No. of Neutrons	No. of Electrons	Charge
Example:				
$^{40}_{20}$Ca	20	20	20	0
$^{23}_{11}$Na	11	_____	11	0
$^{32}_{16}$S^{2-}	16	16	_____	2−
_____	8	8	8	0
$^{24}_{12}$Mg^{2+}	_____	12	_____	2+
_____	19	20	18	_____

2.38 Fill in the blanks:

Atomic Symbol	No. of Protons	No. of Neutrons	No. of Electrons	Charge
Example:				
$^{27}_{13}$Al	13	14	13	0
$^{39}_{19}$K	19	_____	19	0
$^{31}_{15}$P^{3-}	15	16	_____	_____
_____	29	34	27	2+
$^{55}_{26}$Fe^{2+}	_____	29	_____	2+
_____	8	8	10	_____

Development of Atomic Theory

2.39 What are the major postulates of Dalton's atomic theory?

2.40 What points of the Bohr theory are no longer current?

2.41 Note the major accomplishment of each of the following:
  a. Chadwick
  b. deBroglie
  c. Geiger
  d. Bohr

2.42 Note the major accomplishment of each of the following:
  a. Dalton
  b. Crookes
  c. Thomson
  d. Rutherford

2.43 Describe the experiment that provided the basis for our understanding of the nucleus.

2.44 Describe the process that occurs when electrical energy is applied to a sample of hydrogen gas.

2.45 What are the most important points of the Bohr theory?

2.46 Give two reasons why the Bohr theory did not stand the test of time.

2.47 Describe the series of experiments that characterized the electron.

2.48 List at least three properties of the electron.

2.49 What is a cathode ray? Which subatomic particle is detected?

2.50 Pictured is a cathode ray tube. Show the path that an electron would follow in the tube.

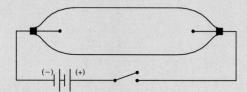

2.51 Fill in the blanks:
  a. An isotope of an element differs in mass because the atom has a different number of _____.
  b. The atomic number gives the number of _____ in the nucleus.
  c. The mass number of an atom is due to the number of _____ and _____ in the nucleus.
  d. A charged atom is called a(n) _____.
  e. Electrons surround the _____ and have a _____ charge.

2.52 Label each of the following statements as true or false:
  a. An atom with an atomic number of 7 and a mass of 14 is identical to an atom with an atomic number of 6 and a mass of 14.
  b. Neutral atoms have the same number of electrons as protons.
  c. The mass of an atom is due to the sum of the number of protons, neutrons, and electrons.
  d. Energy is required to promote an electron from a lower energy level to a higher energy level.
  e. The speed of light is directly proportional to its wavelength.

2.53 Rank the various regions of the electromagnetic spectrum in order of increasing wavelength.

2.54 Rank the various regions of the electromagnetic spectrum in order of increasing energy.

Modern Atomic Theory

2.55 Describe the meaning of the deBroglie hypothesis.

2.56 Describe the meaning of Heisenberg's Uncertainty Principle.

2.57 What was the major contribution of Bohr's atomic model?

2.58 What was the major deficiency of Bohr's atomic model?

Critical Thinking Problems

1. A natural sample of chromium, taken from the ground, will contain four isotopes: Cr-50, Cr-52, Cr-53 and Cr-54. Predict which isotope is in greatest abundance. Explain your reasoning.

2. Copper, silver, and gold are termed coinage metals; they are often used to mint coins (see Figure 5.1). List all the chemical and physical properties that you believe would make these elements suitable for coins. Now, choose three metals that would not be suitable for coinage, and explain why.

3. Rutherford's theory of the nucleus was based on the measurement of the results of a series of interactions. Explain how the process of reading this page involves similar principles.

4. Crookes's cathode ray tube experiment inadvertently supplied the basic science for a number of modern high-tech devices. List a few of these devices and describe how they involve one or more aspects of this historic experiment.

3

Elements, Atoms, Ions, and the Periodic Table

Organization and understanding go hand-in-hand.

Learning Goals

1 Recognize the important subdivisions of the periodic table: periods, groups (families), metals, and nonmetals.

2 Use the periodic table to obtain information about an element.

3 Describe the relationship between the electronic structure of an element and its position in the periodic table.

4 Write electron configurations for atoms of the most commonly-occurring elements.

5 Know the meaning of the octet rule and its predictive usefulness.

6 Use the octet rule to predict the charge of common cations and anions.

7 Utilize the periodic table and its predictive power to estimate the relative sizes of atoms and ions, as well as relative magnitudes of ionization energy and electron affinity.

8 Use values of ionization energies and electron affinities to predict ion formation.

Managing Mountains of Information

Recall for a moment the first time that you sat down in front of a computer. Perhaps it was connected to the Internet; somewhere in its memory was a word processor program, a spreadsheet, a few games, and many other features with strange-sounding names. Your challenge, very simply, was to use this device to access and organize information. Several manuals, all containing hundreds of pages of bewilderment, were your only help. How did you overcome this seemingly impossible task?

We are quite sure that you did not succeed without doing some reading and talking to people who had experience with computers. Also, you did not attempt to memorize every single word in each manual.

Success with a computer or any other storehouse of information results from developing an overall understanding of the way in which the system is organized. Certain facts must be memorized, but seeing patterns and using these relationships allows us to accomplish a wide variety of tasks that involve similar logic.

The study of chemistry is much like "real life." Just as it is impossible to memorize every single fact that will allow you to run a computer or drive an automobile in traffic, it is equally impossible to learn every fact in chemistry. Knowing the organization and logic of a process, along with a few key facts, makes a task manageable.

One powerful organizational device in chemistry is the periodic table. Its use in organizing and predicting the behavior of all of the known elements (and many of the compounds formed from these elements) is the subject of this chapter.

Introduction

We discussed some of the early experiments that established the existence of fundamental atomic particles (protons, neutrons, and electrons) and their relationship within the atom in Chapter 2. Let us now consider the relationships among the elements themselves. The unifying concept is called the *periodic law,* and it gives rise to an organized "map" of the elements that relates their structure to their chemical and physical properties. This "map" is the *periodic table.*

As we study the periodic law and periodic table, we shall see that the chemical and physical properties of elements follow directly from the electronic structure of the atoms that make up these elements. A thorough familiarity with the arrangement of the periodic table is vital to the study of chemistry. It not only allows us to predict the structure and properties of the various elements, but it also serves as the basis for developing an understanding of chemical bonding, or the process of forming molecules. Additionally, the properties and behavior of these larger units on a macroscopic scale (bulk properties) are fundamentally related to the properties of the atoms that comprise them.

3.1 The Periodic Law and the Periodic Table

Learning Goal

In 1869, Dmitri Mendeleev, a Russian, and Lothar Meyer, a German, working independently, found ways of arranging elements in order of increasing atomic mass such that elements with similar properties were grouped together in a *table of elements.* The **periodic law** is embodied by Mendeleev's statement, "the elements if arranged according to their atomic weights (masses), show a distinct *periodicity* (regular variation) of their properties." The *periodic table* (Figure 3.1) is a visual representation of the periodic law.

Chemical and physical properties of elements correlate with the electronic structure of the atoms that make up these elements. In turn, the electronic structure correlates with position on the periodic table.

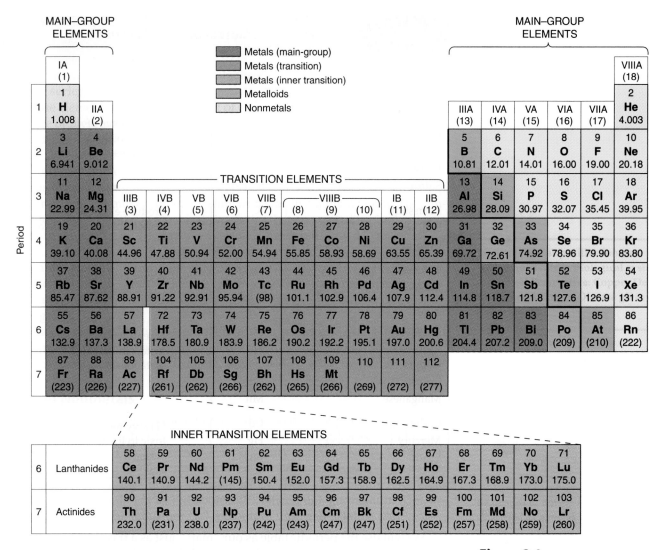

Figure 3.1

Classification of the elements: the periodic table.

A thorough familiarity with the arrangement of the periodic table allows us to predict electronic structure and physical and chemical properties of the various elements. It also serves as the basis for understanding chemical bonding.

The concept of "periodicity" may be illustrated by examining a portion of the modern periodic table (see Figure 3.1). The elements in the second row (beginning with lithium, Li, and proceeding to the right) show a marked difference in properties. However, sodium (Na) has properties similar to those of lithium, and sodium is therefore placed below lithium; once sodium is fixed in this position, the elements Mg through Ar have properties remarkably similar (though not identical) to those of the elements just above them. The same is true throughout the complete periodic table.

Mendeleev arranged the elements in his original periodic table in order of increasing atomic mass. However, as our knowledge of atomic structure increased, atomic numbers became the basis for the organization of the table.

The modern periodic law states that *the physical and chemical properties of the elements are periodic functions of their atomic numbers.* If we arrange the elements in order of increasing number of protons, the properties of the elements repeat at regular intervals.

Not all of the elements are of equal importance to an introductory study of chemistry. Table 3.1 lists 20 of the elements that are most important to biological systems, along with their symbols and a brief description of their functions.

A MEDICAL PERSPECTIVE

Copper Deficiency and Wilson's Disease

An old adage tells us that we should consume all things in moderation. This is very true of many of the trace minerals, such as copper. Too much copper in the diet causes toxicity and too little copper results in a serious deficiency disease.

Copper is extremely important for the proper functioning of the body. It aids in the absorption of iron from the intestine and facilitates iron metabolism. It is critical for the formation of hemoglobin and red blood cells in the bone marrow. Copper is also necessary for the synthesis of collagen, a protein that is a major component of the connective tissue. It is essential to the central nervous system in two important ways. First, copper is needed for the synthesis of norepinephrine and dopamine, two chemicals that are necessary for the transmission of nerve signals. Second, it is required for the deposition of the myelin sheath (a layer of insulation) around nerve cells. Release of cholesterol from the liver depends on copper, as does bone development and proper function of the immune and blood clotting systems.

The estimated safe and adequate daily dietary intake (ESADDI) for adults is 1.5–3.0 mg. Meats, cocoa, nuts, legumes, and whole grains provide significant amounts of copper. The accompanying table shows the amount of copper in some common foods.

Although getting enough copper in the diet would appear to be relatively simple, it is estimated that Americans often ingest only marginal levels of copper, and we absorb only 25–40% of that dietary copper. Despite these facts, it appears that copper deficiency is not a serious problem in the United States.

Individuals who are at risk for copper deficiency include people who are recovering from abdominal surgery, which causes decreased absorption of copper from the intestine. Others at risk are premature babies and people who are sustained solely by intravenous feedings that are deficient in copper. In addition, people who ingest high doses of antacids or take excessive supplements of zinc, iron, or vitamin C can develop copper deficiency because of reduced copper absorption. Because copper is involved in so many processes in the body, it is not surprising that the symptoms of copper deficiency are many and diverse. They include anemia; decreased red and white blood cell counts; heart disease; increased levels of serum cholesterol; loss of bone; defects in the nervous system, immune system, and connective tissue; and abnormal hair.

Some of these symptoms are seen among people who suffer from the rare genetic disease known as Menkes' kinky hair syndrome. The symptoms of this disease, which is caused by a defect in the ability to absorb copper from the intestine, include very low copper levels in the serum, kinky white hair, slowed growth, and degeneration of the brain.

Just as too little copper causes serious problems, so does an excess of copper. At doses greater than about 15 mg, copper causes toxicity that results in vomiting. The effects of extended exposure to excess copper are apparent when we look at Wil-

Copper in One-Cup Portions of Food

Food	Mass of copper (mg)
Sesame seeds	5.88
Cashews	3.04
Oysters	2.88
Sunflower seeds	2.52
Peanuts, roasted	1.85
Crabmeat	1.71
Walnuts	1.28
Almonds	1.22
Cereal, All Bran	0.98
Tuna fish	0.93
Wheat germ	0.70
Prunes	0.69
Kidney beans	0.56
Dried apricots	0.56
Lentils, cooked	0.54
Sweet potato, cooked	0.53
Dates	0.51
Whole milk	0.50
Raisins	0.45
Cereal, C. W. Post, Raisins	0.40
Grape Nuts	0.38
Whole-wheat bread	0.34
Cooked cereal, Roman Meal	0.32

Source: From David C. Nieman, Diane E. Butterworth, and Catherine N. Nieman, *Nutrition*, Revised First Edition. Copyright © 1992 Wm. C. Brown Communications, Inc., Dubuque, Iowa. All Rights Reserved. Reprinted by permission.

son's disease. This is a genetic disorder in which excess copper cannot be removed from the body and accumulates in the cornea of the eye, liver, kidneys, and brain. The symptoms include a greenish ring around the cornea, cirrhosis of the liver, copper in the urine, dementia and paranoia, drooling, and progressive tremors. As a result of the condition, the victim generally dies in early adolescence. Wilson's disease can be treated with moderate success if it is recognized early, before permanent damage has occurred to any tissues. The diet is modified to reduce the intake of copper; for instance, such foods as chocolate are avoided. In addition, the drug penicillamine is administered. This compound is related to the antibiotic penicillin but has no antibacterial properties; rather it has the ability to bind to copper in the blood and enhance its excretion by the kidneys into the urine. In this way the brain degeneration and tissue damage that are normally seen with the disease can be lessened.

Refer to the periodic table (Figure 3.1) and find the following information:

a. the symbol of the noble gas in period 3
b. the lightest element in Group IVA
c. the only metalloid in Group IIIA
d. the element whose atoms contain 18 protons

For each of the following element symbols, give the name of the element, its atomic number, and its atomic mass:

a. He
b. F
c. Mn

For each of the following element symbols, give the name of the element, its atomic number, and its atomic mass:

a. Mg
b. Ne
c. Se

3.2 Electron Arrangement and the Periodic Table

A primary objective of studying chemistry is to understand the way in which atoms join together to form chemical compounds. The most important factor in this *bonding process* is the arrangement of the electrons in the atoms that are combining. The **electronic configuration** describes the arrangement of electrons in atoms. The periodic table is helpful because it provides us with a great deal of information about the electron arrangement or electronic configuration of atoms.

Learning Goal

Valence Electrons

If we picture two spherical objects that we wish to join together, perhaps with glue, the glue can be applied to the surface, and the two objects can then be brought into contact. We can extend this analogy to two atoms that are modeled as spherical objects. Although this is not a perfect analogy, it is apparent that the surface interaction is of primary importance. Although the positively charged nucleus and "interior" electrons certainly play a role in bonding, we can most easily understand the process by considering only the outermost electrons. We refer to these as *valance electrons.* **Valence electrons** are the outermost electrons in an atom, which are involved, or have the potential to become involved in the bonding process.

For representative elements the number of valence electrons in an atom corresponds to the number of the *group* or *family* in which the atom is found. For example, elements such as hydrogen and sodium (in fact, all alkali metals, Group IA or 1) have a valence of 1 (or one valence electron). From left to right in period 2, beryllium, Be (Group IIA or 2), has two valence electrons; boron, B (Group IIIA or 3), has three; carbon, C (Group IVA or 4), has four; and so forth.

We have seen that an atom may have several energy levels, or regions where electrons are located. These energy levels are symbolized by n, the lowest energy level being assigned a value of $n = 1$. Each energy level may contain up to a fixed maximum number of electrons. For example, the $n = 1$ energy level may contain a maximum of two electrons. Thus hydrogen (atomic number = 1) has one electron and helium (atomic number = 2) has two electrons in the $n = 1$ level. Only these elements have electrons *exclusively* in the first energy level:

Metals tend to have fewer valence electrons, and nonmetals tend to have more valence electrons.

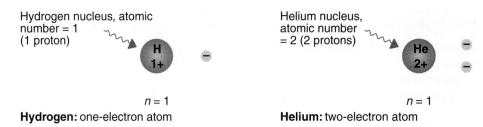

Hydrogen: one-electron atom

Helium: two-electron atom

These two elements make up the first period of the periodic table. Period 1 contains all elements whose *maximum* energy level is $n = 1$. In other words, the $n = 1$ level is the *outermost* electron region for hydrogen and helium. Hydrogen has one electron and helium has two electrons in the $n = 1$ level.

The valence electrons of elements in the second period are in the $n = 2$ energy level. (Remember that you must fill the $n = 1$ level with two electrons before adding electrons to the next level.) The third electron of lithium (Li) and the remaining electrons of the second period elements must be in the $n = 2$ level and are considered the valence electrons for lithium and the remaining second period elements.

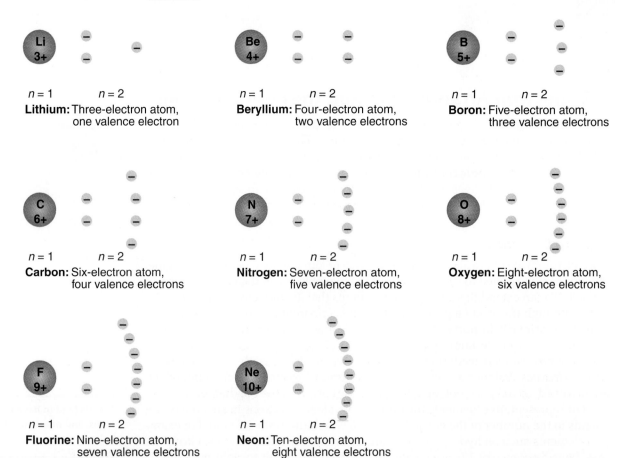

Lithium: Three-electron atom, one valence electron

Beryllium: Four-electron atom, two valence electrons

Boron: Five-electron atom, three valence electrons

Carbon: Six-electron atom, four valence electrons

Nitrogen: Seven-electron atom, five valence electrons

Oxygen: Eight-electron atom, six valence electrons

Fluorine: Nine-electron atom, seven valence electrons

Neon: Ten-electron atom, eight valence electrons

The electron distribution (arrangement) of the first twenty elements of the periodic table is given in Table 3.2.

Two general rules of electron distribution are based on the periodic law:

RULE 1:   The number of valence electrons in an atom equals the *group* number for all representative (A group) elements. ∎

RULE 2:   The energy level ($n = 1, 2$, etc.) in which the valence electrons are located corresponds to the *period* in which the element may be found. ∎

Table 3.2	The Electron Distribution for the First Twenty Elements of the Periodic Table				
Element Symbol and Name	Total Number of Electrons (Valence Electrons)	Electrons in $n = 1$	Electrons in $n = 2$	Electrons in $n = 3$	Electrons in $n = 4$
H, hydrogen	1 (1)	1	0	0	0
He, helium	2 (2)	2	0	0	0
Li, lithium	3 (1)	2	1	0	0
Be, beryllium	4 (2)	2	2	0	0
B, boron	5 (3)	2	3	0	0
C, carbon	6 (4)	2	4	0	0
N, nitrogen	7 (5)	2	5	0	0
O, oxygen	8 (6)	2	6	0	0
F, fluorine	9 (7)	2	7	0	0
Ne, neon	10 (8)	2	8	0	0
Na, sodium	11 (1)	2	8	1	0
Mg, magnesium	12 (2)	2	8	2	0
Al, aluminum	13 (3)	2	8	3	0
Si, silicon	14 (4)	2	8	4	0
P, phosphorus	15 (5)	2	8	5	0
S, sulfur	16 (6)	2	8	6	0
Cl, chlorine	17 (7)	2	8	7	0
Ar, argon	18 (8)	2	8	8	0
K, potassium	19 (1)	2	8	8	1
Ca, calcium	20 (2)	2	8	8	2

For example,

Group IA	*Group IIA*	*Group IIIA*	*Group VIIA*
Li	Ca	Al	Br
one valence electron in $n = 2$ energy level period 2	two valence electrons in $n = 4$ energy level period 4	three valence electrons in $n = 3$ energy level period 3	seven valence electrons in $n = 4$ energy level period 4

EXAMPLE 3.1

Determining Electron Arrangement

Provide the total number of electrons, total number of valence electrons, and energy level in which the valence electrons are found for the silicon (Si) atom.

Solution

Step 1. Determine the position of silicon in the periodic table. Silicon is found in Group IVA and period 3 of the table. Silicon has an atomic number of 14.

> ***Step 2.*** The atomic number provides the number of electrons in an atom. Silicon therefore has 14 electrons.
>
> ***Step 3.*** Because silicon is in Group IV, only four of the 14 electrons are valence electrons.
>
> ***Step 4.*** Silicon has two electrons in $n = 1$, eight electrons in $n = 2$, and four electrons in the $n = 3$ level.

Question 3.5

For each of the following elements, provide the *total* number of electrons and *valence* electrons in its atom:

a. Na   d. Cl
b. Mg   e. Ar
c. S

Question 3.6

For each of the following elements, provide the *total* number of electrons and *valence* electrons in its atom:

a. K   d. O
b. F   e. Ca
c. P

The Quantum Mechanical Atom

As we noted at the end of Chapter 2, the success of Bohr's theory was short-lived. Emission spectra of multi-electron atoms (recall that the hydrogen atom has only one electron) could not be explained by Bohr's theory. DeBroglie's statement that electrons have wave properties served to intensify the problem. Bohr stated that electrons in atoms had very specific locations. The very nature of waves, spread out in space, defies such an exact model of electrons in atoms. Furthermore, the exact model is contradictory to Heisenberg's Uncertainty Principle.

The basic concept of the Bohr theory, that the energy of an electron in an atom is quantized, was refined and expanded by an Austrian physicist, Erwin Schröedinger. He described electrons in atoms in probability terms, developing equations that emphasize the wavelike character of electrons. Although Schröedinger's approach was founded on complex mathematics, we can readily use models of electron probability regions that enable us to gain a reasonable insight into atomic structure without the need to understand the underlying mathematics.

Schröedinger's theory, often described as quantum mechanics, incorporates Bohr's principal energy levels ($n = 1, 2$, and so forth); however, it is proposed that each of these levels is made up of one or more sublevels. Each sublevel, in turn, contains one or more atomic orbitals. In the following section we shall look at each of these regions in more detail and learn how to predict the way that electrons are arranged in stable atoms.

Energy Levels and Sublevels

Principal Energy Levels

The principal energy levels are designated $n = 1, 2, 3$, and so forth. The number of possible sublevels in a principal energy level is also equal to n. When $n = 1$, there can be only one sublevel; $n = 2$ allows two sublevels, and so forth.

The total electron capacity of a principal level is $2(n)^2$. For example:

$n = 1$ $2(1)^2$ Capacity = $2e^-$

$n = 2$ $2(2)^2$ Capacity = $8e^-$

$n = 3$ $2(3)^2$ Capacity = $18e^-$

Sublevels

The sublevels, or subshells, are symbolized as s, p, d, f, and so forth; they increase in energy in the following order:

$$s < p < d < f$$

We specify both the principal energy level and type of sublevel when describing the location of an electron—for example, 1s, 2s, 2p. Energy level designations for the first four principal energy levels follow:

- The first principal energy level ($n = 1$) has one possible sublevel: 1s.
- The second principal energy level ($n = 2$) has two possible sublevels: 2s and 2p.
- The third principal energy level ($n = 3$) has three possible sublevels: 3s, 3p, and 3d.
- The fourth principal energy level ($n = 4$) has four possible sublevels: 4s, 4p, 4d, and 4f.

Orbitals

An **orbital** is a specific region of a sublevel containing a maximum of two electrons.

Figure 3.2 depicts a model of an s orbital. It is spherically symmetrical, much like a Ping-Pong ball. Its volume represents a region where there is a high probability of finding electrons of similar energy. A close inspection of Figure 3.2 shows that this probability decreases as we approach the outer surface (the decreasing color density in the model represents a decrease in the electron density). The nucleus is at the center of the s orbital. At that point the probability of finding the electron is zero; electrons cannot reside in the nucleus. Only one s orbital can be found in any n level. Atoms with many electrons, occupying a number of n levels, have an s orbital in each n level. Consequently 1s, 2s, 3s, and so forth are possible orbitals.

Figure 3.3 describes the shapes of the three possible p orbitals within a given level. Each has the same shape, and that shape appears much like a dumbbell; these three orbitals differ only in the direction they extend into space. Imaginary coordinates x, y, and z are superimposed on these models to emphasize this fact. These three orbitals, termed p_x, p_y, and p_z, may coexist in a single atom. Their arrangement is shown in Figure 3.4.

In a similar fashion, five possible d orbitals and seven possible f orbitals exist. d Orbitals exist only in $n = 3$ and higher principal energy levels; f orbitals exist only in $n = 4$ and higher principal energy levels. Because of their complexity, we will not discuss the shapes of d and f orbitals.

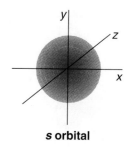

s orbital

Figure 3.2
Representation of an s orbital.

The shape and orientation of atomic orbitals strongly influence the structure and properties of compounds.

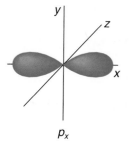

p_x

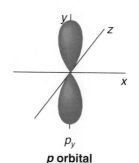

p_y

p orbital

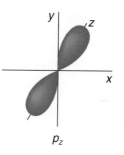

p_z

Figure 3.3
Representation of the three p orbitals, p_x, p_y, and p_z.

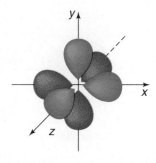

Figure 3.4
The orientation of three *p* orbitals in space.

Section 2.3 discusses the properties of electrons demonstrated by Thomson.

Electrons in Sublevels

We can deduce the maximum electron capacity of each sublevel based on the information just given.

For the *s* sublevel:

$$1 \text{ orbital} \times \frac{2e^- \text{ capacity}}{\text{orbital}} = 2e^- \text{ capacity}$$

For the *p* sublevel:

$$3 \text{ orbitals} \times \frac{2e^- \text{ capacity}}{\text{orbital}} = 6e^- \text{ capacity}$$

For the *d* sublevel:

$$5 \text{ orbitals} \times \frac{2e^- \text{ capacity}}{\text{orbital}} = 10e^- \text{ capacity}$$

For the *f* sublevel:

$$7 \text{ orbitals} \times \frac{2e^- \text{ capacity}}{\text{orbital}} = 14e^- \text{ capacity}$$

Electron Spin

As we have noted, each atomic orbital has a maximum capacity of two electrons. The electrons are perceived to *spin* on an imaginary axis, and the two electrons in the same orbital must have opposite spins: clockwise and counterclockwise. Their behavior is analogous to two ends of a magnet. Remember, electrons have magnetic properties. The electrons exhibit sufficient magnetic attraction to hold themselves together despite the natural repulsion that they "feel" for each other, owing to their similar charge (remember, like charges repel). Electrons must therefore have opposite spins to coexist in an orbital. A pair of electrons in one orbital that possess opposite spins are referred to as *paired* electrons.

Electron Configuration and the Aufbau Principle

Learning Goal

The arrangement of electrons in atomic orbitals is referred to as the atom's electron configuration. The *aufbau*, or building up, *principle* helps us to represent the electron configuration of atoms of various elements. According to this principle, electrons fill the lowest-energy orbital that is available first. We should also recall that the maximum capacity of an *s* level is two, that of a *p* level is six, that of a *d* level is ten, and that of an *f* level is fourteen electrons. Consider the following guidelines for writing electron configurations:

Rules for Writing Electron Configurations

- Obtain the total number of electrons in the atoms from the atomic number found on the periodic table.
- Electrons in atoms occupy the lowest energy orbitals that are available, beginning with 1*s*.
- Each principal energy level, *n*, can contain only *n* subshells.
- Each sublevel is composed of one (*s*) or more (three *p*, five *d*, seven *f*) orbitals.
- No more than two electrons can be placed in any orbital.
- The maximum number of electrons in any principal energy level is $2(n)^2$.
- The theoretical order of orbital filling is depicted in Figure 3.5.

Now let us look at several elements:

Hydrogen: Hydrogen is the simplest atom; it has only one electron. That electron must be in the lowest principal energy level ($n = 1$) and the lowest orbital (*s*). We indicate the number of electrons in a region with a *superscript*, so we write $1s^1$.

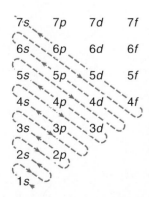

Figure 3.5
A useful way to remember the filling order for electrons in atoms.

Helium: Helium has two electrons, which will fill the lowest energy level. The ground state (lowest energy) electron configuration for helium is $1s^2$.

Lithium: Lithium has three electrons. The first two are configured as helium. The third must go into the orbital of the lowest energy in the second principal energy level; therefore the configuration is $1s^2\, 2s^1$.

Beryllium Through Neon: The second principal energy level can contain eight electrons $[2(2)^2]$, two in the s level and six in the p level. The "building up" process results in

Be	$1s^2\, 2s^2$
B	$1s^2\, 2s^2\, 2p^1$
C	$1s^2\, 2s^2\, 2p^2$
N	$1s^2\, 2s^2\, 2p^3$
O	$1s^2\, 2s^2\, 2p^4$
F	$1s^2\, 2s^2\, 2p^5$
Ne	$1s^2\, 2s^2\, 2p^6$

Sodium Through Argon: Electrons in these elements retain the basic $1s^2\, 2s^2\, 2p^6$ arrangement of the preceding element, neon; new electrons enter the third principal energy level:

Na	$1s^2\, 2s^2\, 2p^6\, 3s^1$
Mg	$1s^2\, 2s^2\, 2p^6\, 3s^2$
Al	$1s^2\, 2s^2\, 2p^6\, 3s^2\, 3p^1$
Si	$1s^2\, 2s^2\, 2p^6\, 3s^2\, 3p^2$
P	$1s^2\, 2s^2\, 2p^6\, 3s^2\, 3p^3$
S	$1s^2\, 2s^2\, 2p^6\, 3s^2\, 3p^4$
Cl	$1s^2\, 2s^2\, 2p^6\, 3s^2\, 3p^5$
Ar	$1s^2\, 2s^2\, 2p^6\, 3s^2\, 3p^6$

By knowing the order of filling of atomic orbitals, lowest to highest energy, you may write the electron configuration for any element. The order of orbital filling can be represented by the diagram in Figure 3.5. Such a diagram provides an easy way of predicting the electron configuration of the elements. Remember that the diagram is based on an energy scale, with the lowest energy orbital at the beginning of the "path" and the highest energy orbital at the end of the "path." An alternative way of representing orbital energies is through the use of an energy level diagram, such as the one in Figure 3.6.

EXAMPLE 3.2

Writing the Electron Configuration of Tin

Tin, Sn, has an atomic number of 50; thus we must place fifty electrons in atomic orbitals. We must also remember the total electron capacities of orbital types: s, 2; p, 6; d, 10; and f, 14. The electron configuration is as follows:

$$1s^2\, 2s^2\, 2p^6\, 3s^2\, 3p^6\, 4s^2\, 3d^{10}\, 4p^6\, 5s^2\, 4d^{10}\, 5p^2$$

As a check, count electrons in the electron configuration to see that we have accounted for all fifty electrons of the Sn atom.

Question 3.7

Give the electron configuration for an atom of:
a. sulfur
b. calcium

Question 3.8

Give the electron configuration for an atom of:
a. potassium
b. phosphorus

Recall that the prefix *iso* (Greek *isos*) means equal.

These ions are particularly stable. Each ion is **isoelectronic** (that is, it has the same number of electrons) with its nearest noble gas neighbor and has an octet of electrons in its outermost energy level.

Sodium is typical of each element in its group. Knowing that sodium forms a 1+ ion leads to the prediction that H, Li, K, Rb, Cs, and Fr also will form 1+ ions. Furthermore, magnesium, which forms a 2+ ion, is typical of each element in its group; Be^{2+}, Ca^{2+}, Sr^{2+}, and so forth are the resulting ions.

Nonmetallic elements, located at the right of the periodic table, tend to gain electrons to become isoelectronic with the nearest noble gas element, forming negative ions called **anions.**

Section 4.2 discusses the naming of ions.

Consider:

The ion of fluorine is the *fluoride ion;* the ion of oxygen is the *oxide ion;* and the ion of nitrogen is the *nitride ion.*

$$F + 1e^- \longrightarrow F^- \quad \text{(isoelectronic with Ne, 10e}^-\text{)}$$

Fluorine atom ($9e^-$)   Fluoride ion ($10e^-$)

$$O + 2e^- \longrightarrow O^{2-} \quad \text{(isoelectronic with Ne, 10e}^-\text{)}$$

Oxygen atom ($8e^-$)   Oxide ion ($10e^-$)

$$N + 3e^- \longrightarrow N^{3-} \quad \text{(isoelectronic with Ne, 10e}^-\text{)}$$

Nitrogen atom ($7e^-$)   Nitride ion ($10e^-$)

As in the case of positive ion formation, each of these negative ions has an octet of electrons in its outermost energy level.

The element fluorine, forming F^-, indicates that the other halogens, Cl, Br, and I, behave as a true family and form Cl^-, Br^-, and I^- ions. Also, oxygen and the other nonmetals in its group form $2-$ ions; nitrogen and phosphorus form $3-$ ions.

Question 3.11

Give the charge of the most probable ion resulting from each of the following elements. With what element is the ion isoelectronic?
a. Ca   d. Mg
b. Sr   e. P
c. S

Question 3.12

Which of the following pairs of atoms and ions are isoelectronic?
a. Cl^-, Ar   d. Li^+, Ne
b. Na^+, Ne   e. O^{2-}, F^-
c. Mg^{2+}, Na^+

The transition metals tend to form positive ions by losing electrons, just like the representative metals. Metals, whether representative or transition, share this characteristic. However, the transition elements are characterized as "variable valence" elements; depending on the type of substance with which they react, they may form more than one stable ion. For example, iron has two stable ionic forms:

$$Fe^{2+} \text{ and } Fe^{3+}$$

copper can exist as

$$Cu^+ \text{ and } Cu^{2+}$$

and elements such as vanadium, V, and manganese, Mn, each can form four different stable ions.

A CLINICAL PERSPECTIVE

Dietary Calcium

"**D**rink your milk!" "Eat all of your vegetables!" These imperatives are almost universal memories from our childhood. Our parents knew that calcium, present in abundance in these foods, was an essential element for the development of strong bones and healthy teeth.

Recent studies, spanning the fields of biology, chemistry, and nutrition science indicate that the benefits of calcium go far beyond bones and teeth. This element has been found to play a role in the prevention of disease throughout our bodies.

Calcium is the most abundant mineral (metal) in the body. It is ingested as the calcium ion (Ca^{2+}) either in its "free" state or "combined," as a part of a larger compound; calcium dietary supplements often contain ions in the form of calcium carbonate. The acid naturally present in the stomach produces the calcium ion:

$$CaCO_3 + 2H^+ \longrightarrow Ca^{2+} + H_2O + CO_2$$

calcium + stomach ⟶ calcium + water + carbon
carbonate   acid   ion   dioxide

Vitamin D serves as the body's regulator of calcium ion uptake, release, and transport in the body (see Appendix E.3).

Calcium is responsible for a variety of body functions including:

- transmission of nerve impulses
- release of "messenger compounds" that enable communication among nerves
- blood clotting
- hormone secretion
- growth of living cells throughout the body

The body's storehouse of calcium is bone tissue. When the supply of calcium from external sources, the diet, is insufficient, the body uses a mechanism to compensate for this shortage. With vitamin D in a critical role, this mechanism removes calcium from bone to enable other functions to continue to take place. It is evident then that prolonged dietary calcium deficiency can weaken the bone structure. Unfortunately, current studies show that as many as 75% of the American population may not be consuming sufficient amounts of calcium. Develop-

ing an understanding of the role of calcium in premenstrual syndrome, cancer, and blood pressure regulation is the goal of three current research areas.

Calcium and premenstrual syndrome (PMS). Dr. Susan Thys-Jacobs, a gynecologist at St. Luke's-Roosevelt Hospital Center in New York City, and colleagues at eleven other medical centers are conducting a study of calcium's ability to relieve the discomfort of PMS. They believe that women with chronic PMS have calcium blood levels that are normal only because calcium is continually being removed from the bone to maintain an adequate supply in the blood. To complicate the situation, vitamin D levels in many young women are very low (as much as 80% of a person's vitamin D is made in the skin, upon exposure to sunlight; many of us now minimize our exposure to the sun because of concerns about ultraviolet radiation and skin cancer). Because vitamin D plays an essential role in calcium metabolism, even if sufficient calcium is consumed, it may not be used efficiently in the body.

Colon cancer. The colon is lined with a type of cell (epithelial cell) that is similar to those that form the outer layers of skin. Various studies have indicated that by-products of a high-fat diet are irritants to these epithelial cells and produce abnormal cell growth in the colon. Dr. Martin Lipkin, Rockefeller University in New York, and his colleagues have shown that calcium ions may bind with these irritants, reducing their undesirable effects. It is believed that a calcium-rich diet, low in fat, and perhaps use of a calcium supplement can prevent or reverse this abnormal colon cell growth, delaying or preventing the onset of colon cancer.

Blood pressure regulation. Dr. David McCarron, a blood pressure specialist at the Oregon Health Sciences University, believes that dietary calcium levels may have a significant influence on hypertension (high blood pressure). Preliminary studies show that a diet rich in low-fat dairy products, fruits, and vegetables, all high in calcium, may produce a significant lowering of blood pressure in adults with mild hypertension.

The take-home lesson appears clear: a high calcium, low fat diet promotes good health in many ways. Once again, our parents were right!

Predicting the charge of an ion or the various possible ions for a given transition metal is not an easy task. Energy differences between valence electrons of transition metals are small and not easily predicted from the position of the element in the periodic table. In fact, in contrast to representative metals, the transition metals show great similarities within a *period* as well as within a *group*.

3.4 Trends in the Periodic Table

If our model of the atom is a tiny sphere whose radius is determined by the distance between the center of the nucleus and the boundary of the region where the

The radius of an atom is traditionally defined as one-half of the distance between atoms in a covalent bond. The covalent bond is discussed in Section 4.1.

from the atom, and this, in part, accounts for the extreme stability and nonreactivity of the noble gases.

Electron Affinity

The energy released when a single electron is added to an isolated atom is the **electron affinity.** If we consider ionization energy in relation to positive ion formation (remember that the magnitude of the ionization energy tells us the ease of *removal* of an electron, hence the ease of forming positive ions), then electron affinity provides a measure of the ease of forming negative ions. A large electron affinity (energy released) indicates that the atom becomes more stable as it becomes a negative ion (through gaining an electron). Consider the gain of an electron by a bromine atom:

$$\text{Br} + \text{e}^- \longrightarrow \text{Br}^- + \text{energy}$$

Electron affinity

Periodic trends for electron affinity are as follows:

- Electron affinities generally decrease down a group.
- Electron affinities generally increase across a period.

Remember these trends are not absolute. Exceptions exist, as seen in Figure 3.10.

Question 3.13

Rank Be, N, and F in order of increasing
a. atomic size
b. ionization energy
c. electron affinity

Question 3.14

Rank Cl, Br, I, and F in order of increasing
a. atomic size
b. ionization energy
c. electron affinity

Figure 3.10

The periodic variation of electron affinity. Note the very low values for the noble gases and the elements on the far left of the periodic table. These elements do not form negative ions. In contrast, F, Cl, and Br readily form negative ions.

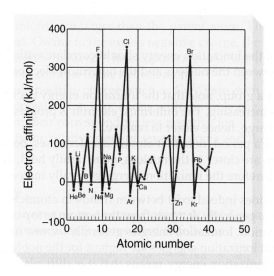

Summary

3.1 The Periodic Law and the Periodic Table

The *periodic law* is an organized "map" of the elements that relates their structure to their chemical and physical properties. It states that the elements, when arranged according to their atomic numbers, show a distinct periodicity (regular variation) of their properties. The periodic table is the result of the periodic law.

The modern periodic table exists in several forms. The most important variation is in group numbering. The tables in this text use the two most commonly accepted numbering systems.

A horizontal row of elements in the periodic table is referred to as a *period.* The periodic table consists of seven periods. The *lanthanide series* is a part of period 6; the *actinide series* is a part of period 7.

The columns of elements in the periodic table are called *groups* or *families.* The elements of a particular family share many similarities in physical and chemical properties because of the similarities in electronic structure. Some of the most important groups are named; for example, the *alkali metals* (IA or 1), *alkaline earth metals* (IIA or 2), the *halogens* (VIIA or 17), and the *noble gases* (VIII or 18).

Group A elements are called *representative elements;* Group B elements are *transition elements.* A bold zigzag line runs from top to bottom of the table, beginning to the left of boron (B) and ending between polonium (Po) and astatine (At). This line acts as the boundary between *metals* to the left and *nonmetals* to the right. Elements straddling the boundary, *metalloids,* have properties intermediate between those of metals and nonmetals.

3.2 Electron Arrangement and the Periodic Table

The outermost electrons in an atom are *valence electrons.* For representative elements the number of valence electrons in an atom corresponds to the group or family number (old system). Metals tend to have fewer valence electrons than nonmetals.

Electron configuration of the elements is predictable, using the aufbau principle. Knowing the electron configuration, we can identify valence electrons and begin to predict the kinds of reactions that the elements will undergo.

Elements in the last family, the noble gases, have either two valence electrons (helium) or eight valence electrons (neon, argon, krypton, xenon, and radon). Their most important properties are their extreme stability and lack of reactivity. A full valence level is responsible for this unique stability.

3.3 The Octet Rule

The *octet rule* tells us that in chemical reactions, elements will gain, lose, or share the minimum number of electrons necessary to achieve the electron configuration of the nearest noble gas.

Metallic elements tend to form cations. The ion is *isoelectronic* with its nearest noble gas neighbor and has a stable octet of electrons in its outermost energy level. Nonmetallic elements tend to gain electrons to become isoelectronic with the nearest noble gas element, forming anions.

3.4 Trends in the Periodic Table

Atomic size decreases from left to right and from bottom to top in the periodic table. *Cations* are smaller than the parent atom. *Anions* are larger than the parent atom. Ions with multiple positive charge are even smaller than their corresponding monopositive ion; ions with multiple negative charge are larger than their corresponding less negative ion.

The energy required to remove an electron from the atom is the *ionization energy.* Down a group, the ionization energy generally decreases. Across a period, the ionization energy generally increases.

The energy released when a single electron is added to a neutral atom in the gaseous state is known as the *electron affinity.* Electron affinities generally decrease proceeding down a group and increase proceeding across a period.

Key Terms

actinide series (3.1)
alkali metal (3.1)
alkaline earth metal (3.1)
anion (3.3)
cation (3.3)
electron affinity (3.4)
electronic configuration (3.2)
group (3.1)
halogen (3.1)
ionization energy (3.4)
isoelectronic (3.3)
lanthanide series (3.1)

metal (3.1)
metalloid (3.1)
noble gas (3.1)
nonmetal (3.1)
octet rule (3.3)
orbital (3.2)
period (3.1)
periodic law (3.1)
representative element (3.1)
transition element (3.1)
valence electron (3.2)

Questions and Problems

The Periodic Law and the Periodic Table

3.15 Define each of the following terms:
    a. periodic law
    b. period

c. group
d. ion

3.16 Define each of the following terms:
a. electron configuration
b. octet rule
c. ionization energy
d. isoelectronic

3.17 Label each of the following statements as true or false:
a. Elements of the same group have similar properties.
b. Atomic size decreases from left to right across a period.

3.18 Label each of the following statements as true or false:
a. Ionization energy increases from top to bottom within a group.
b. Representative metals are located on the left in the periodic table.

3.19 For each of the elements Na, Ni, Al, P, Cl, and Ar, provide the following information:
a. Which are metals?
b. Which are representative metals?
c. Which tend to form positive ions?
d. Which are inert or noble gases?

3.20 For each of the elements Ca, K, Cu, Zn, Br, and Kr provide the following information:
a. Which are metals?
b. Which are representative metals?
c. Which tend to form positive ions?
d. Which are inert or noble gases?

3.21 Provide the name of the element represented by each of the following symbols:
a. Na
b. K
c. Mg

3.22 Provide the name of the element represented by each of the following symbols:
a. Ca
b. Cu
c. Co

3.23 Which group of the periodic table is known as the alkali metals? List them.

3.24 Which group of the periodic table is known as the alkaline earth metals? List them.

3.25 Which group of the periodic table is known as the halogens? List them.

3.26 Which group of the periodic table is known as the noble gases? List them.

3.27 What are the major differences between the early and modern periodic tables?

3.28 Provide the name of the element represented by each of the following symbols:
a. B
b. Si
c. As

3.29 What is meant by the term *metalloid*?

3.30 Give three examples of elements that are:
a. metals
b. metalloids
c. nonmetals

Electron Arrangement and the Periodic Table

3.31 How many valence electrons are found in an atom of each of the following elements?
a. H   d. F
b. Na   e. Ne
c. B   f. He

3.32 How many valence electrons are found in an atom of each of the following elements?
a. Mg   d. Br
b. K   e. Ar
c. C   f. Xe

3.33 What is the common feature of the electron configurations of elements in Group IA (1)?

3.34 What is the common feature of the electron configurations of elements in Group VIIIA (18)?

3.35 How do we calculate the electron capacity of a principal energy level?

3.36 What sublevels would be found in each of the following principal energy levels?
a. $n = 1$ c. $n = 3$
b. $n = 2$ d. $n = 4$

3.37 Distinguish between a principal energy level and a sublevel.

3.38 Distinguish between a sublevel and an orbital.

3.39 Sketch a diagram and describe our current model of an s orbital.

3.40 How is a $2s$ orbital different from a $1s$ orbital?

3.41 How many p orbitals can exist in a given principal energy level?

3.42 Sketch diagrams of a set of p orbitals. How does a p_x orbital differ from a p_y orbital? From a p_z orbital?

3.43 How does a $3p$ orbital differ from a $2p$ orbital?

3.44 What is the maximum number of electrons that an orbital can hold?

3.45 What is the maximum number of electrons in each of the following energy levels?
a. $n = 1$
b. $n = 2$
c. $n = 3$

3.46 a. What is the maximum number of s electrons that can exist in any one principal energy level?
b. How many p electrons?
c. How many d electrons?
d. How many f electrons?

3.47 In which orbital is the highest-energy electron located in each of the following elements?
a. Al   d. Ca
b. Na   e. Fe
c. Sc   f. Cl

3.48 Using only the periodic table or list of elements, write the electron configuration of each of the following atoms:
a. B   d. V
b. S   e. Cd
c. Ar   f. Te

3.49 Which of the following electron configurations are not possible? Why?
a. $1s^2 1p^2$
b. $1s^2 2s^2 2p^2$
c. $1s^2, 2s^2, 2p^6, 2d^1$
d. $1s^2, 2s^3$

3.50 For each incorrect electron configuration in Question 3.49, assume that the number of electrons is correct, identify the element, and write the correct electron configuration.

The Octet Rule

3.51 Give the most probable ion formed from each of the following elements:
a. Li   d. Br
b. O   e. S
c. Ca   f. Al

3.52 Using only the periodic table or list of elements, write the electron configuration of each of the following ions:
   a. I^-
   b. Ba^{2+}
   c. Se^{2-}
   d. Al^{3+}

3.53 Which of the following pairs of atoms and/or ions are isoelectronic with one another?
   a. O^{2-}, Ne
   b. S^{2-}, Cl^-

3.54 Which of the following pairs of atoms and/or ions are isoelectronic with one another?
   a. F^-, Cl^-
   b. K^+, Ar

3.55 Why do Group IA (1) metals form only one ion (1+)? Does the same hold true for Group IIA (2): Can they form only a 2+ ion?

3.56 Why are noble gases so nonreactive?

3.57 Which species in each of the following groups would you expect to find in nature?
   a. Na, Na^+, Na^-
   b. S^{2-}, S^-, S^+
   c. Cl, Cl^-, Cl^+

3.58 Which atom or ion in each of the following groups would you expect to find in nature?
   a. K, K^+, K^-
   b. O^{2-}, O, O^{2+}
   c. Br, Br^-, Br^+

3.59 Write the electron configuration of each of the following biologically important ions:
   a. Ca^{2+}
   b. Mg^{2+}

3.60 Write the electron configuration of each of the following biologically important ions:
   a. K^+
   b. Cl^-

Trends in the Periodic Table

3.61 Arrange each of the following lists of elements in order of increasing atomic size:
   a. N, O, F
   b. Li, K, Cs
   c. Cl, Br, I

3.62 Arrange each of the following lists of elements in order of increasing atomic size:
   a. Al, Si, P, Cl, S
   b. In, Ga, Al, B, Tl
   c. Sr, Ca, Ba, Mg, Be
   d. P, N, Sb, Bi, As

3.63 Which of the elements has the highest electron affinity?

3.64 Which of the elements has the highest ionization energy?

3.65 Arrange each of the following lists of elements in order of increasing ionization energy:
   a. N, O, F
   b. Li, K, Cs
   c. Cl, Br, I

3.66 Arrange each of the following lists of elements in order of decreasing electron affinity:
   a. Na, Li, K
   b. Br, F, Cl
   c. S, O, Se

3.67 Explain why a positive ion is always smaller than its parent atom.

3.68 Explain why a negative ion is always larger than its parent atom.

3.69 Explain why a fluoride ion is commonly found in nature but a fluorine atom is not.

3.70 Explain why a sodium ion is commonly found in nature but a sodium atom is not.

Critical Thinking Problems

1. Name five elements that you came in contact with today. Were they in combined form or did they exist in the form of atoms? Were they present in pure form or in mixtures? If mixtures, were they heterogeneous or homogeneous? Locate each in the periodic table by providing the group and period designation, for example: Group IIA (2), period 3.

2. The periodic table is incomplete. It is possible that new elements will be discovered from experiments using high-energy particle accelerators. Predict as many properties as you can that might characterize the recently discovered element that has an atomic number of 118. Can you suggest an appropriate name for this element?

3. The element titanium is now being used as a structural material for bone and socket replacement (shoulders, knees). Predict properties that you would expect for such applications; go to the library and look up the properties of titanium and evaluate your answer.

4. Imagine that you have undertaken a voyage to an alternate universe. Using your chemical skills, you find a collection of elements quite different than those found here on earth. After measuring their properties and assigning symbols for each, you wish to organize them as Mendeleev did for our elements. Design a periodic table using the information you have gathered:

Symbol	Mass (amu)	Reactivity	Electrical Conductivity
A	2.0	High	High
B	4.0	High	High
C	6.0	Moderate	Trace
D	8.0	Low	0
E	10.0	Low	0
F	12.0	High	High
G	14.0	High	High
H	16.0	Moderate	Trace
I	18.0	Low	0
J	20.0	None	0
K	22.0	High	High
L	24.0	High	High

Predict the reactivity and conductivity of an element with a mass of 30.0 amu. What element in our universe does this element most closely resemble?

5. Why does the octet rule not work well for compounds of lanthanide and actinide elements? Suggest a number other than eight that may be more suitable.

4

Structure and Properties of Ionic and Covalent Compounds

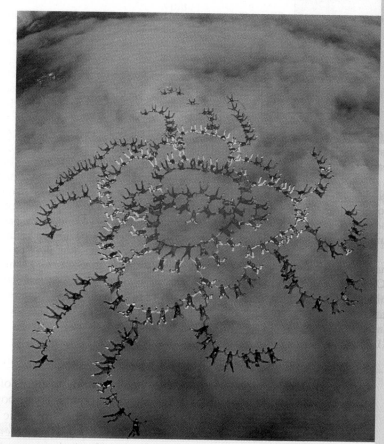

The pattern formed depends on the courage and skill of the individuals.

Learning Goals

1 Classify compounds as having ionic, covalent, or polar covalent bonds.

2 Name common inorganic compounds using standard conventions and recognize the common names of frequently used substances.

3 Write the formulas of compounds when provided with the name of the compound.

4 Predict the differences in physical state, melting and boiling points, solid-state structure, and solution chemistry that result from differences in bonding.

5 Draw Lewis structures for covalent compounds and complex inorganic ions.

6 Describe the relationship between stability and bond energy.

7 Predict the geometry of molecules and ions using the octet rule and Lewis structure.

8 Understand the role that molecular geometry plays in determining the solubility and melting and boiling points of compounds.

9 Use the principles of VSEPR theory and molecular geometry to predict relative melting points, boiling points, and solubilities of compounds.

(higher electron density) hence, more negative. The other end of the bond (in this case, the H atom) is less electron rich (lower electron density) hence, more positive. These two ends, one somewhat positive and the other somewhat negative may be described as electronic poles, hence the term polar covalent bonds.

Once again, we can use the predictive power of the periodic table to help us determine whether a particular bond is polar or nonpolar covalent. We already know that elements that tend to form negative ions (by gaining electrons) are found to the right of the table whereas positive ion formers (that may lose electrons) are located on the left side of the table. Elements whose atoms strongly attract electrons are described as electronegative elements. Linus Pauling, a chemist noted for his theories on chemical bonding, developed a scale of relative electronegativities that correlates reasonably well with the positions of the elements in the periodic table.

Electronegativity

Electronegativity (E_n) is a measure of the ability of an atom to attract electrons in a chemical bond. Elements with high electronegativity have a greater ability to attract electrons than do elements with low electronegativity. Pauling developed a method to assign values of electronegativity to many of the elements in the periodic table. These values range from a low of 0.7 to a high of 4.0, 4.0 being the most electronegative element.

Figure 4.2 shows that the most electronegative elements (excluding the nonreactive noble gas elements) are located in the upper right corner of the periodic table, whereas the least electronegative elements are found in the lower left corner of the table. In general, electronegativity values increase as we proceed left to right and bottom to top of the table. Like other periodic trends, numerous exceptions occur.

If we picture the covalent bond as a competition for electrons between two positive centers, it is the difference in electronegativity, ΔE_n, that determines the extent of polarity. Consider: H_2 or H—H

$$\Delta E_n = \frac{\text{Electronegativity}}{\text{of hydrogen}} - \frac{\text{Electronegativity}}{\text{of hydrogen}}$$

$$\Delta E_n = 2.1 - 2.1 = 0$$

The bond in H_2 is nonpolar covalent. Bonds between *identical* atoms are *always* nonpolar covalent. Also, Cl_2 or Cl—Cl

$$\Delta E_n = \frac{\text{Electronegativity}}{\text{of chlorine}} - \frac{\text{Electronegativity}}{\text{of chlorine}}$$

$$\Delta E_n = 3.0 - 3.0 = 0$$

The bond in Cl_2 is nonpolar covalent. Now consider HCl or H—Cl

$$\Delta E_n = \frac{\text{Electronegativity}}{\text{of chlorine}} - \frac{\text{Electronegativity}}{\text{of hydrogen}}$$

$$\Delta E_n = 3.0 - 2.1 = 0.9$$

The bond in HCl is polar covalent.

4.2 Naming Compounds and Writing Formulas of Compounds

Nomenclature is the assignment of a correct and unambiguous name to each and every chemical compound. Assignment of a name to a structure or deducing the structure from a name is a necessary first step in any discussion of these compounds.

Linus Pauling is the only person to receive two Nobel Prizes in very unrelated fields; the chemistry award in 1954 and eight years later, the Nobel Peace Prize. His career is a model of interdisciplinary science, with important contributions ranging from chemical physics to molecular biology.

By convention, the electronegativity difference is calculated by subtracting the less electronegative element's value from the value for the more electronegative element. In this way, negative numbers are avoided.

A ΔE_n value of 1.9 is generally accepted as the boundary between a polar covalent and an ionic compound.

IA (1)												IIIA (13)	IVA (14)	VA (15)	VIA (16)	VIIA (17)
H 2.1	IIA (2)															
Li 1.0	**Be** 1.5											**B** 2.0	**C** 2.5	**N** 3.0	**O** 3.5	**F** 4.0
Na 0.9	**Mg** 1.2	IIIB (3)	IVB (4)	VB (5)	VIB (6)	VIIB (7)	(8) VIIIB (9)	(10)	IB (11)	IIA (12)		**Al** 1.5	**Si** 1.8	**P** 2.1	**S** 2.5	**Cl** 3.0
K 0.8	**Ca** 1.0	**Sc** 1.3	**Ti** 1.5	**V** 1.6	**Cr** 1.6	**Mn** 1.5	**Fe** 1.8	**Co** 1.9 **Ni** 1.9	**Cu** 1.9	**Zn** 1.6	**Ga** 1.6	**Ge** 1.8	**As** 2.0	**Se** 2.4	**Br** 2.8
Rb 0.8	**Sr** 1.0	**Y** 1.2	**Zr** 1.4	**Nb** 1.6	**Mo** 1.8	**Tc** 1.9	**Ru** 2.2	**Rh** 2.2 **Pd** 2.2	**Ag** 1.9	**Cd** 1.7	**In** 1.7	**Sn** 1.8	**Sb** 1.9	**Te** 2.1	**I** 2.5
Cs 0.7	**Ba** 0.9	**La*** 1.1	**Hf** 1.3	**Ta** 1.5	**W** 1.7	**Re** 1.9	**Os** 2.2	**Ir** 2.2 **Pt** 2.2	**Au** 2.4	**Hg** 1.9	**Tl** 1.8	**Pb** 1.9	**Bi** 1.9	**Po** 2.0	**At** 2.2
Fr 0.7	**Ra** 0.9	**Ac** 1.1														

Below 1.0

1.0–3.0

Above 3.0

*Lathanides: 1.1 — 1.3
Actinides: 1.3 — 1.5

Alkali metals

Figure 4.2
Electronegativities of the elements.

Ionic Compounds

The "shorthand" symbol for a compound is its formula—for example,

NaCl   and   MgBr$_2$

The formula identifies the number and type of the various atoms that make up the compound. The number of like atoms in the unit is shown by the use of a subscript. The presence of only one atom is understood when no subscript is present.

The formula NaCl indicates that each ion pair consists of one sodium cation (Na^+) and one chloride anion (Cl^-). Similarly, the formula MgBr$_2$ indicates that one magnesium ion and two bromide ions combine to form the compound.

In Chapter 3 we learned that positive ions are formed from elements that

- are located at the left of the periodic table,
- are referred to as *metals*, and
- have low ionization energies, low electron affinities, and hence easily *lose* electrons.

Elements that form negative ions, on the other hand,

- are located at the right of the periodic table (but exclude the noble gases),
- are referred to as *nonmetals*, and
- have high ionization energies, high electron affinities, and hence easily *gain* electrons.

In short, metals and nonmetals usually react to produce ionic compounds resulting from the transfer of one or more electrons from the metal to the nonmetal.

Although ionic compounds are sometimes referred to as ion pairs, in the solid state these ion pairs do not actually exist as individual units. The positive ions exert attractive forces on several negative ions, and the negative ions are attracted to several positive centers. Positive and negative ions arrange themselves in a regular three-dimensional repeating array to produce a stable arrangement known as a **crystal lattice.** The lattice structure for sodium chloride is shown from two different perspectives in Figure 4.3. The **formula** of an ionic compound is the smallest whole-number ratio of ions in the crystal.

Writing Formulas of Ionic Compounds from the Identities of the Component Ions

It is important to be able to write the formula of an ionic compound when provided with the identities of the ions that make up the compound. The charge of each ion can usually be determined from the group (family) of the periodic table in which the parent element is found. The cations and anions must combine in such a way that the resulting formula unit has a net charge of zero.

Consider the following examples.

EXAMPLE 4.1

Predicting the Formula of an Ionic Compound

Predict the formula of the ionic compound formed from the reaction of sodium and oxygen atoms.

Solution

Sodium is in group IA (or 1); it has *one* valence electron. Loss of this electron produces Na^+. Oxygen is in group VIA (or 16); it has *six* valence electrons. A gain of two electrons (to create a stable octet) produces O^{2-}. Two positive charges are necessary to counterbalance two negative charges on the oxygen anion. Because each sodium ion carries a 1+ charge, two sodium ions are needed for each O^{2-}. The subscript 2 is used to indicate that the formula unit contains two sodium ions. Thus the formula of the compound is Na_2O.

EXAMPLE 4.2

Predicting the Formula of an Ionic Compound

Predict the formula of the compound formed by the reaction of aluminum and oxygen atoms.

Solution

Aluminum is in group IIIA (or 13) of the periodic table; we predict that it has three valence electrons. Loss of these electrons produces Al^{3+}. Oxygen is in group VIA (or 16) of the periodic table and has six valence electrons. A gain of two electrons (to create a stable octet) produces O^{2-}. How can we combine Al^{3+} and O^{2-} to yield a unit of zero charge? It is necessary that *both* the cation and anion be multiplied by factors that will result in a zero net charge:

$$2 \times (+3) = +6 \quad \text{and} \quad 3 \times (-2) = -6$$
$$2 \times Al^{3+} = +6 \quad \text{and} \quad 3 \times O^{2-} = -6$$

Hence the formula is Al_2O_3.

Question 4.1

Predict the formulas of the compounds formed from the combination of ions of the following elements:
a. lithium and bromine
b. calcium and bromine
c. calcium and nitrogen

Predict the formulas of the compounds formed from the combination of ions of the following elements:

a. potassium and chlorine
b. magnesium and bromine
c. magnesium and nitrogen

Writing Names of Ionic Compounds from the Formula of the Compound

Nomenclature, the way in which compounds are named, is based on their formulas. The name of the cation appears first, followed by the name of the anion. The positive ion has the name of the element; the negative ion is named by using the *stem* of the name of the element joined to the suffix *-ide*. Some examples follow.

Learning Goal

2

Formula	*+ion*	*and*	*−ion stem*	+	*ide*	=	*Compound name*
NaCl	sodium		chlor	+	ide		sodium chloride
Na_2O	sodium		ox	+	ide		sodium oxide
Li_2S	lithium		sulf	+	ide		lithium sulfide
$AlBr_3$	aluminum		brom	+	ide		aluminum bromide
CaO	calcium		ox	+	ide		calcium oxide

If the cation and anion exist in only one common charged form, there is no ambiguity between formula and name. Sodium chloride *must be* NaCl, and lithium sulfide *must be* Li_2S, so that the sum of positive and negative charges is zero. With many elements, such as the transition metals, several ions of different charge may exist. Fe^{2+}, Fe^{3+} and Cu^+, Cu^{2+} are two common examples. Clearly, an ambiguity exists if we use the name iron for both Fe^{2+} and Fe^{3+} or copper for both Cu^+ and Cu^{2+}. Two systems have been developed to avoid this problem: the *Stock system* and the *common nomenclature system*.

In the Stock system (systematic name), a Roman numeral indicates the magnitude of the cation's charge. In the older common nomenclature system the suffix *-ous* indicates the lower ionic charge, and the suffix *-ic* indicates the higher ionic charge. Consider the examples in Table 4.1.

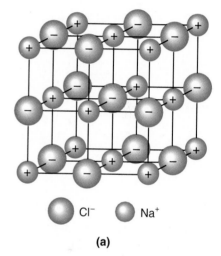

Cl^- Na^+

(a)

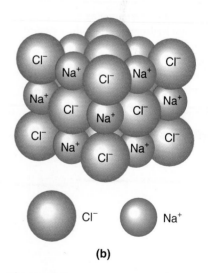

(b)

Figure 4.3

The arrangement of ions in a crystal of NaCl (sodium chloride, table salt). (a) Microscopic arrangement of ions as point charges. (b) Microscopic arrangement of the spherical ions in crystal lattice.

Table 4.1	Systematic (Stock) and Common Names for Iron and Copper Ions

For systematic name:

Formula	*+ Ion Charge*	*Cation Name*	*Compound Name*
$FeCl_2$	2+	Iron(II)	Iron(II) chloride
$FeCl_3$	3+	Iron(III)	Iron(III) chloride
Cu_2O	1+	Copper(I)	Copper(I) oxide
CuO	2+	Copper(II)	Copper(II) oxide

For common nomenclature:

Formula	*+ Ion Charge*	*Cation Name*	*Common -ous/ic Name*
$FeCl_2$	2+	Ferrous	Ferrous chloride
$FeCl_3$	3+	Ferric	Ferric chloride
Cu_2O	1+	Cuprous	Cuprous oxide
CuO	2+	Cupric	Cupric oxide

Table 4.2	Common Monatomic Cations and Anions		
Cation	Name	Anion	Name
H^+	Hydrogen ion	H^-	Hydride ion
Li^+	Lithium ion	F^-	Fluoride ion
Na^+	Sodium ion	Cl^-	Chloride ion
K^+	Potassium ion	Br^-	Bromide ion
Cs^+	Cesium ion	I^-	Iodide ion
Be^{2+}	Beryllium ion	O^{2-}	Oxide ion
Mg^{2+}	Magnesium ion	S^{2-}	Sulfide ion
Ca^{2+}	Calcium ion	N^{3-}	Nitride ion
Ba^{2+}	Barium ion	P^{3-}	Phosphide ion
Al^{3+}	Aluminum ion		
Ag^+	Silver ion		

The ions of principal importance are highlighted in blue.

Systematic names are easier and less ambiguous than common names. Whenever possible, we will use this system of nomenclature. The older, common names (-ous, -ic) are less specific; furthermore, they often use the Latin names of the elements (for example, iron compounds use *ferr-*, from *ferrum*, the Latin word for iron).

Monatomic ions are ions consisting of a single atom. Common monatomic ions are listed in Table 4.2. The ions that are particularly important in biological systems are highlighted in blue.

Polyatomic ions, such as the hydroxide ion, OH^-, are composed of two or more atoms bonded together. These ions, although bonded to other ions with ionic bonds, are themselves held together by covalent bonds.

The polyatomic ion has an *overall* positive or negative charge. Some common polyatomic ions are listed in Table 4.3. The formulas, charges, and names of these polyatomic ions, especially those highlighted in blue, should be memorized.

The following examples are formulas of several compounds containing polyatomic ions.

Formula	Cation	Anion	Name
NH_4Cl	NH_4^+	Cl^-	ammonium chloride
$Ca(OH)_2$	Ca^{2+}	OH^-	calcium hydroxide
Na_2SO_4	Na^+	SO_4^{2-}	sodium sulfate
$NaHCO_3$	Na^+	HCO_3^-	sodium bicarbonate

> Sodium bicarbonate may also be named sodium hydrogen carbonate, a preferred and less ambiguous name. Likewise, Na_2HPO_4 is named sodium hydrogen phosphate, and other ionic compounds are named similarly.

Question 4.3

Name each of the following compounds:
a. KCN
b. MgS
c. $Mg(CH_3COO)_2$

Question 4.4

Name each of the following compounds:
a. Li_2CO_3
b. $FeBr_2$
c. $CuSO_4$

Learning Goal

Writing Formulas of Ionic Compounds from the Name of the Compound

It is also important to be able to write the correct formula when given the compound name. To do this, we must be able to predict the charge of monatomic ions and

Table 4.3	Common Polyatomic Cations and Anions
Ion	**Name**
NH_4^+	Ammonium
NO_2^-	Nitrite
NO_3^-	Nitrate
SO_3^{2-}	Sulfite
SO_4^{2-}	Sulfate
HSO_4^-	Hydrogen sulfate
OH^-	Hydroxide
CN^-	Cyanide
PO_4^{3-}	Phosphate
HPO_4^{2-}	Hydrogen phosphate
$H_2PO_4^-$	Dihydrogen phosphate
CO_3^{2-}	Carbonate
HCO_3^-	Bicarbonate
ClO^-	Hypochlorite
ClO_2^-	Chlorite
ClO_3^-	Chlorate
ClO_4^-	Perchlorate
CH_3COO^- (or $C_2H_3O_2^-$)	Acetate
MnO_4^-	Permanganate
$Cr_2O_7^{2-}$	Dichromate
CrO_4^{2-}	Chromate
O_2^{2-}	Peroxide

The most commonly encountered ions are highlighted in blue.

remember the charge and formula of polyatomic ions. Equally important, the relative number of positive and negative ions in the unit must result in a net (compound) charge of zero. The compounds are electrically neutral. Two examples follow.

EXAMPLE 4.3

Writing a Formula When Given the Name of the Compound

Write the formula of sodium sulfate.

Solution

Step 1. The sodium ion is Na^+, a group I (or 1) element. The sulfate ion is SO_4^{2-} (from Table 4.3).

Step 2. Two positive charges, two sodium ions, are needed to cancel the charge on one sulfate ion (two negative charges).

Hence the formula is Na_2SO_4.

EXAMPLE 4.4

Writing a Formula When Given the Name of the Compound

Write the formula of ammonium sulfide.

Solution

Step 1. The ammonium ion is NH_4^+ (from Table 4.3). The sulfide ion is S^{2-} (from its position on the periodic table).

> **Step 2.** Two positive charges are necessary to cancel the charge on one sulfide ion (two negative charges).
>
> Hence the formula is $(NH_4)_2S$.
>   Note that parentheses must be used whenever a subscript accompanies a polyatomic ion.

Question 4.5

Write the formula for each of the following compounds:
a. calcium carbonate
b. sodium bicarbonate
c. copper(I) sulfate

Question 4.6

Write the formula for each of the following compounds:
a. sodium phosphate
b. potassium bromide
c. iron(II) nitrate

Covalent Compounds

Naming Covalent Compounds

Learning Goal

Most covalent compounds are formed by the reaction of nonmetals. **Molecules** are compounds characterized by covalent bonding. We saw earlier that ionic compounds are not composed of single units but are a part of a massive three-dimensional crystal structure in the solid state. Covalent compounds exist as discrete molecules in the solid, liquid, and gas states. This is a distinctive feature of covalently bonded substances.

The conventions for naming covalent compounds follow:

1. The names of the elements are written in the order in which they appear in the formula.
2. A prefix (Table 4.4) indicating the number of each kind of atom found in the unit is placed before the name of the element.
3. If only one atom of a particular kind is present in the molecule, the prefix mono- is usually omitted from the first element.
4. The stem of the name of the last element is used with the suffix -ide.
5. The final vowel in a prefix is often dropped before a vowel in the stem name.

By convention the prefix *mono-* is often omitted from the second element as well (dinitrogen oxide, not dinitrogen monoxide). In other cases, common usage retains the prefix (carbon monoxide, not carbon oxide).

Table 4.4	Prefixes Used to Denote Numbers of Atoms in a Compound
Prefix	**Number of Atoms**
Mono-	1
Di-	2
Tri-	3
Tetra-	4
Penta-	5
Hexa-	6
Hepta-	7
Octa-	8
Nona-	9
Deca-	10

EXAMPLE 4.5

Naming a Covalent Compound

Name the covalent compound N_2O_4.

Solution

Step 1. two nitrogen atoms   four oxygen atoms

Step 2. di-   tetra-

Step 3. dinitrogen   tetr(a)oxide

The name is dinitrogen tetroxide.

The following are examples of other covalent compounds.

Formula	Name
N_2O	dinitrogen monoxide
NO_2	nitrogen dioxide
SiO_2	silicon dioxide
CO_2	carbon dioxide
CO	carbon monoxide

Question 4.7

Name each of the following compounds:

a. B_2O_3 c. ICl
b. NO d. PCl_3

Question 4.8

Name each of the following compounds:

a. H_2S c. PCl_5
b. CS_2 d. P_2O_5

Writing Formulas of Covalent Compounds

Many compounds are so familiar to us that their *common names* are generally used. For example, H_2O is water, NH_3 is ammonia, C_2H_5OH (ethanol) is alcohol, and $C_6H_{12}O_6$ is glucose. It is useful to be able to correlate both systematic and common names with the corresponding molecular formula and vice versa.

When common names are used, formulas of covalent compounds can be written *only* from memory. You *must* remember that water is H_2O, ammonia is NH_3, and so forth. This is the major disadvantage of common names. Because of their widespread use, however, they cannot be avoided and must be memorized.

Compounds named by using Greek prefixes are easily converted to formulas. Consider the following examples.

Learning Goal 3

EXAMPLE 4.6

Writing the Formula of a Covalent Compound

Write the formula of nitrogen monoxide.

Solution

Nitrogen has no prefix; one is understood. Oxide has the prefix *mono*—one oxygen. Hence the formula is NO.

EXAMPLE 4.7

Writing the Formula of a Covalent Compound

Write the formula of dinitrogen tetroxide.

Solution

Nitrogen has the prefix *di*—two nitrogen atoms. Oxygen has the prefix *tetr(a)*—four oxygen atoms. Hence the formula is N_2O_4.

Question 4.9

Write the formula of each of the following compounds:
a. diphosphorus pentoxide
b. silicon dioxide

Question 4.10

Write the formula of each of the following compounds:
a. nitrogen trifluoride
b. carbon monoxide

4.3 Properties of Ionic and Covalent Compounds

Learning Goal

The differences in ionic and covalent bonding result in markedly different properties for ionic and covalent compounds. Because covalent molecules are distinct units, they have less tendency to form an extended structure in the solid state. Ionic compounds, with ions joined by electrostatic attraction, do not have definable units but form a crystal lattice composed of enormous numbers of positive and negative ions in an extended three-dimensional network. The effects of this basic structural difference are summarized in this section.

Physical State

All ionic compounds (for example, NaCl, KCl, and $NaNO_3$) are solids at room temperature; covalent compounds may be solids (sugar), liquids (H_2O, ethanol), or gases (carbon monoxide, carbon dioxide). The three-dimensional crystal structure that is characteristic of ionic compounds holds them in a rigid, solid arrangement, whereas molecules of covalent compounds may be fixed, as in a solid, or more mobile, a characteristic of liquids and gases.

Melting and Boiling Points

The **melting point** is the temperature at which a solid is converted to a liquid and the **boiling point** is the temperature at which a liquid is converted to a gas at a specified pressure. Considerable energy is required to break apart an ionic crystal lattice with uncountable numbers of ionic interactions and convert the ionic substance to a liquid or a gas. As a result, the melting and boiling temperatures for ionic compounds are generally higher than those of covalent compounds, whose molecules interact less strongly in the solid state. A typical ionic compound, sodium chloride, has a melting point of 801°C; methane, a covalent compound, melts at −182°C. Exceptions to this general rule do exist; diamond, a covalent solid with an extremely high melting point, is a well-known example.

Structure of Compounds in the Solid State

Ionic solids are *crystalline,* characterized by a regular structure, whereas covalent solids may either be crystalline or have no regular structure. In the latter case they are said to be *amorphous.*

A HUMAN PERSPECTIVE

How the Elements Came into Being

The current, most widely held theory of the origin of the universe is the "big bang" theory. An explosion of very dense matter was followed by expansion into space of the fragments resulting from this explosion. This is one of the scenarios that have been created by scientists fascinated by the origins of matter, the stars and planets, and life as we know it today.

The first fragments, or particles, were protons and neutrons moving with tremendous velocity and possessing large amounts of energy. Collisions involving these high-energy protons and neutrons formed deuterium atoms (^{2}H), which are isotopes of hydrogen. As the universe expanded and cooled, tritium (^{3}H), another hydrogen isotope, formed as a result of collisions of neutrons with deuterium atoms. Subsequent capture of a proton produced helium (He). Scientists theorize that a universe that was principally composed of hydrogen and helium persisted for perhaps 100,000 years until the temperature decreased sufficiently to allow the formation of a simple molecule, hydrogen, two atoms of hydrogen bonded together (H_2).

Many millions of years later, the effect of gravity caused these small units to coalesce, first into clouds and eventually into stars, with temperatures of millions of degrees. In this setting, these small collections of protons and neutrons combined to form larger atoms such as carbon (C) and oxygen (O), then sodium (Na), neon (Ne), magnesium (Mg), silicon (Si), and so forth. Subsequent explosions of stars provided the conditions that formed many larger atoms. These fragments, gathered together by the force of gravity, are the most probable origin of the planets in our own solar system.

The reactions that formed the elements as we know them today were a result of a series of *fusion reactions*, the joining of nuclei to produce larger atoms at very high temperatures (millions of degrees Celsius). These fusion reactions are similar to processes that are currently being studied as a possible alternative source of nuclear power. We shall study such nuclear processes in more detail in Chapter 10.

Nuclear reactions of this type do not naturally occur on the earth today. The temperature is simply too low. As a result we have, for the most part, a collection of stable elements existing as chemical compounds, atoms joined together by chemical bonds while retaining their identity even in the combined state. Silicon exists all around us as sand and soil in a combined form, silicon dioxide; most metals exist as a part of a chemical compound, such as iron ore. We are learning more about the structure and properties of these compounds in this chapter.

Solutions of Ionic and Covalent Compounds

In Chapter 2 we saw that mixtures are either heterogeneous or homogeneous. A homogeneous mixture is a solution. Many ionic solids dissolve in solvents, such as water. An ionic solid, if soluble, will form positive and negative ions in solution by **dissociation.**

Because ions in water are capable of carrying (conducting) a current of electricity, we refer to these compounds as **electrolytes,** and the solution is termed an **electrolytic solution.** Covalent solids dissolved in solution usually retain their neutral (molecular) character and are **nonelectrolytes.** The solution is not an electrical conductor.

The role of the solvent in the dissolution of solids is discussed in Section 4.5.

4.4 Drawing Lewis Structures of Molecules and Polyatomic Ions

Lewis Structures of Molecules

In Section 4.1, we used Lewis structures of individual atoms to help us understand the bonding process. To begin to explain the relationship between molecular structure and molecular properties, we will first need a set of guidelines to help us write Lewis structures for more complex molecules.

1. *Use chemical symbols for the various elements to write the skeletal structure of the compound.* To accomplish this, place the bonded atoms next to one another. This is relatively easy for simple compounds; however, as the number of atoms in

Learning Goal

The skeletal structure indicates only the relative positions of atoms in the molecule or ion. Bonding information results from the Lewis structure.

A CLINICAL PERSPECTIVE

Blood Pressure and the Sodium Ion/Potassium Ion Ratio

When you have a physical exam, the physician measures your blood pressure. This indicates the pressure of blood against the walls of the blood vessels each time the heart pumps. A blood pressure reading is always characterized by two numbers. With every heartbeat there is an increase in pressure; this is the systolic blood pressure. When the heart relaxes between contractions, the pressure drops; this is the diastolic pressure. Thus the blood pressure is expressed as two values—for instance, 117/72—measured in millimeters of mercury. Hypertension is simply defined as high blood pressure. To the body it means that the heart must work too hard to pump blood, and this can lead to heart failure or heart disease.

Heart disease accounts for 50% of all deaths in the United States. Epidemiological studies correlate the following major risk factors with heart disease: heredity, sex, race, age, diabetes, cigarette smoking, high blood cholesterol, and hypertension. Obviously, we can do little about our age, sex, and genetic heritage, but we can stop smoking, limit dietary cholesterol, and maintain a normal blood pressure.

The number of Americans with hypertension is alarmingly high: 60 million adults and children. More than 10 million of these individuals take medication to control blood pressure, at a cost of nearly $2.5 billion each year. In many cases, blood pressure can be controlled without medication by increasing physical activity, losing weight, decreasing consumption of alcohol, and limiting intake of sodium.

It has been estimated that the average American ingests 7.5–10 g of salt (NaCl) each day. Because NaCl is about 40% (by mass) sodium ions, this amounts to 3–4 g of sodium daily. Until 1989 the Food and Nutrition Board of the National Academy of Sciences National Research Council's defined *estimated safe and adequate daily dietary intake* (ESADDI) of sodium ion was 1.1–3.3 g. Clearly, Americans exceed this recommendation.

Recently, studies have shown that excess sodium is not the sole consideration in the control of blood pressure. More important is the sodium ion/potassium ion (Na^+/K^+) ratio. That ratio should be about 0.6; in other words, our diet should contain about 67% more potassium than sodium. Does the typical American diet fall within this limit? Definitely not! Young American males (25–30 years old) consume a diet with a $Na^+/K^+ = 1.07$, and the diet of females of the same age range has a $Na^+/K^+ = 1.04$. It is little wonder that so many Americans suffer from hypertension.

How can we restrict sodium in the diet, while increasing the potassium? The following table lists a variety of foods that are low in sodium and high in potassium. These include fresh fruits and vegetables and fruit juices, a variety of cereals, unsalted nuts, and cooked dried beans (legumes). The table also notes some high-sodium, low-potassium foods. Notice that most of these are processed or prepared foods. This points out how difficult it can be to control sodium in the diet. The majority of the sodium that we ingest comes from commercially prepared foods. The consumer must read the nutritional information printed on cans and packages to determine whether the sodium levels are within acceptable limits.

Low Sodium Ion, High Potassium Ion Foods

Food Category	Examples
Fruit and fruit juices	Pineapple, grapefruit, pears, strawberries, watermelon, raisins, bananas, apricots, oranges
Low-sodium cereals	Oatmeal (unsalted), Roman Meal Hot Cereal, shredded wheat
Nuts (unsalted)	Hazelnuts, macadamia nuts, almonds, peanuts, cashews, coconut
Vegetables	Summer squash, zucchini, eggplant, cucumber, onions, lettuce, green beans, broccoli
Beans (dry, cooked)	Great Northern beans, lentils, lima beans, red kidney beans

High Sodium Ion, Low Potassium Ion Foods

Food Category	Examples
Fats	Butter, margarine, salad dressings
Soups	Onion, mushroom, chicken noodle, tomato, split pea
Breakfast cereals	Many varieties; consult the label for specific nutritional information.
Breads	Most varieties
Processed meats	Most varieties
Cheese	Most varieties

the compound increases, the possible number of arrangements increases dramatically. We may be told the pattern of arrangement of the atoms in advance; if not, we can make an intelligent guess and see if a reasonable Lewis structure can be constructed. Three considerations are very important here:

- the least electronegative atom will be placed in the central position (the central atom),
- hydrogen and fluorine (and the other halogens) often occupy terminal positions,
- carbon often forms chains of carbon-carbon covalent bonds.

2. *Determine the number of valence electrons associated with each atom; combine them to determine the total number of valence electrons in the compound.* However, if we are representing polyatomic cations or anions, we must account for the charge on the ion. Specifically:

   - for polyatomic cations, subtract one electron for each unit of positive charge. This accounts for the fact that the positive charge arises from electron loss.
   - for polyatomic anions, add one electron for each unit of negative charge. This accounts for excess negative charge resulting from electron gain.

3. *Connect the central atom to each of the surrounding atoms using electron pairs.* Then complete the octets of all of the atoms bonded to the central atom. Recall that hydrogen needs only two electrons to complete its valence shell. Electrons not involved in bonding must be represented as lone pairs and the total number of electrons in the structure must equal the number of valence electrons computed in our second step.

4. *If the octet rule is not satisfied for the central atom, move one or more electron pairs* from the surrounding atoms to create double or triple bonds until all atoms have an octet.

5. *After you are satisfied with the Lewis structure that you have constructed, perform a final electron count* verifying that the total number of electrons **and** the number around each atom are correct.

Now, let us see how these guidelines are applied in the examples that follow.

> The central atom is often the element farthest to the left and/or lowest in the periodic table.
> The central atom is often the element in the compound for which there is only one atom.
> Hydrogen is *never* the central atom.

Drawing Lewis Structures of Covalent Compounds

EXAMPLE 4.8

Draw the Lewis structure of carbon dioxide, CO_2.

Solution

Draw a skeletal structure of the molecule, arranging the atoms in their most probable order.

For CO_2, two possibilities exist:

$$C—O—O \quad \text{and} \quad O—C—O$$

Referring to Figure 4.2, we find that the electronegativity of oxygen is 3.5 whereas that of carbon is 2.5. Our strategy dictates that the least electronegative atom, in this case carbon, is the central atom. Hence the skeletal structure O—C—O may be presumed correct.

Next, we want to determine the number of valence electrons on each atom and add them to arrive at the total for the compound.

For CO_2,

$$\begin{array}{l} 1\text{ C atom} \times 4 \text{ valence electrons} = 4\text{ e}^- \\ 2\text{ O atoms} \times 6 \text{ valence electrons} = 12\text{ e}^- \\ \hline \phantom{2\text{ O atoms} \times 6 \text{ valence electrons} = } 16\text{ e}^-\text{ total} \end{array}$$

Now, use electron pairs to connect the central atom, C, to each oxygen with a single bond.

$$O:C:O$$

Distribute the electrons around the atoms (in pairs if possible) in an attempt to satisfy the octet rule, eight electrons around each element.

$$:\ddot{O}:C:\ddot{O}:$$

This structure satisfies the octet rule for each oxygen atom, but not the carbon atom (only four electrons surround the carbon).

However, when this structure is modified by moving two electrons from each oxygen atom to a position between C and O, each oxygen and carbon atom is surrounded by eight electrons. The octet rule is satisfied, and the structure below is the most probable Lewis structure for CO_2.

$$\ddot{O}::C::\ddot{O}$$

In this structure, four electrons (two electron pairs) are located between C and each O, and these electrons are shared in covalent bonds. Because a **single bond** is composed of two electrons (one electron pair) and because four electrons "bond" the carbon atom to each oxygen atom in this structure, there must be two bonds between each oxygen atom and the carbon atom, a **double bond:**

The notation for a single bond $:$ is equivalent to — (one pair of electrons).

The notation for a double bond $::$ is equivalent to $=$ (two pairs of electrons).

We may write CO_2 as shown above or, replacing $:$ with —,

$$\overline{O}=C=\overline{O}$$

As a final step, let us do some "electron accounting." There are eight dashes, or electron pairs, and they correspond to sixteen valence electrons (8 pair × 2e$^-$/pair). Furthermore, there are eight electrons around each atom and the octet rule is satisfied. Therefore

$$\overline{O}=C=\overline{O}$$

is a satisfactory way to depict the structure of CO_2

EXAMPLE 4.9

Drawing Lewis Structures of Covalent Compounds

Draw the Lewis structure of ammonia, NH_3.

Solution

When trying to implement the first step in our strategy we may be tempted to make H our central atom because it is less electronegative than N. But, remember the margin note in this section:
   "Hydrogen is *never* the central atom"
Hence:

$$\begin{array}{c} H \\ | \\ H-N-H \end{array}$$

is our skeletal structure.
   Applying our strategy to determine the total valence electrons for the molecule, we find that there are five valence electrons in nitrogen and one in each of the three hydrogens, for a total of eight valence electrons.

Applying our strategy for distribution of valence electrons results in the following Lewis diagram:

$$H : \overset{..}{\underset{..}{N}} : H$$ with H above N

This satisfies the octet rule for nitrogen (eight electrons around N) and hydrogen (two electrons around each H) and is an acceptable structure for ammonia. Ammonia may also be written:

$$H—\underset{\_}{N}—H$$ with H above N

Note the pair of nonbonding electrons on the nitrogen atom. These are often called a **lone pair,** or *unshared* pair, of electrons. As we will see later in this section, lone pair electrons have a profound effect on molecular geometry. The geometry, in turn, affects the reactivity of the molecule.

Draw a Lewis structure for each of the following covalent compounds:

a. H_2O (water)
b. CH_4 (methane)

Draw a Lewis structure for each of the following covalent compounds:

a. C_2H_6 (ethane)
b. N_2 (nitrogen gas)

Lewis Structures of Polyatomic Ions

The strategies for writing the Lewis structures of polyatomic ions are similar to those for neutral compounds. There is, however, one major difference: The charge on the ion must be accounted for when computing the total number of valence electrons.

Learning Goal

5

EXAMPLE 4.10

Drawing Lewis Structures of Polyatomic Cations

Draw the Lewis structure of the ammonium ion, NH_4^+.

Solution

The ammonium ion has the following skeletal structure and charge:

$$\left[H—N—H \right]^+$$ with H above and H below N

The total number of valence electrons is determined by subtracting one electron for each unit of positive charge.

Solution

Nitrogen is less electronegative than oxygen; therefore, nitrogen is the central atom and the skeletal structure is:

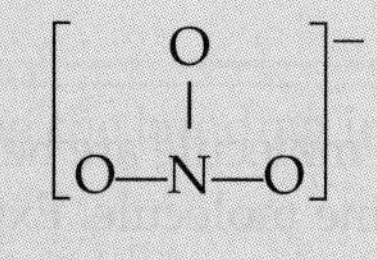

$$\left[\begin{array}{c} O \\ | \\ O{-}N{-}O \end{array} \right]^{-}$$

The pool of valence electrons for anions is determined by adding one electron for each unit of negative charge:

$$\begin{array}{l} 1 \text{ N atom} \times 5 \text{ valence electrons} = 5 \text{ e}^{-} \\ 3 \text{ O atoms} \times 6 \text{ valence electrons} = 18 \text{ e}^{-} \\ \underline{+ 1 \text{ negative charge} = 1 \text{ e}^{-}} \\ 24 \text{ e}^{-} \text{ total} \end{array}$$

Distributing the electrons throughout the structure results in the legitimate Lewis structures:

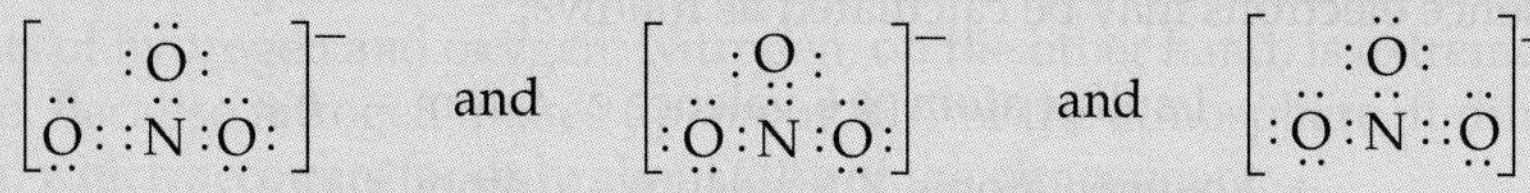

$$\left[\begin{array}{c} :\ddot{O}: \\ \ddot{O}::\ddot{N}:\ddot{O}: \end{array} \right]^{-} \text{ and } \left[\begin{array}{c} :\ddot{O}: \\ :\ddot{O}:\ddot{N}:\ddot{O}: \end{array} \right]^{-} \text{ and } \left[\begin{array}{c} :\ddot{O}: \\ :\ddot{O}:\ddot{N}::\ddot{O} \end{array} \right]^{-}$$

All contribute to the true structure of the nitrate ion, represented as a resonance hybrid.

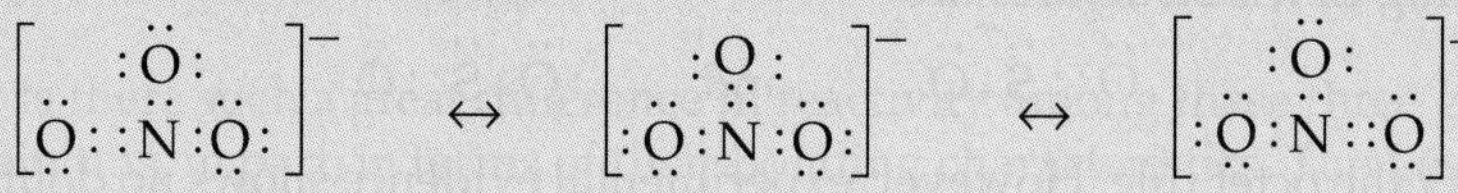

$$\left[\begin{array}{c} :\ddot{O}: \\ \ddot{O}::\ddot{N}:\ddot{O}: \end{array} \right]^{-} \leftrightarrow \left[\begin{array}{c} :\ddot{O}: \\ :\ddot{O}:\ddot{N}:\ddot{O}: \end{array} \right]^{-} \leftrightarrow \left[\begin{array}{c} :\ddot{O}: \\ :\ddot{O}:\ddot{N}::\ddot{O} \end{array} \right]^{-}$$

Question 4.19

SeO_2, like SO_2, has two resonance forms. Draw their Lewis structures.

Question 4.20

Explain any similarities between the structures for SeO_2 and SO_2 (in Question 4.19) in light of periodic relationships.

Lewis Structures and Exceptions to the Octet Rule

The octet rule is remarkable in its ability to realistically model bonding and structure in covalent compounds. But, like any model, it does not adequately describe all systems. Beryllium, boron, and aluminum, in particular, tend to form compounds in which they are surrounded by fewer than eight electrons. This situation is termed an *incomplete octet*. Other molecules, such as nitric oxide:

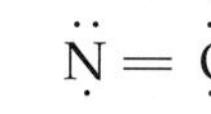

$$\dot{N} = \ddot{O}$$

are termed *odd electron* molecules. Note that it is impossible to pair all electrons to achieve an octet simply because the compound contains an odd number of valence electrons. Elements in the third period and beyond may involve *d* orbitals and form an *expanded octet*, with ten or even twelve electrons surrounding the central atom. Examples 4.14 and 4.15 illustrate common exceptions to the octet rule.

EXAMPLE 4.14

Drawing Lewis Structures of Covalently Bonded Compounds that are Exceptions to the Octet Rule

Draw the Lewis structure of beryllium hydride, BeH_2.

Solution

A reasonable skeletal structure of BeH_2 is:

$$H—Be—H$$

The total number of valence electrons in BeH_2 is:

$$1 \text{ beryllium atom } \times 2 \text{ valence e}^-/\text{atom} = 2 \text{ e}^-$$
$$\underline{2 \text{ hydrogen atoms } \times 1 \text{ valence e}^-/\text{atom} = 2 \text{ e}^-}$$
$$4 \text{ e}^- \text{ total}$$

The resulting Lewis structure must be:

$$H:Be:H \quad \text{ or } \quad H—Be—H$$

It is apparent that there is no way to satisfy the octet rule for Be in this compound. Consequently, BeH_2 is an exception to the octet rule. It contains an incomplete octet.

EXAMPLE 4.15

Drawing Lewis Structures of Covalently Bonded Compounds that are Exceptions to the Octet Rule

Draw the Lewis structure of phosphorus pentafluoride.

Solution

A reasonable skeletal structure of PF_5 is:

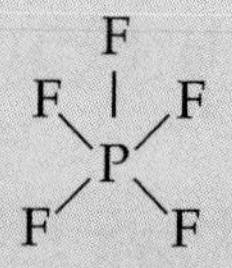

Phosphorus is a third-period element; it may have an expanded octet. The total number of valence electrons is:

$$1 \text{ phosphorus atom } \times 5 \text{ valence e}^-/\text{atom} = 5 \text{ e}^-$$
$$\underline{5 \text{ fluorine atoms } \times 7 \text{ valence e}^-/\text{atom} = 35 \text{ e}^-}$$
$$40 \text{ e}^- \text{ total}$$

Distributing the electrons around each F in the skeletal structure results in the Lewis structure:

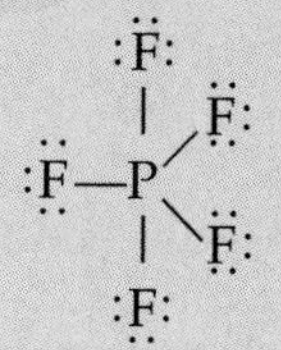

PF_5 is an example of a compound with an expanded octet.

CHEMISTRY CONNECTION

The Chemistry of Automobile Air Bags

Each year, thousands of individuals are killed or seriously in-jured in automobile accidents. Perhaps most serious is the front-end collision. The car decelerates or stops virtually on im-pact; the momentum of the passengers, however, does not stop, and the driver and passengers are thrown forward toward the dashboard and the front window. Suddenly, passive parts of the automobile, such as control knobs, the rearview mirror, the steering wheel, the dashboard, and the windshield, become lethal weapons.

Automobile engineers have been aware of these problems for a long time. They have made a series of design improve-ments to lessen the potential problems associated with front-end impact. Smooth switches rather than knobs, recessed hardware, and padded dashboards are examples. These changes, coupled with the use of lap and shoulder belts, which help to immobilize occupants of the car, have decreased the fre-quency and severity of the impact and lowered the death rate for this type of accident.

An almost ideal protection would be a soft, fluffy pillow, providing a cushion against impact. Such a device, an air bag inflated only on impact, is now standard equipment for the protection of the driver and front-seat passenger.

How does it work? Ideally, it inflates only when severe front-end impact occurs; it inflates very rapidly (in approxi-mately 40 milliseconds), then deflates to provide a steady deceleration, cushioning the occupants from impact. A remark-ably simple chemical reaction makes this a reality.

When solid sodium azide (NaN_3) is detonated by mechani-cal energy produced by an electric current, it decomposes to form solid sodium and nitrogen gas:

$$2NaN_3(s) \longrightarrow 2Na(s) + 3N_2(g)$$

The nitrogen gas inflates the air bag, cushioning the driver and front-seat passenger.

The solid sodium azide has a high density (characteristic of solids) and thus occupies a small volume. It can easily be stored in the center of a steering wheel or in the dashboard. The rate of the detonation is very rapid. In milliseconds it produces three moles of N_2 gas for every two moles of NaN_3. The N_2 gas occu-pies a relatively large volume because its density is low. This is a general property of gases.

Figuring out how much sodium azide is needed to produce enough nitrogen to properly inflate the bag is an example of a practical application of the chemical arithmetic that we are learning in this chapter.

Introduction

The calculation of chemical quantities based on chemical equations, termed *stoichiometry*, is the application of logic and arithmetic to chemical systems to an-swer questions such as the following:

A pharmaceutical company wishes to manufacture 1000 kg of a product next year. How much of each of the starting materials must be ordered? If the starting materials cost \$20/g, how much money must be budgeted for the project?

We often need to predict the quantity of a product produced from the reaction of a given amount of material. This calculation is possible. It is equally possible to calculate how much of a material would be necessary to produce a desired amount of product. One of many examples was shown in the preceding Chemistry Con-nection: the need to solve a very practical problem.

What is required is a recipe: a procedure to follow. The basis for our recipe is the *chemical equation*. A properly written chemical equation provides all of the nec-essary information for the chemical calculation. That critical information is the *combining ratio* of elements or compounds that must occur to produce a certain amount of product or products.

In this chapter we define the mole, the fundamental unit of measure of chem-ical arithmetic, learn to write and balance chemical equations, and use these tools to perform calculations of chemical quantities.

5.1 The Mole Concept and Atoms

Atoms are exceedingly small, yet their masses have been experimentally determined for each of the elements. The unit of measurement for these determinations is the **atomic mass unit,** abbreviated amu:

$$1 \text{ amu} = 1.661 \times 10^{-24} \text{ g}$$

The Mole and Avogadro's Number

The exact value of the atomic mass unit is defined in relation to a standard, just as the units of the metric system represent defined quantities. The carbon-12 isotope has been chosen and is assigned a mass of exactly 12 atomic mass units. Hence this standard reference point defines an atomic mass unit as exactly one-twelfth the mass of a carbon-12 atom.

The periodic table provides atomic weights in atomic mass units. These atomic weights are average values, based on the contribution of all the isotopes of the particular element. For example, the average mass of a carbon atom is 12.01 amu and

$$\frac{12.01 \text{ amu C}}{\text{C atom}} \times \frac{1.661 \times 10^{-24} \text{ g C}}{1 \text{ amu C}} = 1.995 \times \frac{10^{-23} \text{ g C}}{\text{C atom}}$$

The average mass of a helium atom is 4.003 amu and

$$\frac{4.003 \text{ amu He}}{\text{He atom}} \times \frac{1.661 \times 10^{-24} \text{ g He}}{1 \text{ amu He}} = 6.649 \times \frac{10^{-24} \text{ g He}}{\text{He atom}}$$

In everyday work, chemists use much larger quantities of matter (typically, grams or kilograms). A more practical unit for defining a "collection" of atoms is the **mole:**

$$1 \text{ mol of atoms} = 6.022 \times 10^{23} \text{ atoms of an element}$$

This number is **Avogadro's number.** Amedeo Avogadro, a nineteenth-century scientist, conducted a series of experiments that provided the basis for the mole concept.

The practice of defining a unit for a quantity of small objects is common; a *dozen* eggs, a *ream* of paper, and a *gross* of pencils are well-known examples. Similarly, a mole is 6.022×10^{23} individual units of anything. We could, if we desired, speak of a mole of eggs or a mole of pencils. However, in chemistry we use the mole to represent a specific quantity of atoms, ions, or molecules.

The mole (mol) and the atomic mass unit (amu) are related. The atomic mass of an element corresponds to the average mass of a single atom in amu *and* the mass of a mole of atoms in grams.

The mass of 1 mol of atoms, in grams, is defined as the **molar mass.** Consider this relationship for sodium in Example 5.1.

Learning Goal
1

The term atomic weight is not correct but is a fixture in common usage. Just remember that atomic weight is really "average atomic mass."

Learning Goal
2

EXAMPLE 5.1

Relating Avogadro's Number to Molar Mass

Calculate the mass, in grams, of Avogadro's number of sodium atoms.

Solution

The average mass of one sodium atom is 22.99 amu. This may be formatted as the conversion factor:

$$\frac{22.99 \text{ amu Na}}{1 \text{ atom Na}}$$

As previously noted, 1 amu is 1.661×10^{-24} g, and 6.022×10^{23} atoms of sodium is Avogadro's number. Similarly, these relationships may be formatted as:

$$1.661 \times 10^{-24} \frac{\text{g Na}}{\text{amu}} \text{ and } 6.022 \times 10^{23} \frac{\text{atoms Na}}{\text{mol Na}}$$

Formatting this information as a series of conversion factors, using the factor-label method, we have

Section 1.3 discusses the use of conversion factors.

$$22.99 \frac{\text{amu Na}}{\text{atom Na}} \times 1.661 \times 10^{-24} \frac{\text{g Na}}{\text{amu Na}} \times 6.022 \times 10^{23} \frac{\text{atoms Na}}{\text{mol Na}} = 22.99 \frac{\text{g Na}}{\text{mol Na}}$$

The average mass of one *atom* of sodium, in units of amu, is *numerically identical* to the mass of *Avogadro's number of atoms*, expressed in units of grams. Hence the molar mass of sodium is 22.99 g Na/mol.

The sodium example is not unique. The relationship holds for every element in the periodic table.

Because Avogadro's number of particles (atoms) is 1 mol, it follows that

the average mass of one atom of hydrogen is 1.008 amu

and

the mass of 1 mol of hydrogen atoms is 1.008 g

or

the average mass of one atom of carbon is 12.01 amu

and

the mass of 1 mol of carbon atoms is 12.01 g

and so forth. One mole of atoms of *any element* contains the same number, Avogadro's number, of atoms, 6.022×10^{23} atoms.

The difference in mass of a mole of two different elements can be quite striking (Figure 5.1). For example, a mole of hydrogen atoms is 1.008 g, and a mole of lead atoms is 207.19 g.

Figure 5.1

The comparison of approximately one mole each of silver (as Morgan and Peace dollars), gold (as Canadian Maple Leaf coins), and copper (as pennies) shows the considerable difference in mass (as well as economic value) of equivalent moles of different substances.

Question 5.1

Calculate the mass, in grams, of Avogadro's number of aluminum atoms.

Question 5.2

Calculate the mass, in grams, of Avogadro's number of mercury atoms.

Calculating Atoms, Moles, and Mass

Learning Goal

Performing calculations based on a chemical equation requires a facility for relating the number of atoms of an element to a corresponding number of moles of that element and ultimately to their mass in grams. Such calculations involve the use of conversion factors. This type of calculation was first described in Chapter 1. Some examples follow.

EXAMPLE 5.2

Converting Moles to Atoms

How many iron atoms are present in 3.0 mol of iron metal?

Solution

The calculation is based on choosing the appropriate conversion factor. The relationship

$$\frac{6.022 \times 10^{23} \text{ atoms Fe}}{1 \text{ mol Fe}}$$

follows directly from

$$1 \text{ mol Fe} = 6.022 \times 10^{23} \text{ atoms Fe}$$

Using this conversion factor, we have

$$\text{number of atoms of Fe} = 3.0 \text{ mol Fe} \times \frac{6.022 \times 10^{23} \text{ atoms Fe}}{1 \text{ mol Fe}}$$

$$= 18 \times 10^{23} \text{ atoms of Fe, or}$$

$$= 1.8 \times 10^{24} \text{ atoms of Fe}$$

Converting Atoms to Moles

EXAMPLE 5.3

Calculate the number of moles of sulfur represented by 1.81×10^{24} atoms of sulfur.

Solution

$$1.81 \times 10^{24} \text{ atoms S} \times \frac{1 \text{ mol S}}{6.022 \times 10^{23} \text{ atoms S}} = 3.01 \text{ mol S}$$

Note that this conversion factor is the inverse of that used in Example 5.2. Remember, the conversion factor must cancel units that should not appear in the final answer.

Converting Moles of a Substance to Mass in Grams

EXAMPLE 5.4

What is the mass, in grams, of 3.01 mol of sulfur?

Solution

We know from the periodic table that 1 mol of sulfur has a mass of 32.06 g. Setting up a suitable conversion factor between grams and moles results in

$$3.01 \text{ mol S} \times \frac{32.06 \text{ g S}}{1 \text{ mol S}} = 96.5 \text{ g S}$$

Converting Kilograms to Moles

EXAMPLE 5.5

Calculate the number of moles of sulfur in 1.00 kg of sulfur.

Solution

$$1.00 \text{ kg S} \times \frac{10^3 \text{ g S}}{1 \text{ kg S}} \times \frac{1 \text{ mol S}}{32.06 \text{ g S}} = 31.2 \text{ mol S}$$

- heat or light absorbed or emitted as the result of a reaction,
- changes in the way the sample behaves in an electric or magnetic field before and after a reaction, and
- changes in electrical properties before and after a reaction.

Whether we use our senses or a $100,000 computerized instrument, the "bottom line" is the same: We are measuring a change in one or more chemical or physical properties in an effort to understand the changes taking place in a chemical system.

Disease can be described as a chemical system (actually a biochemical system) gone awry. Here, too, the underlying changes may not be obvious. Just as technology has helped chemists see subtle chemical changes in the laboratory, medical diagnosis has been revolutionized in our lifetimes using very similar technology. Some of these techniques are described in the Clinical Perspective, Magnetic Resonance Imaging, in Chapter 10.

5.5 Balancing Chemical Equations

Learning Goal

The chemical equation shows the *molar quantity* of reactants needed to produce a certain *molar quantity* of products.

The relative number of moles of each product and reactant is indicated by placing a whole-number *coefficient* before the formula of each substance in the chemical equation. A coefficient of 2 (for example, 2NaCl) indicates that 2 mol of sodium chloride are involved in the reaction. Also, $3NH_3$ signifies 3 mol of ammonia; it means that 3 mol of nitrogen atoms and 3×3, or 9, mol of hydrogen atoms are involved in the reaction. The coefficient 1 is understood, not written. Therefore H_2SO_4 would be interpreted as 1 mol of sulfuric acid, or 2 mol of hydrogen atoms, 1 mol of sulfur atoms, and 4 mol of oxygen atoms.

The equation

$$CaCO_3(s) \xrightarrow{\Delta} CaO(s) + CO_2(g)$$

is balanced as written. On the reactant side we have

1 mol Ca
1 mol C
3 mol O

On the product side there are

1 mol Ca
1 mol C
3 mol O

The coefficients indicate *relative* numbers of moles: 10 mol of $CaCO_3$ produce 10 mol of CaO; 0.5 mol of $CaCO_3$ produce 0.5 mol of CaO; and so forth.

Therefore the law of conservation of mass is obeyed, and the equation is balanced as written.

Now consider the reaction of aqueous hydrogen chloride with solid calcium metal in aqueous solution:

$$HCl(aq) + Ca(s) \longrightarrow CaCl_2(aq) + H_2(g)$$

The equation, as written, is not balanced.

Reactants	Products
1 mol H atoms	2 mol H atoms
1 mol Cl atoms	2 mol Cl atoms
1 mol Ca atoms	1 mol Ca atoms

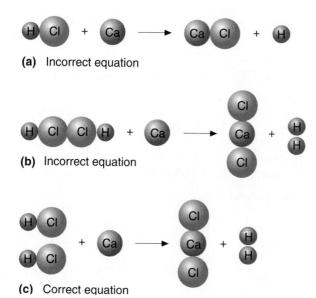

(a) Incorrect equation

(b) Incorrect equation

(c) Correct equation

Figure 5.5
Balancing the equation HCl + Ca →
CaCl$_2$ + H$_2$. (a) Neither product is the
correct chemical species. (b) The reactant,
HCl, is incorrectly represented as H$_2$Cl$_2$.
(c) This equation is correct; all species are
correct, and the law of conservation of
mass is obeyed.

We need 2 mol of both H and Cl on the left, or reactant, side. An *incorrect* way of balancing the equation is as follows:

$$H_2Cl_2(aq) + Ca(s) \longrightarrow CaCl_2(aq) + H_2(g)$$

Not a correct equation

The equation satisfies the law of conservation of mass; however, we have altered one of the reacting species. Hydrogen chloride is HCl, not H$_2$Cl$_2$. We must remember that *we cannot alter any chemical substance in the process of balancing the equation.* We can *only* introduce coefficients into the equation. Changing subscripts changes the identity of the chemicals involved, and that is not permitted. The equation must represent the reaction accurately. The correct equation is

$$2HCl(aq) + Ca(s) \longrightarrow CaCl_2(aq) + H_2(g)$$

Correct equation

This process is illustrated in Figure 5.5.

Many equations are balanced by trial and error. After the identity of the products and reactants, the physical state, and the reaction conditions are known, the following steps provide a method for correctly balancing a chemical equation:

Step 1. Count the number of moles of atoms of each element on both product and reactant side.

Step 2. Determine which elements are not balanced.

Step 3. Balance one element at a time using coefficients.

Step 4. After you believe that you have successfully balanced the equation, check, as in Step 1, to be certain that mass conservation has been achieved.

Let us apply these steps to the reaction of calcium with aqueous hydrogen chloride:

$$HCl(aq) + Ca(s) \longrightarrow CaCl_2(aq) + H_2(g)$$

Step 1. **Reactants**   **Products**
1 mol H atoms   2 mol H atoms
1 mol Cl atoms   2 mol Cl atoms
1 mol Ca atoms   1 mol Ca atoms

Coefficients placed in front of the formula indicate the relative number of moles of compound (represented by the formula) that are involved in the reaction. Subscripts placed to the lower right of the atomic symbol indicate the relative number of atoms in the compound.

Water (H$_2$O) and hydrogen peroxide (H$_2$O$_2$) illustrate the effect a subscript can have. The two compounds show marked differences in physical and chemical properties.

Step 2. The numbers of moles of H and Cl are not balanced.

Step 3. Insertion of a 2 before HCl on the reactant side should balance the equation:

$$2HCl(aq) + Ca(s) \longrightarrow CaCl_2(aq) + H_2(g)$$

Step 4. Check for mass balance:

Reactants	Products
2 mol H atoms	2 mol H atoms
2 mol Cl atoms	2 mol Cl atoms
1 mol Ca atoms	1 mol Ca atoms

Hence the equation is balanced.

EXAMPLE 5.10

Balancing Equations

Balance the following equation: Hydrogen gas and oxygen gas react explosively to produce gaseous water.

Solution

Recall that hydrogen and oxygen are diatomic gases; therefore

$$H_2(g) + O_2(g) \longrightarrow H_2O(g)$$

Note that the moles of hydrogen atoms are balanced but that the moles of oxygen atoms are not; therefore we must first balance the moles of oxygen atoms:

$$H_2(g) + O_2(g) \longrightarrow 2H_2O(g)$$

Balancing moles of oxygen atoms creates an imbalance in the number of moles of hydrogen atoms, so

$$2H_2(g) + O_2(g) \longrightarrow 2H_2O(g)$$

The equation is balanced, with 4 mol of hydrogen atoms and 2 mol of oxygen atoms on each side of the reaction arrow.

EXAMPLE 5.11

Balancing Equations

Balance the following equation: Propane gas, C_3H_8, a fuel, reacts with oxygen gas to produce carbon dioxide and water vapor. The reaction is

$$C_3H_8(g) + O_2(g) \longrightarrow CO_2(g) + H_2O(g)$$

Solution

First, balance the carbon atoms; there are 3 mol of carbon atoms on the left and only 1 mol of carbon atoms on the right. We need $3CO_2$ on the right side of the equation:

$$C_3H_8(g) + O_2(g) \longrightarrow 3CO_2(g) + H_2O(g)$$

Next, balance the hydrogen atoms; there are 2 mol of hydrogen atoms on the right and 8 mol of hydrogen atoms on the left. We need $4H_2O$ on the right:

$$C_3H_8(g) + O_2(g) \longrightarrow 3CO_2(g) + 4H_2O(g)$$

There are now 10 mol of oxygen atoms on the right and 2 mol of oxygen atoms on the left. To balance, we must have $5O_2$ on the left side of the equation:

$$C_3H_8(g) + 5O_2(g) \longrightarrow 3CO_2(g) + 4H_2O(g)$$

Remember: In every case, be sure to check the final equation for mass balance.

Balancing Equations

Balance the following equation: Butane gas, C_4H_{10}, a fuel used in pocket lighters, reacts with oxygen gas to produce carbon dioxide and water vapor. The reaction is

$$C_4H_{10}(g) + O_2(g) \longrightarrow CO_2(g) + H_2O(g)$$

Solution

First, balance the carbon atoms; there are 4 mol of carbon atoms on the left and only 1 mol of carbon atoms on the right:

$$C_4H_{10}(g) + O_2(g) \longrightarrow 4CO_2(g) + H_2O(g)$$

Next, balance hydrogen atoms; there are 10 mol of hydrogen atoms on the left and only 2 mol of hydrogen atoms on the right:

$$C_4H_{10}(g) + O_2(g) \longrightarrow 4CO_2(g) + 5H_2O(g)$$

There are now 13 mol of oxygen atoms on the right and only 2 mol of oxygen atoms on the left. Therefore a coefficient of 6.5 is necessary for O_2.

$$C_4H_{10}(g) + 6.5O_2(g) \longrightarrow 4CO_2(g) + 5H_2O(g)$$

Fractional or decimal coefficients are often needed and used. However, the preferred form requires all integer coefficients. Multiplying each term in the equation by a suitable integer (2, in this case) satisfies this requirement. Hence

$$2C_4H_{10}(g) + 13O_2(g) \longrightarrow 8CO_2(g) + 10H_2O(g)$$

The equation is balanced, with 8 mol of carbon atoms, 20 mol of hydrogen atoms, and 26 mol of oxygen atoms on each side of the reaction arrow.

EXAMPLE 5.12

When balancing equations, we find that it is often most efficient to begin by balancing the atoms in the most complicated formulas.

Balancing Equations

Balance the following equation: Aqueous ammonium sulfate reacts with aqueous lead nitrate to produce aqueous ammonium nitrate and solid lead sulfate. The reaction is

$$(NH_4)_2SO_4(aq) + Pb(NO_3)_2(aq) \longrightarrow NH_4NO_3(aq) + PbSO_4(s)$$

Solution

In this case the polyatomic ions remain as intact units. Therefore we can balance them as we would balance molecules rather than as atoms.

EXAMPLE 5.13

There are two ammonium ions on the left and only one ammonium ion on the right. Hence

$$(NH_4)_2SO_4(aq) + Pb(NO_3)_2(aq) \longrightarrow 2NH_4NO_3(aq) + PbSO_4(s)$$

No further steps are necessary. The equation is now balanced. There are two ammonium ions, two nitrate ions, one lead ion, and one sulfate ion on each side of the reaction arrow.

Question 5.9

Balance each of the following chemical equations:
a. $Fe(s) + O_2(g) \longrightarrow Fe_2O_3(s)$
b. $C_6H_6(l) + O_2(g) \longrightarrow CO_2(g) + H_2O(g)$

Question 5.10

Balance each of the following chemical equations:
a. $S_2Cl_2(s) + NH_3(g) \longrightarrow N_4S_4(s) + NH_4Cl(s) + S_8(s)$
b. $C_2H_5OH(l) + O_2(g) \longrightarrow CO_2(g) + H_2O(g)$

5.6 The Extent of Chemical Reactions

All of the reactions discussed thus far go to completion; the reactants are, for all practical purposes, completely converted to products. However, many reactions do not completely convert reactant to product. The explanation is simply that some tendency exists for the reverse reaction to occur; product converting back to the reactant. Reactions that do not go to completion are easily recognized; a double arrow ⇆ symbolizes the tendency for simultaneous forward and reverse reactions. For example, a mixture of N_2 and H_2 will not completely react to produce NH_3, and we represent this incomplete reaction as:

$$N_2(g) + 3H_2(g) \rightleftharpoons 2NH_3(g)$$

Such reactions are termed **equilibrium reactions.** After some time, a mixture of N_2, H_2, and NH_3 remains, and the amount of each component no longer changes. However, if we could peer into this mixture and see the individual molecules, we would observe a beehive of activity. NH_3 molecules would be decomposing to form H_2 and N_2. At the same time, these H_2 and N_2 molecules would be colliding and re-forming NH_3. One process would occur as fast as the other, so no net change in amounts would be apparent.

The process that we have just described is termed a **dynamic equilibrium.** A dynamic equilibrium is characterized by constancy amid change. The measured quantities of reactants and products in the container do not change when equilibrium is reached, although the individual molecules are continuously interconverting.

Section 7.3 and Section 7.6 discuss the properties of solutions.

In Chapter 8 we shall investigate the process of equilibrium in detail. In Section 6.2 we will discover that equilibrium processes govern physical as well as chemical change. Evaporation and boiling of liquids can be explained using equilibrium principles. Solubility of solids and gases in liquids is an equilibrium process as well. Explanations of events ranging from salt melting ice on a cold winter day to cellular transport in living systems rely on an equilibrium model.

In the next section, we will learn how to calculate quantities of reactants and products. It is important for you to remember that these calculations are valid only for complete reactions (single arrow). Other approaches, introduced in Chapter 8, are needed for incomplete, equilibrium processes (double arrow).

Carbon Monoxide Poisoning: A Case of Combining Ratios

A fuel, such as methane, CH_4, burned in an excess of oxygen produces carbon dioxide and water:

$$CH_4(g) + 2O_2(g) \longrightarrow CO_2(g) + 2H_2O(g)$$

The same combustion in the presence of insufficient oxygen produces carbon monoxide and water:

$$2CH_4(g) + 3O_2(g) \longrightarrow 2CO(g) + 4H_2O(g)$$

The combustion of methane, repeated over and over in millions of gas furnaces, is responsible for heating many of our homes in the winter. The furnace is designed to operate under conditions that favor the first reaction and minimize the second; excess oxygen is available from the surrounding atmosphere. Furthermore, the vast majority of exhaust gases (containing principally CO, CO_2, H_2O, and unburned fuel) are removed from the home through the chimney. However, if the chimney becomes obstructed, or the burner malfunctions, carbon monoxide levels within the home can rapidly reach hazardous levels.

Why is exposure to carbon monoxide hazardous? Hemoglobin, an iron-containing compound, binds with O_2 and trans-

ports it throughout the body. Carbon monoxide also combines with hemoglobin, thereby blocking oxygen transport. The binding affinity of hemoglobin for carbon monoxide is about 200 times as great as for O_2. Therefore, to maintain O_2 binding and transport capability, our exposure to carbon monoxide must be minimal. Proper ventilation and suitable oxygen-to-fuel ratio are essential for any combustion process in the home, automobile, or workplace. In recent years carbon monoxide sensors have been developed. These sensors sound an alarm when toxic levels of CO are reached. These warning devices have helped to create a safer indoor environment.

The example we have chosen is an illustration of what is termed the *law of multiple proportions*. This law states that identical reactants may produce different products, depending on their combining ratio. The experimental conditions (in this case, the quantity of available oxygen) determine the preferred path of the chemical reaction. In Section 5.7 we will learn how to use a properly balanced equation, representing the chemical change occurring, to calculate quantities of reactants consumed or products produced.

5.7 Calculations Using the Chemical Equation

General Principles

The calculation of quantities of products and reactants based on a balanced chemical equation is important in many fields. The synthesis of drugs and other complex molecules on a large scale is conducted on the basis of a balanced equation. This minimizes the waste of expensive chemical compounds used in these reactions. Similarly, the ratio of fuel and air in a home furnace or automobile must be adjusted carefully, according to their combining ratio, to maximize energy conversion, minimize fuel consumption, and minimize pollution.

Learning Goal **7**

In carrying out chemical calculations we apply the following guidelines.

1. The chemical formulas of all reactants and products must be known.
2. The basis for the calculations is a balanced equation because the conservation of mass must be obeyed. If the equation is not properly balanced, the calculation is meaningless.
3. The calculations are performed in terms of moles. The coefficients in the balanced equation represent the relative number of moles of products and reactants.

We have seen that the number of moles of products and reactants often differs in a balanced equation. For example,

$$C(s) + O_2(g) \longrightarrow CO_2(g)$$

is a balanced equation. Two moles of reactants combine to produce one mole of product:

$$1 \text{ mol C} + 1 \text{ mol O}_2 \longrightarrow 1 \text{ mol CO}_2$$

However, 1 mol of C *atoms* and 2 mol of O *atoms* produce 1 mol of C *atoms* and 2 mol of O *atoms*. In other words, the number of moles of reactants and products may differ, but the number of moles of atoms cannot. The formation of CO_2 from C and O_2 may be described as follows:

$$C(s) + O_2(g) \longrightarrow CO_2(g)$$

$$1 \text{ mol C} + 1 \text{ mol } O_2 \longrightarrow 1 \text{ mol } CO_2$$

$$12.0 \text{ g C} + 32.0 \text{ g } O_2 \longrightarrow 44.0 \text{ g } CO_2$$

The mole is the basis of our calculations. However, moles are generally measured in grams (or kilograms). A facility for interconversion of moles and grams is fundamental to chemical arithmetic (see Figure 5.2). These calculations are reviewed in Example 5.14.

Use of Conversion Factors

Conversion between Moles and Grams

Conversion from moles to grams, and vice versa, requires only the formula weight of the compound of interest. Consider the following examples.

EXAMPLE 5.14

Converting between Moles and Grams

a. Convert 1.00 mol of oxygen gas, O_2, to grams.

Solution

Use the following path:

$$\boxed{\text{moles of oxygen}} \longrightarrow \boxed{\text{grams of oxygen}}$$

The molar mass of oxygen (O_2) is

$$\frac{32.0 \text{ g } O_2}{1 \text{ mol } O_2}$$

Therefore

$$1.00 \text{ mol } O_2 \times \frac{32.0 \text{ g } O_2}{1 \text{ mol } O_2} = 32.0 \text{ g } O_2$$

b. How many grams of carbon dioxide are contained in 10.0 mol of carbon dioxide?

Solution

Use the following path:

$$\boxed{\text{moles of carbon dioxide}} \longrightarrow \boxed{\text{grams of carbon dioxide}}$$

The formula weight of CO_2 is

$$\frac{44.0 \text{ g } CO_2}{1 \text{ mol } CO_2}$$

and

$$10.0 \text{ mol } CO_2 \times \frac{44.0 \text{ g } CO_2}{1 \text{ mol } CO_2} = 4.40 \times 10^2 \text{ g } CO_2$$

c. How many moles of sodium are contained in 1 lb (454 g) of sodium metal?

Solution

Use the following path:

$$\text{grams of sodium} \longrightarrow \text{moles of sodium}$$

The number of moles of sodium atoms is

$$454 \text{ g Na} \times \frac{1 \text{ mol Na}}{22.99 \text{ g Na}} = 19.7 \text{ mol Na}$$

Note that each factor can be inverted producing a second possible factor. Only one will allow the appropriate unit cancellation.

Perform each of the following conversions:
a. 5.00 mol of water to grams of water
b. 25.0 g of LiCl to moles of LiCl

Question **5.11**

Perform each of the following conversions:
a. 1.00×10^{-5} mol of $C_6H_{12}O_6$ to micrograms of $C_6H_{12}O_6$
b. 35.0 g of $MgCl_2$ to moles of $MgCl_2$

Question **5.12**

Conversion of Moles of Reactants to Moles of Products

In Example 5.11 we balanced the equation for the reaction of propane and oxygen as follows:

$$C_3H_8(g) + 5O_2(g) \longrightarrow 3CO_2(g) + 4H_2O(g)$$

In this reaction, 1 mol of C_3H_8 corresponds to, or results in,

5 mol of O_2 being consumed and

3 mol of CO_2 being formed and

4 mol of H_2O being formed.

This information may be written in the form of a conversion factor or ratio:

$$1 \text{ mol } C_3H_8/5 \text{ mol } O_2$$

Translated: One mole of C_3H_8 reacts with five moles of O_2.

$$1 \text{ mol } C_3H_8/3 \text{ mol } CO_2$$

Translated: One mole of C_3H_8 produces three moles of CO_2.

$$1 \text{ mol } C_3H_8/4 \text{ mol } H_2O$$

Translated: One mole of C_3H_8 produces four moles of H_2O.

Conversion factors, based on the chemical equation, permit us to perform a variety of calculations.

Let us look at a few examples, based on the combustion of propane and the equation that we balanced in Example 5.11.

EXAMPLE 5.15

Calculating Reacting Quantities

Calculate the number of grams of O_2 that will react with 1.00 mol of C_3H_8.

Solution

Two conversion factors are necessary to solve this problem:

1. conversion from moles of C_3H_8 to moles of O_2 and
2. conversion of moles of O_2 to grams of O_2.

Therefore our path is

$$\text{moles} \atop C_3H_8 \longrightarrow \text{moles} \atop O_2 \longrightarrow \text{grams} \atop O_2$$

and

$$1.00 \ \cancel{\text{mol } C_3H_8} \times \frac{5 \ \cancel{\text{mol } O_2}}{1 \ \cancel{\text{mol } C_3H_8}} \times \frac{32.0 \ \text{g } O_2}{1 \ \cancel{\text{mol } O_2}} = 1.60 \times 10^2 \ \text{g } O_2$$

EXAMPLE 5.16

Calculating Grams of Product from Moles of Reactant

Calculate the number of grams of CO_2 produced from the combustion of 1.00 mol of C_3H_8.

Solution

Employ logic similar to that used in Example 5.15 and use the following path:

$$\text{moles} \atop C_3H_8 \longrightarrow \text{moles} \atop CO_2 \longrightarrow \text{grams} \atop CO_2$$

Then

$$1.00 \ \cancel{\text{mol } C_3H_8} \times \frac{3 \ \cancel{\text{mol } CO_2}}{1 \ \cancel{\text{mol } C_3H_8}} \times \frac{44.0 \ \text{g } CO_2}{1 \ \cancel{\text{mol } CO_2}} = 132 \ \text{g } CO_2$$

EXAMPLE 5.17

Relating Masses of Reactants and Products

Calculate the number of grams of C_3H_8 required to produce 36.0 g of H_2O.

Solution

It is necessary to convert

1. grams of H_2O to moles of H_2O,
2. moles of H_2O to moles of C_3H_8, and
3. moles of C_3H_8 to grams of C_3H_8.

Use the following path:

$$\begin{array}{ccccccc} \text{grams} & \longrightarrow & \text{moles} & \longrightarrow & \text{moles} & \longrightarrow & \text{grams} \\ H_2O & & H_2O & & C_3H_8 & & C_3H_8 \end{array}$$

Then

$$36.0 \text{ g } H_2O \times \frac{1 \text{ mol } H_2O}{18.0 \text{ g } H_2O} \times \frac{1 \text{ mol } C_3H_8}{4 \text{ mol } H_2O} \times \frac{44.0 \text{ g } C_3H_8}{1 \text{ mol } C_3H_8} = 22.0 \text{ g } C_3H_8$$

Question **5.13**

The balanced equation for the combustion of ethanol (ethyl alcohol) is:

$$C_2H_5OH(l) + 3O_2(g) \longrightarrow 2CO_2(g) + 3H_2O(g)$$

a. How many moles of O_2 will react with 1 mol of ethanol?
b. How many grams of O_2 will react with 1 mol of ethanol?

Question **5.14**

How many grams of CO_2 will be produced by the combustion of 1 mol of ethanol? (See Question 5.13.)

Let's consider an example that requires us to write and balance the chemical equation, use conversion factors, and calculate the amount of a reactant consumed in the chemical reaction.

EXAMPLE 5.18

Calculating a Quantity of Reactant

Calcium hydroxide may be used to neutralize (completely react with) aqueous hydrochloric acid. Calculate the number of grams of hydrochloric acid that would be neutralized by 0.500 mol of solid calcium hydroxide.

The reaction between an acid and a base produces a salt and water (Chapter 9).

Solution

The formula for calcium hydroxide is $Ca(OH)_2$ and that for hydrochloric acid is HCl. The unbalanced equation produces calcium chloride and water as products:

$$Ca(OH)_2(s) + HCl(aq) \longrightarrow CaCl_2(aq) + H_2O(l)$$

First, balance the equation:

$$Ca(OH)_2(s) + 2HCl(aq) \longrightarrow CaCl_2(aq) + 2H_2O(l)$$

Next, determine the necessary conversion:

1. moles of $Ca(OH)_2$ to moles of HCl and
2. moles of HCl to grams of HCl.

Use the following path:

$$\begin{array}{ccccc} \text{moles} & \longrightarrow & \text{moles} & \longrightarrow & \text{grams} \\ Ca(OH)_2 & & HCl & & HCl \end{array}$$

$$0.500 \text{ mol } Ca(OH)_2 \times \frac{2 \text{ mol } HCl}{1 \text{ mol } Ca(OH)_2} \times \frac{36.5 \text{ g } HCl}{1 \text{ mol } HCl} = 36.5 \text{ g } HCl$$

This reaction is illustrated in Figure 5.6.

Figure 5.6

An illustration of the law of conservation of mass. In this example, 1 mol of calcium hydroxide and 2 mol of hydrogen chloride react to produce 3 mol of product (2 mol of water and 1 mol of calcium chloride). The total mass, in grams, of reactant(s) consumed is equal to the total mass, in grams, of product(s) formed. *Note:* In reality, HCl does not exist as discrete molecules in water. The HCl separates to form H^+ and Cl^-. Ionization in water will be discussed with the chemistry of acids and bases in Chapter 9.

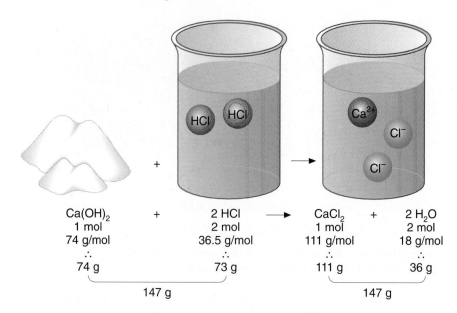

$Ca(OH)_2$	+	2 HCl	$\longrightarrow$	$CaCl_2$	+	$2 H_2O$
1 mol		2 mol		1 mol		2 mol
74 g/mol		36.5 g/mol		111 g/mol		18 g/mol
∴		∴		∴		∴
74 g		73 g		111 g		36 g

$$\underbrace{\qquad\qquad\qquad}_{147\text{ g}} \qquad\qquad \underbrace{\qquad\qquad\qquad}_{147\text{ g}}$$

EXAMPLE 5.19

Mass is a laboratory unit, whereas moles is a calculation unit. The laboratory balance is calibrated in units of mass (grams). Although moles are essential for calculation, often the starting point and objective are in mass units. As a result, our path is often grams → moles → grams.

Calculating Reactant Quantities

What mass of sodium hydroxide, NaOH, would be required to produce 8.00 g of the antacid milk of magnesia, $Mg(OH)_2$, by the reaction of $MgCl_2$ with NaOH?

Solution

$$MgCl_2(aq) + 2NaOH(aq) \longrightarrow Mg(OH)_2(s) + 2NaCl(aq)$$

The equation tells us that 2 mol of NaOH form 1 mol of $Mg(OH)_2$. If we calculate the number of moles of $Mg(OH)_2$ in 8.00 g of $Mg(OH)_2$, we can determine the number of moles of NaOH necessary and then the mass of NaOH required:

$$\boxed{\text{mass } Mg(OH)_2} \longrightarrow \boxed{\text{moles } Mg(OH)_2} \longrightarrow \boxed{\text{moles } NaOH} \longrightarrow \boxed{\text{mass } NaOH}$$

$$58.3 \text{ g } Mg(OH)_2 = 1 \text{ mol } Mg(OH)_2$$

Therefore

$$8.00 \text{ g } Mg(OH)_2 \times \frac{1 \text{ mol } Mg(OH)_2}{58.3 \text{ g } Mg(OH)_2} = 0.137 \text{ mol } Mg(OH)_2$$

Two moles of NaOH react to give one mole of $Mg(OH)_2$. Therefore

$$0.137 \text{ mol } Mg(OH)_2 \times \frac{2 \text{ mol } NaOH}{1 \text{ mol } Mg(OH)_2} = 0.274 \text{ mol } NaOH$$

40.0 g of NaOH = 1 mol of NaOH. Therefore

$$0.274 \text{ mol } NaOH \times \frac{40.0 \text{ g } NaOH}{1 \text{ mol } NaOH} = 11.0 \text{ g } NaOH$$

The calculation may be done in a single step:

$$8.00 \text{ g Mg(OH)}_2 \times \frac{1 \text{ mol Mg(OH)}_2}{58.3 \text{ g Mg(OH)}_2} \times \frac{2 \text{ mol NaOH}}{1 \text{ mol Mg(OH)}_2} \times \frac{40.0 \text{ g NaOH}}{1 \text{ mol NaOH}} = 11.0 \text{ g NaOH}$$

Note once again that we have followed a logical and predictable path to the solution:

$$\begin{array}{ccccccc}
\text{grams} & \longrightarrow & \text{moles} & \longrightarrow & \text{moles} & \longrightarrow & \text{grams} \\
\text{Mg(OH)}_2 & & \text{Mg(OH)}_2 & & \text{NaOH} & & \text{NaOH}
\end{array}$$

A general problem-solving strategy is summarized in Figure 5.7. By systematically applying this strategy, you will be able to solve virtually any problem requiring calculations based on the chemical equation.

Metallic iron reacts with O_2 gas to produce iron(III) oxide.
a. Write and balance the equation.
b. Calculate the number of grams of iron needed to produce 5.00 g of product.

Question 5.15

Barium carbonate decomposes upon heating to barium oxide and carbon dioxide.
a. Write and balance the equation.
b. Calculate the number of grams of carbon dioxide produced by heating 50.0 g of barium carbonate.

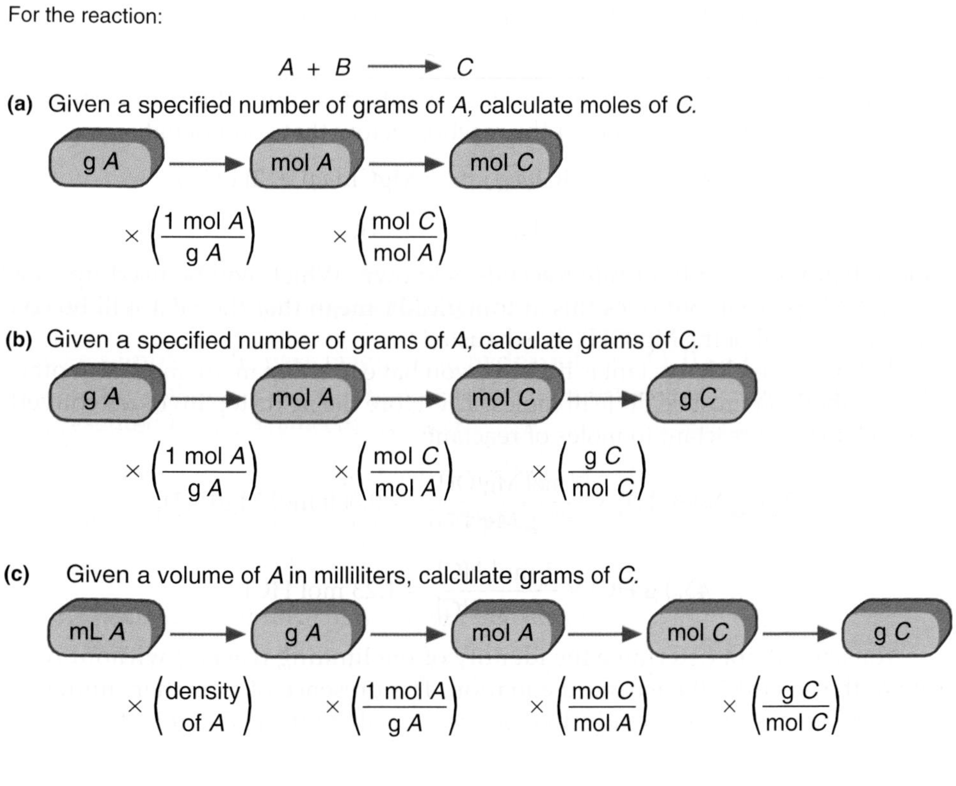

Question 5.16

For the reaction:

$$A + B \longrightarrow C$$

(a) Given a specified number of grams of A, calculate moles of C.

$$\boxed{\text{g } A} \longrightarrow \boxed{\text{mol } A} \longrightarrow \boxed{\text{mol } C}$$
$$\times \left(\frac{1 \text{ mol } A}{\text{g } A}\right) \qquad \times \left(\frac{\text{mol } C}{\text{mol } A}\right)$$

(b) Given a specified number of grams of A, calculate grams of C.

$$\boxed{\text{g } A} \longrightarrow \boxed{\text{mol } A} \longrightarrow \boxed{\text{mol } C} \longrightarrow \boxed{\text{g } C}$$
$$\times \left(\frac{1 \text{ mol } A}{\text{g } A}\right) \qquad \times \left(\frac{\text{mol } C}{\text{mol } A}\right) \qquad \times \left(\frac{\text{g } C}{\text{mol } C}\right)$$

(c) Given a volume of A in milliliters, calculate grams of C.

$$\boxed{\text{mL } A} \longrightarrow \boxed{\text{g } A} \longrightarrow \boxed{\text{mol } A} \longrightarrow \boxed{\text{mol } C} \longrightarrow \boxed{\text{g } C}$$
$$\times \left(\begin{array}{c}\text{density} \\ \text{of } A\end{array}\right) \quad \times \left(\frac{1 \text{ mol } A}{\text{g } A}\right) \quad \times \left(\frac{\text{mol } C}{\text{mol } A}\right) \quad \times \left(\frac{\text{g } C}{\text{mol } C}\right)$$

Figure 5.7
A general problem-solving strategy, using molar quantities.

then

$$V = \frac{(3.13 \times 10^3 \text{ mol})(0.0821 \text{ L-atm K}^{-1} \text{ mol}^{-1})(293 \text{ K})}{1.00 \text{ atm}}$$

$$= 7.53 \times 10^4 \text{ L}$$

Question 6.11

What volume is occupied by 10.0 g N_2 at 30.°C and a pressure of 750 torr?

Question 6.12

A 20.0-L gas cylinder contains 4.80 g H_2 at 25°C. What is the pressure of this gas?

Question 6.13

How many moles of N_2 gas will occupy a 5.00-L container at standard temperature and pressure?

Question 6.14

At what temperature will 2.00 mol He fill a 2.00-L container at standard pressure?

Dalton's Law of Partial Pressures

Learning Goal

Our discussion of gases so far has presumed that we are working with a single pure gas. A *mixture* of gases exerts a pressure that is the *sum* of the pressures that each gas would exert if it were present alone under the same conditions. This is known as **Dalton's law** of partial pressures.

Stated another way, the total pressure of a mixture of gases is the sum of the **partial pressures.** That is,

$$P_t = p_1 + p_2 + p_3 + \dots$$

in which P_t = total pressure and $p_1, p_2, p_3, \dots$, are the partial pressures of the component gases. For example, the total pressure of our atmosphere is equal to the sum of the pressures of N_2 and O_2 (the principal components of air):

$$P_{air} = p_{N_2} + p_{O_2}$$

Other gases, such as argon (Ar), carbon dioxide (CO_2), carbon monoxide (CO), and methane (CH_4) are present in the atmosphere at very low partial pressures. However, their presence may result in dramatic consequences; one such gas is carbon dioxide. Classified as a "greenhouse gas," it exerts a significant effect on our climate. Its role is described in "An Environmental Perspective: The Greenhouse Effect and Global Warming."

Kinetic Molecular Theory of Gases

Learning Goal

The kinetic molecular theory of gases provides a reasonable explanation of the behavior of gases that we have studied in this chapter. The macroscopic properties result from the action of the individual molecules comprising the gas.

The **kinetic molecular theory** can be summarized as follows:

1. Gases are made up of small atoms or molecules that are in constant, random motion.
2. The distance of separation among these atoms or molecules is very large in comparison to the size of the individual atoms or molecules. In other words, a gas is mostly empty space.

AN ENVIRONMENTAL PERSPECTIVE

The Greenhouse Effect and Global Warming

A greenhouse is a bright, warm, and humid environment for growing plants, vegetables, and flowers even during the cold winter months. It functions as a closed system in which the concentration of water vapor is elevated and visible light streams through the windows; this creates an ideal climate for plant growth.

Some of the visible light is absorbed by plants and soil in the greenhouse and radiated as infrared radiation. This radiated energy is blocked by the glass or absorbed by water vapor and carbon dioxide (CO_2). This trapped energy warms the greenhouse and is a form of solar heating: light energy is converted to heat energy.

On a global scale, the same process takes place. Although more than half of the sunlight that strikes the earth's surface is reflected back into space, the fraction of light that is absorbed produces sufficient heat to sustain life. How does this happen? Greenhouse gases, such as CO_2, trap energy radiated from the earth's surface and store it in the atmosphere. This moderates our climate. The earth's surface would be much colder and more inhospitable if the atmosphere was not able to capture some reasonable amount of solar energy.

Can we have too much of a good thing? It appears so. Since 1900 the atmospheric concentration of CO_2 has increased from 296 parts per million (ppm) to over 350 ppm (approximately 17% increase). The energy demands of technological and population growth have caused massive increases in the combustion of organic matter and carbon-based fuels (coal, oil, and natural gas), adding over 50 billion tons of CO_2 to that already present in the atmosphere. Photosynthesis naturally removes CO_2 from the atmosphere. However, the removal of forestland to create living space and cropland decreases the amount of vegetation available to consume atmospheric CO_2 through photosynthesis. The rapid destruction of the Amazon rainforest is just the latest of many examples.

If our greenhouse model is correct, an increase in CO_2 levels should produce global warming, perhaps changing our climate in unforeseen and undesirable ways.

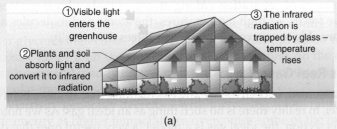

① Visible light enters the greenhouse

② Plants and soil absorb light and convert it to infrared radiation

③ The infrared radiation is trapped by glass – temperature rises

(a)

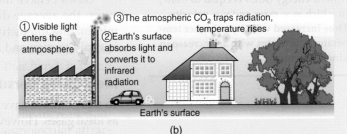

① Visible light enters the atmposphere

② Earth's surface absorbs light and converts it to infrared radiation

③ The atmospheric CO_2 traps radiation, temperature rises

Earth's surface

(b)

(a) A greenhouse traps solar radiation as heat. (b) Our atmosphere also acts as a solar collector. Carbon dioxide, like the windows of a greenhouse, allows the visible light to enter and traps the heat.

3. All of the atoms and molecules behave independently. No attractive or repulsive forces exist between atoms or molecules in a gas.
4. Atoms and molecules collide with each other and with the walls of the container without *losing* energy. The energy is *transferred* from one atom or molecule to another.
5. The average kinetic energy of the atoms or molecules increases or decreases in proportion to absolute temperature.

We know that gases are easily *compressible*. The reason is that a gas is mostly empty space, providing space for the particles to be pushed closer together.

Gases will *expand* to fill any available volume because they move freely with sufficient energy to overcome their attractive forces.

Gases have a *low density*. Density is defined as mass per volume. Because gases are mostly empty space, they have a low mass per volume.

Gases readily *diffuse* through each other simply because they are in continuous motion and paths are readily available because of the large space between adjacent atoms or molecules. Light molecules diffuse rapidly; heavier molecules diffuse more slowly (Figure 6.5).

Learning Goal

4

7

Reactions and Solutions

Formation of a precipitate by mixing two solutions.

Learning Goals

1. Classify chemical reactions by type: combination, decomposition, or replacement.

2. Recognize the various classes of chemical reactions: precipitation, reactions with oxygen, acid–base, and oxidation–reduction.

3. Distinguish among the terms *solution, solute,* and *solvent.*

4. Describe various kinds of solutions, and give examples of each.

5. Describe the relationship between solubility and equilibrium.

6. Calculate solution concentration in weight/volume percent and weight/weight percent.

7. Calculate solution concentration using molarity.

8. Perform dilution calculations.

9. Interconvert molar concentration of ions and milliequivalents/liter.

10. Describe and explain concentration-dependent solution properties.

11. Describe why the chemical and physical properties of water make it a truly unique solvent.

12. Explain the role of electrolytes in blood and their relationship to the process of dialysis.

8

Chemical and Physical Change: Energy, Rate, and Equilibrium

A rapid, exothermic chemical reaction.

Learning Goals

1 Correlate the terms *endothermic* and *exothermic* with heat flow between a *system* and its *surroundings*.

2 State the meaning of the terms *enthalpy, entropy,* and *free energy* and know their implications.

3 Describe experiments that yield thermochemical information and calculate fuel values based on experimental data.

4 Describe the concept of reaction rate and the role of kinetics in chemical and physical change.

5 Describe the importance of *activation energy* and the *activated complex* in determining reaction rate.

6 Predict the way reactant structure, concentration, temperature, and catalysis affect the rate of a chemical reaction.

7 Write rate equations for elementary processes.

8 Recognize and describe equilibrium situations.

9 Write equilibrium-constant expressions and use these expressions to calculate equilibrium constants.

10 Use LeChatelier's principle to predict changes in equilibrium position.

205

CHEMISTRY CONNECTION

The Cost of Energy? More Than You Imagine

When we purchase gasoline for our automobiles or oil for the furnace, we are certainly buying matter. That matter is only a storage device; we are really purchasing the energy stored in the chemical bonds. Combustion, burning in oxygen, releases the stored potential energy in a form suited to its function: mechanical energy to power a vehicle or heat energy to warm a home.

Energy release is a consequence of change. In fuel combustion, this change results in the production of waste products that may be detrimental to our environment. This necessitates the expenditure of time, money, and *more* energy to clean up our surroundings.

If we are paying a considerable price for our energy supply, it would be nice to believe that we are at least getting full value

for our expenditure. Even that is not the case. Removal of energy from molecules also extracts a price. For example, a properly tuned automobile engine is perhaps 30% efficient. That means that less than one-third of the available energy actually moves the car. The other two-thirds is released into the atmosphere as wasted energy, mostly heat energy. The law of conservation of energy tells us that the energy is not destroyed, but it is certainly not available to us in a useful form.

Can we build a 100% efficient energy transfer system? Is there such a thing as cost-free energy? No, on both counts. It is theoretically impossible, and the laws of thermodynamics, which we discuss in this chapter, tell us why this is so.

Introduction

In Chapter 5 we calculated quantities of matter involved in chemical change, assuming that all of the reacting material was consumed and that only products of the reaction remain at the end of the reaction. Often this is not true. Furthermore, not all chemical reactions take place at the same speed; some occur almost instantaneously (explosions), whereas others may proceed for many years (corrosion).

Two concepts play important roles in determining the extent and speed of a chemical reaction: *thermodynamics*, which deals with energy changes in chemical reactions, and *kinetics*, which describes the rate or speed of a chemical reaction.

Although both thermodynamics and kinetics involve energy, they are two separate considerations. A reaction may be thermodynamically favored but very slow; conversely, a reaction may be very fast because it is kinetically favorable yet produce very little (or no) product because it is thermodynamically unfavorable.

In this chapter we investigate the fundamentals of thermodynamics and kinetics, with an emphasis on the critical role that energy changes play in chemical reactions. We consider physical change and chemical change, including the conversions that take place among the states of matter (solid, liquid, and gas). We use these concepts to explain the behavior of reactions that do not go to completion, *equilibrium reactions*. We develop the *equilibrium-constant expression* and demonstrate how equilibrium composition can be altered using *LeChatelier's principle*.

8.1 Thermodynamics

Thermodynamics is the study of energy, work, and heat. It may be applied to chemical change, such as the calculation of the quantity of heat obtainable from the combustion of one gallon of fuel oil. Similarly, energy released or consumed in physical change, such as the boiling or freezing of water, may be determined.

There are three basic laws of thermodynamics, but only the first two will be of concern here. They help us to understand why some chemical reactions occur readily and others do not. For instance, a mixture of concentrated solutions of hydrochloric acid and sodium hydroxide reacts violently releasing a large quantity of

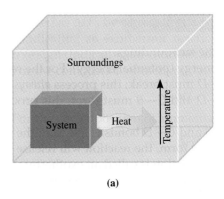
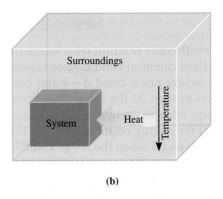

(a) (b)

Figure 8.1

Illustration of heat flow in (a) exothermic, and (b), endothermic reactions.

heat. On the other hand, nitrogen and oxygen have coexisted in the atmosphere for thousands of years with no significant chemical reaction occurring.

The Chemical Reaction and Energy

John Dalton believed that chemical change involved joining, separating, or rearranging atoms. Two centuries later, this statement stands as an accurate description of chemical reactions. However, we now know much more about the nonmaterial energy changes that are an essential part of every reaction.

Throughout the discussion of thermodynamics and kinetics it will be useful to remember the basic ideas of the kinetic molecular theory (Section 6.1):

- molecules and atoms in a reaction mixture are in constant, random motion;
- these molecules and atoms frequently collide with each other;
- only some collisions, those with sufficient energy, will break bonds in molecules; and
- when reactant bonds are broken, new bonds may be formed and products result.

It is worth noting that we cannot measure an absolute value for energy stored in a chemical system. We can only measure the *change* in energy (energy absorbed or released) as a chemical reaction occurs. Also, it is often both convenient and necessary to establish a boundary between the *system* and its *surroundings*.

The **system** is the process under study. The **surroundings** encompass the rest of the universe. Energy is lost from the system to the surroundings or energy may be gained by the system at the expense of the surroundings. This energy change, in the form of heat, may be measured because the temperature of the system or surroundings may change and this property can be measured. This process is illustrated in Figure 8.1.

Consider the combustion of methane in a Bunsen burner, the system. The temperature of the air surrounding the burner increases, indicating that heat energy of the system (methane and oxygen) is being lost to the surroundings.

Now, an exact temperature measurement of the air before and after the reaction is difficult. However, if we could insulate a portion of the surroundings, to isolate and trap the heat, we could calculate a useful quantity, the heat of the reaction. Experimental strategies for measuring temperature change and calculating heats of reactions, termed *calorimetry*, are discussed in Section 8.2.

Exothermic and Endothermic Reactions

The first law of thermodynamics states that the energy of the universe is constant; this is the law of conservation of energy. The study of energy changes that occur in chemical reactions is a very practical application of the first law. Consider, for example, the generalized reaction:

Learning Goal

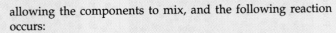

Hot and Cold Packs

Hot packs provide "instant warmth" for hikers and skiers and are used in treatment of injuries such as pulled muscles. Cold packs are in common use today for the treatment of injuries and the reduction of swelling.

These useful items are an excellent example of basic science producing a technologically useful product. (Recall our discussion in Chapter 1 of the relationship of science and technology.)

Both hot and cold packs depend on large energy changes taking place during a chemical reaction. Cold packs rely on an endothermic reaction, and hot packs generate heat energy from an exothermic reaction.

A cold pack is fabricated as two separate compartments within a single package. One compartment contains NH_4NO_3, and the other contains water. When the package is squeezed, the inner boundary between the two compartments ruptures,

allowing the components to mix, and the following reaction occurs:

$$6.7 \text{ kcal/mol} + NH_4NO_3(s) \longrightarrow NH_4^+(aq) + NO_3^-(aq)$$

This reaction is endothermic; heat taken from the surroundings produces the cooling effect.

The design of a hot pack is similar. Here, finely divided iron powder is mixed with oxygen. Production of iron oxide results in the evolution of heat:

$$4Fe + 3O_2 \longrightarrow 2Fe_2O_3 + 198 \text{ kcal/mol}$$

This reaction occurs via an oxidation-reduction mechanism (see Chapter 9). The iron atoms are oxidized, O_2 is reduced. Electrons are transferred from the iron atoms to O_2 and Fe_2O_3 forms exothermically. The rate of the reaction is slow; therefore the heat is liberated gradually over a period of several hours.

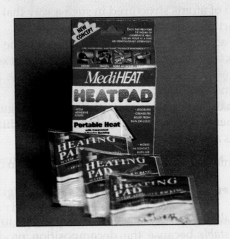

Structure of the Reacting Species

Oppositely charged species often react more rapidly than neutral species. Ions with the same charge do not react, owing to the repulsion of like charges. In contrast, oppositely charged ions attract one another and are often reactive.

Bond strengths certainly play a role in determining reaction rates as well, for the magnitude of the activation energy, or energy barrier, is related to bond strength.

The size and shape of reactant molecules influence the rate of the reaction. Large molecules, containing bulky groups of atoms, may block the reactive part of the molecule from interacting with another reactive substance, causing the reaction to proceed slowly.

The Concentration of Reactants

Concentration is introduced in Section 1.6, and units and calculations are discussed in Sections 7.4 and 7.5.

The rate of a chemical reaction is often a complex function of the concentration of one or more of the reacting substances. The rate will generally *increase* as concentration *increases* simply because a higher concentration means more reactant molecules in a given volume and therefore a greater number of collisions per unit time. If we assume that other variables are held constant, a larger number of collisions leads to a larger number of effective collisions. The explosion (very fast exothermic reaction) of gunpowder is a dramatic example of a rapid rate at high reactant concentration.

The Temperature of Reactants

The rate of a reaction *increases* as the temperature increases, because the kinetic energy of the reacting particles is directly proportional to the Kelvin temperature. Increasing the speed of particles increases the likelihood of collision, and the higher kinetic energy means that a higher percentage of these collisions will result in product formation (effective collisions). A 10°C rise in temperature has often been found to double the reaction rate.

The Physical State of Reactants

The rate of a reaction depends on the physical state of the reactants: solid, liquid, or gas. For a reaction to occur the reactants must collide frequently and have sufficient energy to react. In the solid state, the atoms, ions, or compounds are restricted in their motion. In the gaseous state, the particles are free to move, but the spacing between particles is so great that collisions are relatively infrequent. In the liquid state the particles have both free motion and proximity to each other. Hence reactions tend to be fastest in the liquid state and slowest in the solid state.

These factors were considered in our discussion of the states of matter (Chapter 6).

The Presence of a Catalyst

A **catalyst** is a substance that *increases* the reaction rate. If added to a reaction mixture, the catalytic substance undergoes no net change, nor does it alter the outcome of the reaction. However, the catalyst interacts with the reactants to create an alternative pathway for production of products. This alternative path has a lower activation energy. This makes it easier for the reaction to take place and thus increases the rate. This effect is illustrated in Figure 8.11.

Catalysis is important industrially; it may often make the difference between profit and loss in the sale of a product. For example, catalysis is useful in converting double bonds to single bonds. An important application of this principle involves the process of hydrogenation. Hydrogenation converts one or more of the carbon-carbon double bonds of unsaturated fats (eg: corn oil, olive oil) to single bonds characteristic of saturated fats (such as margarine). The use of a metal catalyst, such as nickel, in contact with the reaction mixture dramatically increases the rate of the reaction.

Sections 12.4 and 18.2 describe the role of catalysis in organic reactions.

Thousands of essential biochemical reactions in our bodies are controlled and speeded up by biological catalysts called *enzymes*.

Sections 20.1 through 20.6 describe enzyme catalysis.

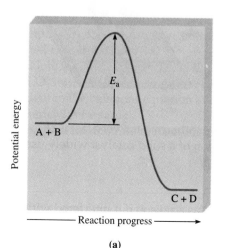

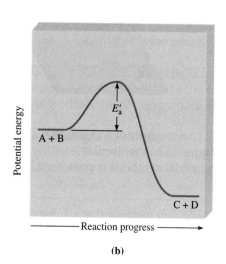

(a) **(b)**

Figure 8.11

The effect of a catalyst on the magnitude of the activation energy of a chemical reaction. Note that the presence of a catalyst decreases the activation energy $(E'_a < E_a)$, thus increasing the rate of the reaction.

$$\text{rate}_r = k_r[NH_3]^{n''}$$

At equilibrium, the forward and reverse rates become equal:

$$\text{rate}_f = \text{rate}_r$$

Consequently,

$$k_f[N_2]^n[H_2]^{n'} = k_r[NH_3]^{n''}$$

Rearranging this equation as we have previously

$$\frac{k_f}{k_r} = \frac{[NH_3]^{n''}}{[N_2]^n[H_2]^{n'}}$$

and using the definition

$$\frac{k_f}{k_r} = K_{eq}$$

yields the equilibrium-constant expression

$$K_{eq} = \frac{[NH_3]^{n''}}{[N_2]^n[H_2]^{n'}}$$

As noted earlier, the exponents n, n', and n'' are experimentally determined. However, for elementary, single-step reactions, the exponents in the rate expression are numerically equal to the coefficients in the balanced chemical equation. Consequently, for this situation, and all other equilibrium situations that we shall encounter in this book, we shall assume that the exponents in the rate expressions are equal to the coefficients in the balanced chemical equation. Hence,

$$K_{eq} = \frac{[NH_3]^2}{[N_2][H_2]^3}$$

It does not matter what initial amounts (concentrations) of reactants or products we choose. When the system reaches equilibrium, the calculated value of K_{eq} will not change. The magnitude of K_{eq} can be altered only by changing the temperature; thus K_{eq} is temperature dependent. The chemical industry uses this fact to advantage by choosing a reaction temperature that will maximize the yield of a desired product.

Question 8.17

How could one determine when a reaction has reached equilibrium?

Question 8.18

Does the attainment of equilibrium imply that no further change is taking place in the system?

Learning Goal

9

The Generalized Equilibrium-Constant Expression for a Chemical Reaction

We write the general form of an equilibrium chemical reaction as

$$aA + bB \rightleftharpoons cC + dD$$

in which A and B represent reactants, C and D represent products, and a, b, c, and d are the coefficients of the balanced equation. The equilibrium constant expression for this general case is

Products **of the overall equilibrium re-action are in the numerator, and** *reactants* **are in the denominator.**

$$K_{eq} = \frac{[C]^c[D]^d}{[A]^a[B]^b}$$

[ ] represents molar concentration, *M*.

Writing Equilibrium-Constant Expressions

An equilibrium-constant expression can be written only after a correct, balanced chemical equation that describes the equilibrium system has been developed. A balanced equation is essential because the *coefficients* in the equation become the *exponents* in the equilibrium-constant expression.

Each chemical reaction has a unique equilibrium constant value at a specified temperature. Equilibrium constants listed in the chemical literature are often reported at 25°C, to allow comparison of one system with any other. For any equilibrium reaction, the value of the equilibrium constant changes with temperature.

The brackets represent molar concentration or molarity; recall that molarity has units of mol/L. Although the equilibrium constant may have units (owing to the units on each concentration term), by convention units are usually not used. In our discussion of equilibrium, all equilibrium constants are shown as *unitless*.

A properly written equilibrium-constant expression may not include all of the terms in the chemical equation upon which it is based. Only the concentration of gases and substances in solution are shown, because their concentrations can change. Concentration terms for pure liquids and solids are *not* shown. The concentration of a pure liquid is constant; a solid also has a fixed concentration and, for solution reactions, is not really a part of the solution. When a solid is formed it exists as a solid phase in contact with a liquid phase (the solution).

> The exponents correspond to the *coefficients* of the balanced equation.

EXAMPLE 8.6

Writing an Equilibrium-Constant Expression

Write an equilibrium-constant expression for the reversible reaction:

$$H_2(g) + F_2(g) \rightleftharpoons 2HF(g)$$

Solution

Inspection of the chemical equation reveals that no solids or pure liquids are present. Hence all reactants and products appear in the equilibrium-constant expression:

The numerator term is the product term $[HF]^2$.
The denominator terms are the reactants $[H_2]$ and $[F_2]$.

Note that each term contains an exponent identical to the corresponding coefficient in the balanced equation. Arranging the numerator and denominator terms as a fraction and setting the fraction equal to K_{eq} yields

$$K_{eq} = \frac{[HF]^2}{[H_2][F_2]}$$

EXAMPLE 8.7

Writing an Equilibrium-Constant Expression

Write an equilibrium-constant expression for the reversible reaction:

$$MnO_2(s) + 4HCl(aq) \rightleftharpoons MnCl_2(aq) + Cl_2(g) + 2H_2O(l)$$

Solution

MnO_2 is a solid and H_2O is a pure liquid. Thus they are not written in the equilibrium-constant expression.

$$\boxed{MnO_2(s)} + 4HCl(aq) \rightleftharpoons MnCl_2(aq) + Cl_2(g) + \boxed{2H_2O(l)}$$

Not a part of the K_{eq} expression

The numerator term includes the remaining products:

$$[MnCl_2] \quad \text{and} \quad [Cl_2]$$

The denominator term includes the remaining reactant:

$$[HCl]^4$$

Note that the exponent is identical to the corresponding coefficient in the chemical equation.

Arranging the numerator and denominator terms as a fraction and setting the fraction equal to K_{eq} yields

$$K_{eq} = \frac{[MnCl_2][Cl_2]}{[HCl]^4}$$

Question 8.19

Write an equilibrium-constant expression for each of the following reversible reactions.

a. $2NO_2(g) \rightleftharpoons N_2(g) + 2O_2(g)$
b. $2H_2O(l) \rightleftharpoons 2H_2(g) + O_2(g)$

Question 8.20

Write an equilibrium-constant expression for each of the following reversible reactions.

a. $2HI(g) \rightleftharpoons H_2(g) + I_2(g)$
b. $PCl_5(s) \rightleftharpoons PCl_3(l) + Cl_2(g)$

Interpreting Equilibrium Constants

What utility does the equilibrium constant have? The reversible arrow in the chemical equation alerts us to the fact that an equilibrium exists. Some measurable quantity of the product and reactant remain. However, there is no indication whether products predominate, reactants predominate, or significant concentrations of both products and reactants are present at equilibrium.

The numerical value of the equilibrium constant provides additional information. It tells us the extent to which reactants have converted to products. This is important information for anyone who wants to manufacture and sell the product. It also is important to anyone who studies the effect of equilibrium reactions on environmental systems and living organisms.

Although an absolute interpretation of the numerical value of the equilibrium constant depends on the form of the equilibrium-constant expression, the following generalizations are useful:

- K_{eq} greater than 1×10^2. A large numerical value of K_{eq} indicates that the numerator (product term) is much larger than the denominator (reactant term) and that at equilibrium mostly product is present.
- K_{eq} less than 1×10^{-2}. A small numerical value of K_{eq} indicates that the numerator (product term) is much smaller than the denominator (reactant term) and that at equilibrium mostly reactant is present.
- K_{eq} between 1×10^{-2} and 1×10^2. In this case the equilibrium mixture contains significant concentrations of both reactants and products.

Question 8.21

At a given temperature, the equilibrium constant for a certain reaction is 1×10^{20}. Does this equilibrium favor products or reactants? Why?

At a given temperature, the equilibrium constant for a certain reaction is 1×10^{-18}. Does this equilibrium favor products or reactants? Why?

Calculating Equilibrium Constants

The magnitude of the equilibrium constant for a chemical reaction is determined experimentally. The reaction under study is allowed to proceed until the composition of products and reactants no longer changes (Figure 8.15). This may be a matter of seconds, minutes, hours, or even months or years, depending on the rate of the reaction. The reaction mixture is then analyzed to determine the molar concentration of each of the products and reactants. These concentrations are substituted in the equilibrium-constant expression and the equilibrium constant is calculated. The following example illustrates this process.

Section 8.3 discusses rates of reaction and Section 7.5 describes molar concentration.

EXAMPLE 8.8

Calculating an Equilibrium Constant

Hydrogen iodide is placed in a sealed container and allowed to come to equilibrium. The equilibrium reaction is:

$$2HI(g) \rightleftharpoons H_2(g) + I_2(g)$$

and the equilibrium concentrations are:

$$[HI] = 0.54 \, M$$
$$[H_2] = 1.72 \, M$$
$$[I_2] = 1.72 \, M$$

Calculate the equilibrium constant.

Solution

First, write the equilibrium-constant expression:

$$K_{eq} = \frac{[H_2][I_2]}{[HI]^2}$$

Then substitute the equilibrium concentrations of products and reactants to obtain

$$K_{eq} = \frac{[1.72][1.72]}{[0.54]^2} = \frac{2.96}{0.29}$$

$$= 10.1 \text{ or } 1.0 \times 10^1 \text{ (two significant figures)}$$

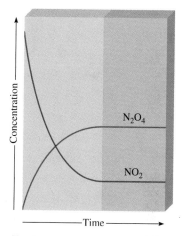

Figure 8.15
The combination reaction of NO_2 molecules produces N_2O_4. Initially, the concentration of reactant (NO_2) diminishes rapidly while the N_2O_4 concentration builds. Eventually, the concentrations of both reactant and product become constant over time (blue area). The equilibrium condition has been attained.

A reaction chamber contains the following mixture at equilibrium:

$$[NH_3] = 0.25 \, M$$
$$[N_2] = 0.11 \, M$$
$$[H_2] = 1.91 \, M$$

If the reaction is:

$$N_2(g) + 3H_2(g) \rightleftharpoons 2NH_3(g)$$

Calculate the equilibrium constant.

Figure 8.17

The effect of heat on equilibrium position. For the reaction:

$CoCl_4^{2-}(aq) + 6H_2O(l) \rightleftharpoons$
(blue)

$\qquad Co(H_2O)_6^{2+}(aq) + 4Cl^-(aq)$
(pink)

Heating the solution favors the blue $CoCl_4^{2-}$ species; cooling favors the pink $Co(H_2O)_6^{2+}$ species.

is unaffected by pressure. The number of moles of gaseous product and reactant are identical. No volume advantage is gained by a shift in equilibrium composition.

In summary:

- Pressure affects the equilibrium composition only of reactions that involve at least one gaseous substance.
- Additionally, the relative number of moles of gaseous products and reactants must differ.
- The equilibrium composition will shift to increase the number of moles of gas when the pressure decreases; it will shift to decrease the number of moles of gas when the pressure increases.

Effect of a Catalyst

A catalyst has no effect on the equilibrium composition. A catalyst increases the rates of both forward and reverse reactions to the same extent. The equilibrium composition *and* equilibrium concentration do not change when a catalyst is used, but the equilibrium composition is achieved in a shorter time. The role of a solid-phase catalyst in the synthesis of ammonia is shown in Figure 8.13.

EXAMPLE 8.9

Predicting Changes in Equilibrium Composition

Earlier in this section we considered the geologically important reaction that occurs in rock and soil.

$$Ca^{2+}(aq) + 2HCO_3^-(aq) \rightleftharpoons CaCO_3(s) + CO_2(aq) + H_2O(l)$$

Predict the effect on the equilibrium composition for each of the following changes.
a. The $[Ca^{2+}]$ is increased.
b. The amount of $CaCO_3$ is increased.
c. The amount of H_2O is increased.
d. The $[HCO_3^-]$ is decreased.
e. A catalyst is added.

Solution

a. The concentration of reactant increases; the equilibrium shifts to the right, and more products are formed.
b. $CaCO_3$ is a solid; solids are not written in the equilibrium-constant expression, so there is no effect on the equilibrium composition.

c. H_2O is a pure liquid; it is not written in the equilibrium expression, so the equilibrium composition is unaffected.
d. The concentration of reactant decreases; the equilibrium shifts to the left, and more reactants are formed.
e. A catalyst has no effect on the equilibrium composition.

For the hypothetical equilibrium reaction

$$A(g) + B(g) \rightleftharpoons C(g) + D(g)$$

predict whether the amount of A in a 5.0-L container would increase, decrease, or remain the same for each of the following changes.
a. Addition of excess B
b. Addition of excess C
c. Removal of some D
d. Addition of a catalyst

For the hypothetical equilibrium reaction

$$A(g) + B(g) \rightleftharpoons C(g) + D(g)$$

predict whether the amount of A in a 5.0-L container would increase, decrease, or remain the same for each of the following changes.
a. Removal of some B
b. Removal of some C
c. Addition of excess D
d. Removal of a catalyst

Summary

8.1 Thermodynamics

Thermodynamics is the study of energy, work, and heat. Thermodynamics can be applied to the study of chemical reactions because we can measure the heat flow (by measuring the temperature change) between the *system* and the *surroundings*. *Exothermic reactions* release energy and products that are lower in energy than the reactants. *Endothermic reactions* require energy input. Heat energy is represented as *enthalpy*, $H°$. The energy gain or loss is the change in enthalpy, $\Delta H°$, and is one factor that is useful in predicting whether a reaction is spontaneous or nonspontaneous.

Entropy, $S°$, is a measure of the randomness of a system. A random, or disordered system has high entropy; a well-ordered system has low entropy. The change in entropy in a chemical reaction, $\Delta S°$, is also a factor in predicting reaction spontaneity.

Free energy, $\Delta G°$, incorporates both factors, enthalpy and entropy; as such, it is an absolute predictor of the spontaneity of a chemical reaction.

8.2 Experimental Determination of Energy Change in Reactions

A *calorimeter* measures heat changes (in calories or joules) that occur in chemical reactions.

The *specific heat* of a substance is the number of calories of heat needed to raise the temperature of 1 g of the substance 1 degree Celsius.

The amount of energy per gram of food is referred to as its *fuel value*. Fuel values are commonly reported in units of *nutritional Calories* (1 nutritional Calorie = 1 kcal). A bomb calorimeter is useful for measurement of the fuel value of foods.

8.3 Kinetics

Chemical *kinetics* is the study of the *rate* or speed of a chemical reaction. Energy for reactions is provided by molecular collisions. If this energy is sufficient, bonds may break, and atoms may recombine in a different arrangement, producing product. A collision producing one or more product molecules is termed an effective collision.

The minimum amount of energy needed for a reaction is the *activation energy*. The reaction proceeds from reactants to products through an intermediate state, the *activated complex*.

Experimental conditions influencing the reaction rate include the structure of the reacting species, the concentration of reactants, the temperature of reactants, the physical state of reactants, and the presence or absence of a catalyst.

A *catalyst* increases the rate of a reaction. The catalytic substance undergoes no net change in the reaction, nor does it alter the outcome of the reaction.

8.4 Equilibrium

Many chemical reactions do not completely convert reactants to products. A mixture of products and reactants exists, and its composition will remain constant until the experimental conditions are changed. This mixture is in a state of *chemical equilibrium*. The reaction continues indefinitely (dynamic), but the concentrations of products and reactants are fixed (equilibrium) because the rates of the forward and reverse reactions are equal. This is a *dynamic equilibrium*.

LeChatelier's principle states that if a stress is placed on an equilibrium system, the system will respond by altering the equilibrium in such a way as to minimize the stress.

Key Terms

activated complex (8.3)
activation energy (8.3)
calorimetry (8.2)
catalyst (8.3)
dynamic equilibrium (8.4)
endothermic reaction (8.1)
enthalpy (8.1)
entropy (8.1)
equilibrium constant (8.4)
equilibrium reaction (8.4)
exothermic reaction (8.1)
free energy (8.1)
fuel value (8.2)

kinetics (8.3)
LeChatelier's principle (8.4)
nutritional Calorie (8.2)
order of the reaction (8.3)
rate constant (8.3)
rate equation (8.3)
rate of chemical reaction (8.3)
reversible reaction (8.4)
specific heat (8.2)
surroundings (8.1)
system (8.1)
thermodynamics (8.1)

Questions and Problems

Energy and Thermodynamics

8.27 Define or explain each of the following terms:
  a. exothermic reaction
  b. endothermic reaction
  c. calorimeter

8.28 Define or explain each of the following terms:
  a. free energy
  b. specific heat
  c. fuel value

8.29 Explain what is meant by the term *enthalpy*.

8.30 Explain what is meant by the term *entropy*.

8.31 5.00 g of octane are burned in a bomb calorimeter containing 2.00×10^2 g H_2O. How much energy, in calories, is released if the water temperature increases 6.00°C?

8.32 0.0500 mol of a nutrient substance is burned in a bomb calorimeter containing 2.00×10^2 g H_2O. If the formula weight of this nutrient substance is 114 g/mol, what is the fuel value (in nutritional Calories) if the temperature of the water increased 5.70°C?

8.33 Calculate the energy released, in joules, in Question 8.31 (recall conversion factors, Chapter 1).

8.34 Calculate the fuel value, in kilojoules, in Question 8.32 (recall conversion factors, Chapter 1).

8.35 Predict whether each of the following processes increases or decreases entropy, and explain your reasoning.
  a. melting of a solid metal
  b. boiling of water

8.36 Predict whether each of the following processes increases or decreases entropy, and explain your reasoning.
  a. burning a log in a fireplace
  b. condensation of water vapor on a cold surface

8.37 Explain why an exothermic reaction produces products that are more stable than the reactants.

8.38 Provide an example of entropy from your own experience.

8.39 Isopropyl alcohol, commonly known as rubbing alcohol, feels cool when applied to the skin. Explain why.

8.40 Energy is required to break chemical bonds during the course of a reaction. When is energy released?

Kinetics

8.41 Define the term *activated complex* and explain its significance in a chemical reaction.

8.42 Define and explain the term *activation energy* as it applies to chemical reactions.

8.43 Sketch a potential energy diagram for a reaction that shows the effect of a catalyst on an exothermic reaction.

8.44 Sketch a potential energy diagram for a reaction that shows the effect of a catalyst on an endothermic reaction.

8.45 Give at least two examples from life sciences in which the rate of a reaction is critically important.

8.46 Give at least two examples from everyday life in which the rate of a reaction is an important consideration.

8.47 Describe how an increase in the concentration of reactants increases the rate of a reaction.

8.48 Describe how an increase in the temperature of reactants increases the rate of a reaction.

8.49 Write the rate expression for the single-step reaction:

$$N_2O_4(g) \rightleftharpoons 2NO_2(g)$$

8.50 Write the rate expression for the single-step reaction:

$$H_2S(aq) + Cl_2(aq) \rightleftharpoons S(s) + 2HCl(aq)$$

8.51 Describe how a catalyst speeds up a chemical reaction.

8.52 Explain how a catalyst can be involved in a chemical reaction without being consumed in the process.

Equilibrium

8.53 Describe the meaning of the term *dynamic equilibrium*.

8.54 What is the relationship between the forward and reverse rates for a reaction at equilibrium?

8.55 Write a valid equilibrium constant for the reaction shown in Question 8.49.

8.56 Write a valid equilibrium constant for the reaction shown in Question 8.50.

8.57 Distinguish between a physical equilibrium and a chemical equilibrium.

8.58 Distinguish between the rate constant and the equilibrium constant for a reaction.

8.59 For the reaction

$$CH_4(g) + Cl_2(g) \rightleftharpoons CH_3Cl(g) + HCl(g) + 26.4 \text{ kcal}$$

predict the effect on the equilibrium (will it shift to the left or to the right, or will there be no change?) for each of the following changes.

a. The temperature is increased.
b. The pressure is increased by decreasing the volume of the container.
c. A catalyst is added.

8.60 For the reaction

$$47 \text{ kcal} + 2SO_3(g) \rightleftharpoons 2SO_2(g) + O_2(g)$$

predict the effect on the equilibrium (will it shift to the left or to the right, or will there be no change?) for each of the following changes.

a. The temperature is increased.
b. The pressure is increased by decreasing the volume of the container.
c. A catalyst is added.

8.61 Label each of the following statements as true or false and explain why.

a. A slow reaction is an incomplete reaction.
b. The rates of forward and reverse reactions are never the same.

8.62 Label each of the following statements as true or false and explain why.

a. A reaction is at equilibrium when no reactants remain.
b. A reaction at equilibrium is undergoing continual change.

8.63 Use LeChatelier's principle to predict whether the amount of PCl_3 in a 1.00-L container is increased, is decreased, or remains the same for the equilibrium

$$PCl_3(g) + Cl_2(g) \rightleftharpoons PCl_5(g) + \text{heat}$$

when each of the following changes is made.

a. PCl_5 is added.   **d.** The temperature is decreased.
b. Cl_2 is added.   **e.** A catalyst is added.
c. PCl_5 is removed.

8.64 Use LeChatelier's principle to predict the effects, if any, of each of the following changes on the equilibrium system, described below, in a closed container.

$$C(s) + 2H_2(g) \rightleftharpoons CH_4(g) + 18 \text{ kcal}$$

a. adding more C.   **d.** increasing the temperature.
b. adding more H_2.   **e.** adding a catalyst.
c. removing CH_4.

1. For the reaction:

$$2NH_3(g) \rightleftharpoons N_2(g) + 3H_2(g)$$

What is the relationship among the equilibrium concentrations of NH_3, N_2, and H_2 for each of the following situations:

• We begin with 2 mol of NH_3 in a 1-L container.
• We begin with 1 mol of N_2 and 3 mol of H_2 in a 1-L container.

Explain your reasoning.

2. Can the following statement ever be true? "Heating a reaction mixture increases the rate of a certain reaction but decreases the yield of product from the reaction." Explain why or why not.

3. Molecules must collide for a reaction to take place. Sketch a model of the orientation and interaction of HI and Cl that is most favorable for the reaction:

$$HI(g) + Cl(g) \longrightarrow HCl(g) + I(g)$$

4. Silver ion reacts with chloride ion to form the precipitate, silver chloride:

$$Ag^+(aq) + Cl^-(aq) \rightleftharpoons AgCl(s)$$

After the reaction reached equilibrium, the chemist filtered 99% of the solid silver chloride from the solution, hoping to shift the equilibrium to the right, to form more product. Critique the chemist's experiment.

5. Human behavior often follows LeChatelier's principle. Provide one example and explain in terms of LeChatelier's principle.

6. A clever device found in some homes is a figurine that is blue on dry, sunny days and pink on damp, rainy days. These figurines are coated with substances containing chemical species that undergo the following equilibrium reaction:

$$Co(H_2O)_6^{2+}(aq) + 4Cl^-(aq) \rightleftharpoons CoCl_4^{2-}(aq) + 6H_2O(l)$$

a. Which substance is blue?
b. Which substance is pink?
c. How is LeChatelier's principle applied here?

7. You have spent the entire morning in a 20°C classroom. As you ride the elevator to the cafeteria, six persons enter the elevator after being outside on a sub-freezing day. You suddenly feel chilled. Explain the heat flow situation in the elevator in thermodynamic terms.

9 Charge-Transfer Reactions: Acids and Bases and Oxidation–Reduction

Solution properties, including color, are often pH dependent.

Learning Goals

1 Identify acids and bases and acid–base reactions.

2 Write equations describing acid–base dissociation and label the conjugate acid–base pairs.

3 Describe the role of the solvent in acid–base reactions, and explain the meaning of the term *pH*.

4 Calculate pH from concentration data.

5 Calculate hydronium and/or hydroxide ion concentration from pH data.

6 Provide examples of the importance of pH in chemical and biochemical systems.

7 Describe the meaning and utility of neutralization reactions.

8 State the meaning of the term *buffer* and describe the applications of buffers to chemical and biochemical systems, particularly blood chemistry.

9 Describe *oxidation* and *reduction*, and describe some practical examples of redox processes.

10 Diagram a voltaic cell and describe its function.

11 Compare and contrast voltaic and electrolytic cells.

Equation balancing is discussed in Chapter 5.

Our objective is to make the balanced equation represent the process actually occurring. We recognize that HCl, NaOH, and NaCl are dissociated in solution:

$$H^+(aq) + Cl^-(aq) + Na^+(aq) + OH^-(aq) \longrightarrow Na^+(aq) + Cl^-(aq) + H_2O(l)$$

We further know that Na^+ and Cl^- are unchanged in the reaction. If we write only those components that actually change, we produce a *net, balanced ionic equation:*

$$H^+(aq) + OH^-(aq) \longrightarrow H_2O(l)$$

If we realize that the H^+ occurs in aqueous solution as the hydronium ion, H_3O^+, the most correct form of the net, balanced ionic equation is

$$H_3O^+(aq) + OH^-(aq) \longrightarrow 2H_2O(l)$$

The equation for any strong acid/strong base neutralization reaction is the same as this equation.

A neutralization reaction may be used to determine the concentration of an unknown acid or base solution. The technique of **titration** involves the addition of measured amounts of a **standard solution** (one whose concentration is known with certainty) to neutralize the second, unknown solution. From the volumes of the two solutions and the concentration of the standard solution the concentration of the unknown solution may be determined. Consider the following application.

EXAMPLE 9.8

Many indicators are naturally occurring substances.

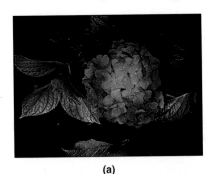

(a)

(b)

Figure 9.4
The color of the petals of the hydrangea is formed by molecules that behave as acid–base indicators. The color is influenced by the pH of the soil in which the hydrangea is grown.

Determining the Concentration of a Solution of Hydrochloric Acid

Step 1. A known volume (perhaps 25.00 mL) of the unknown acid is measured into a flask using a pipet.

Step 2. An **indicator,** a substance that changes color as the solution reaches a certain pH (Figures 9.4 and 9.5), is added to the unknown solution.

Step 3. A solution of sodium hydroxide (perhaps 0.1000 *M*) is carefully added to the unknown solution using a **buret,** which is a long glass tube calibrated in milliliters. A stopcock at the bottom of the buret regulates the amount of liquid dispensed. The standard solution is added until the indicator changes color.

Step 4. At this point, the **equivalence point,** the number of moles of hydroxide ion added is equal to the number of moles of hydronium ion present in the unknown acid.

Step 5. The volume dispensed by the buret (perhaps 35.00 mL) is measured and used in the calculation of the unknown acid concentration.

Step 6. The calculation is as follows:

Pertinent information for this titration includes:

Volume of the unknown acid solution, 25.00 mL

Volume of sodium hydroxide solution added, 35.00 mL

Concentration of the sodium hydroxide solution, 0.1000 *M*

Furthermore, from the balanced equation, we know that HCl and NaOH react in a 1:1 combining ratio.

Using a strategy involving conversion factors

$$35.00 \text{ mL NaOH} \times \frac{1 \text{ L NaOH}}{10^3 \text{ mL NaOH}} \times \frac{0.1000 \text{ mol NaOH}}{\text{L NaOH}} = 3.500 \times 10^{-3} \text{ mol NaOH}$$

Knowing that HCl and NaOH undergo a 1:1 reaction,

$$3.500 \times 10^{-3} \text{ mol NaOH} \times \frac{1 \text{ mol HCl}}{1 \text{ mol NaOH}} = 3.500 \times 10^{-3} \text{ mol HCl}$$

3.500×10^{-3} mol HCl are contained in 25.00 mL of HCl solution. Thus,

$$\frac{3.500 \times 10^{-3} \text{ mol HCl}}{25.00 \text{ mL HCl soln}} \times \frac{10^3 \text{ mL HCl soln}}{1 \text{ L HCl soln}} = 1.400 \times 10^{-1} \text{ mol HCl/L HCl soln}$$

$$= 0.1400 \, M$$

The titration of an acid with a base is depicted in Figure 9.6.

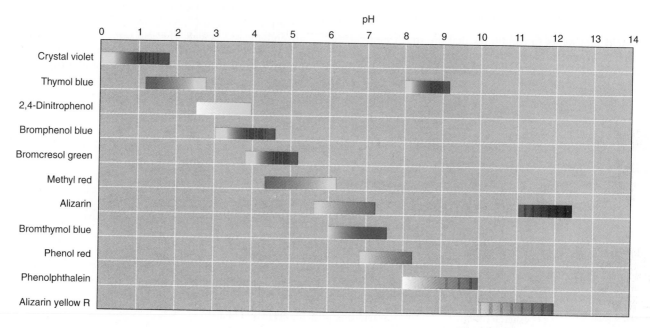

Figure 9.5

The relationship between pH and color of a variety of compounds commonly used as acid–base indicators.

(a)   **(b)**

Figure 9.6

An acid–base titration. (a) An exact volume of a standard solution (in this example, a base) is added to a solution of unknown concentration (in this example, an acid). (b) From the volume (read from the buret) and concentration of the standard solution, coupled with the mass or volume of the unknown, the concentration of the unknown may be calculated.

AN ENVIRONMENTAL PERSPECTIVE

Acid Rain

Acid rain is a global environmental problem that has raised public awareness of the chemicals polluting the air through the activities of our industrial society. Normal rain has a pH of about 5.6 as a result of the chemical reaction between carbon dioxide gas and water in the atmosphere. The following equation shows this reaction:

$$CO_2(g) \quad + \quad H_2O(l) \rightleftharpoons H_2CO_3(aq)$$

Carbon dioxide   Water   Carbonic acid

Acid rain refers to conditions that are much more acidic than this. In upstate New York the rain has as much as 25 times the acidity of normal rainfall. One rainstorm, recorded in West Virginia, produced rainfall that measured 1.5 on the pH scale. This is approximately the pH of stomach acid or about 10,000 times more acidic than "normal rain" (remember that the pH scale is logarithmic).

Acid rain is destroying life in streams and lakes. More than half the highland lakes in the western Adirondack Mountains have no native game fish. In addition to these 300 lakes, 140 lakes in Ontario have suffered a similar fate. It is estimated that 48,000 other lakes in Ontario and countless others in the northeastern and central United States are threatened. Our forests are endangered as well. The acid rain decreases soil pH, which in turn alters the solubility of minerals needed by plants. Studies have shown that about 40% of the red spruce and maple

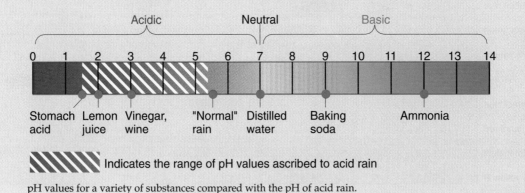

pH values for a variety of substances compared with the pH of acid rain.

Question 9.13

Calculate the molar concentration of a sodium hydroxide solution if 40.00 mL of this solution were required to neutralize 20.00 mL of a 0.2000 M solution of hydrochloric acid.

Question 9.14

Calculate the molar concentration of a sodium hydroxide solution if 36.00 mL of this solution were required to neutralize 25.00 mL of a 0.2000 M solution of hydrochloric acid.

Polyprotic Substances

Not all acid–base reactions occur in a 1:1 combining ratio (as hydrochloric acid and sodium hydroxide in the previous example). Acid–base reactions with other than 1:1 combining ratios occur between what are termed *polyprotic substances.* **Polyprotic substances** donate (as acids) or accept (as bases) more than one proton per formula unit.

trees in New England have died. Increased acidity of rainfall appears to be the major culprit.

What is the cause of this acid rain? The combustion of fossil fuels (gas, oil, and coal) by power plants produces oxides of sulfur and nitrogen. These react with water, as does the CO_2 in normal rain, but the products are strong acids: sulfuric and nitric acids. Let's look at the equations for these processes.

In the atmosphere, nitric oxide (NO) can react with oxygen to produce nitrogen dioxide as shown:

$$2NO(g) \quad + \quad O_2(g) \longrightarrow 2NO_2(g)$$

Nitric oxide   Oxygen Nitrogen dioxide

Nitrogen dioxide (which causes the brown color of smog) then reacts with water to form nitric acid:

$$3NO_2(g) + H_2O(l) \longrightarrow 2HNO_3(aq) + NO(g)$$

A similar chemistry is seen with the sulfur oxides. Coal may contain as much as 3% sulfur. When the coal is burned, the sulfur also burns; this produces choking, acrid sulfur dioxide gas:

$$S(s) + O_2(g) \longrightarrow SO_2(g)$$

By itself, sulfur dioxide can cause serious respiratory problems for people with asthma or other lung diseases, but matters are worsened by the reaction of SO_2 with atmospheric oxygen:

$$2SO_2(g) + O_2(g) \longrightarrow 2SO_3(g)$$

Sulfur trioxide will react with water in the atmosphere:

$$SO_3(g) + H_2O(l) \longrightarrow H_2SO_4(aq)$$

The product, sulfuric acid, is even more irritating to the respiratory tract. When the acid rain created by the reactions shown above falls to earth, the impact is significant.

It is easy to balance these chemical equations, but decades could be required to balance the ecological systems that we have disrupted by our massive consumption of fossil fuels. A sudden decrease of even 25% in the use of fossil fuels would lead to worldwide financial chaos. Development of alternative fuel sources, such as solar energy and safe nuclear power, will help to reduce our dependence on fossil fuels and help us to balance the global equation.

Damage caused by acid rain.

Reactions of Polyprotic Substances

HCl dissociates to produce one H^+ ion for each HCl. For this reason, it is termed a *monoprotic acid*. Its reaction with sodium hydroxide is:

$$HCl(aq) + NaOH(aq) \longrightarrow H_2O(l) + Na^+(aq) + Cl^-(aq)$$

Sulfuric acid, in contrast, is a *diprotic acid*. Each unit of H_2SO_4 produces two H^+ ions (the prefix *di-* indicating two). Its reaction with sodium hydroxide is:

$$H_2SO_4(aq) + 2NaOH(aq) \longrightarrow 2H_2O(l) + 2Na^+(aq) + SO_4^{2-}(aq)$$

Phosphoric acid is a *triprotic acid*. Each unit of H_3PO_4 produces three H^+ ions. Its reaction with sodium hydroxide is:

$$H_3PO_4(aq) + 3NaOH(aq) \longrightarrow 3H_2O(l) + 3Na^+(aq) + PO_4^{3-}(aq)$$

Dissociation of Polyprotic Substances

Sulfuric acid, and other diprotic acids, dissociate in two steps:

Step 1. $H_2SO_4(aq) + H_2O(l) \longrightarrow H_3O^+(aq) + HSO_4^-(aq)$

Step 2. $HSO_4^-(aq) + H_2O(l) \rightleftharpoons H_3O^+(aq) + SO_4^{2-}(aq)$

A CLINICAL PERSPECTIVE

Control of Blood pH

A pH of 7.4 is maintained in blood partly by a carbonic acid–bicarbonate buffer system based on the following equilibrium:

$$H_2CO_3(aq) + H_2O(l) \rightleftharpoons H_3O^+(aq) + HCO_3^-(aq)$$

Carbonic acid   Bicarbonate ion
(weak acid)   (salt)

The regulation process based on LeChatelier's principle is similar to the acetic acid–sodium acetate buffer, which we have already discussed.

Red blood cells transport O_2, bound to hemoglobin, to the cells of body tissue. The metabolic waste product, CO_2, is picked up by the blood and delivered to the lungs.

The CO_2 in the blood also participates in the carbonic acid–bicarbonate buffer equilibrium; carbon dioxide reacts with water in the blood to form carbonic acid:

$$CO_2(aq) + H_2O(l) \rightleftharpoons H_2CO_3(aq)$$

As a result the buffer equilibrium becomes more complex:

$$CO_2(aq) + 2H_2O(l) \rightleftharpoons H_2CO_3(aq) + H_2O(l) \rightleftharpoons H_3O^+(aq) + HCO_3^-(aq)$$

Through this sequence of relationships the concentration of CO_2 in the blood affects the blood pH.

Higher than normal CO_2 concentrations shift the above equilibrium to the right (LeChatelier's principle), increasing $[H_3O^+]$ and lowering the pH. The blood becomes too acidic, leading to numerous medical problems. A situation of high blood CO_2 levels and low pH is termed *acidosis*. Respiratory acidosis results from various diseases (emphysema, pneumonia) that restrict the breathing process, causing the buildup of waste CO_2 in the blood.

Lower than normal CO_2 levels, on the other hand, shift the equilibrium to the left, decreasing $[H_3O^+]$ and making the pH more basic; this condition is termed *alkalosis* (from "alkali," implying basic). Hyperventilation, or rapid breathing, is a common cause of respiratory alkalosis.

Question 9.19

A buffer solution is prepared in such a way that the concentrations of propanoic acid and sodium propanoate are each 2.00×10^{-1} *M*. If the buffer equilibrium is described by

$$\underset{\text{Propanoic acid}}{C_2H_5COOH(aq)} + H_2O(l) \rightleftharpoons H_3O^+(aq) + \underset{\text{Propanoate anion}}{C_2H_5COO^-(aq)}$$

with $K_a = 1.34 \times 10^{-5}$, calculate the pH of the solution.

Question 9.20

Calculate the pH of the buffer solution in Question 9.19 if the concentration of the salt were doubled while the acid concentration remained the same.

9.5 Oxidation–Reduction Processes

Learning Goal

9

Oxidation–reduction processes are responsible for many types of chemical change. Corrosion, the operation of a battery, and biochemical energy-harvesting reactions are a few examples. In this section we explore the basic concepts underlying this class of chemical reactions.

Oxidation and Reduction

Oxidation is defined as a loss of electrons, loss of hydrogen atoms, or gain of oxygen atoms. *Sodium metal*, is, for example, oxidized to a *sodium ion*, losing one electron when it reacts with a nonmetal such as chlorine:

$$Na \longrightarrow Na^+ + e^-$$

A CLINICAL PERSPECTIVE

Oxidizing Agents for Chemical Control of Microbes

Before the twentieth century, hospitals were not particularly sanitary establishments. Refuse, including human waste, was disposed of on hospital grounds. Because many hospitals had no running water, physicians often cleaned their hands and instruments by wiping them on their lab coats and then proceeded to treat the next patient! As you can imagine, many patients died of infections in hospitals.

By the late nineteenth century a few physicians and microbiologists had begun to realize that infectious diseases are transmitted by microbes, including bacteria and viruses. To decrease the number of hospital-acquired infections, physicians like Joseph Lister and Ignatz Semmelweis experimented with chemicals and procedures that were designed to eliminate pathogens from environmental surfaces and from wounds.

Many of the common disinfectants and antiseptics are oxidizing agents. A disinfectant is a chemical that is used to kill or inhibit the growth of pathogens, disease-causing microorganisms, on environmental surfaces. An antiseptic is a milder chemical that is used to destroy pathogens associated with living tissue.

Hydrogen peroxide is an effective antiseptic that is commonly used to cleanse cuts and abrasions. We are all familiar with the furious bubbling that occurs as the enzyme catalase from our body cells catalyzes the breakdown of H_2O_2:

$$2H_2O_2(aq) \longrightarrow 2H_2O(l) + O_2(g)$$

A highly reactive and deadly form of oxygen, the superoxide radical (O_2^-), is produced during this reaction. This molecule inactivates proteins, especially critical enzyme systems.

At higher concentrations (3–6%), H_2O_2 is used as a disinfectant. It is particularly useful for disinfection of soft contact lenses, utensils, and surgical implants because there is no residual toxicity. Concentrations of 6–25% are even used for complete sterilization of environmental surfaces.

Benzoyl peroxide is another powerful oxidizing agent. Ointments containing 5–10% benzoyl peroxide have been used as antibacterial agents to treat acne. The compound is currently found in over-the-counter facial scrubs because it is also an exfoliant, causing sloughing of old skin and replacement with smoother-looking skin. A word of caution is in order; in sensitive individuals, benzoyl peroxide can cause swelling and blistering of tender facial skin.

Chlorine is a very widely used disinfectant and antiseptic. Calcium hypochlorite [$Ca(OCl)_2$] was first used in hospital maternity wards in 1847 by the pioneering Hungarian physician Ignatz Semmelweis. Semmelweis insisted that hospital workers cleanse their hands in a $Ca(OCl)_2$ solution and dramatically reduced the incidence of infection. Today, calcium hypochlorite is more commonly used to disinfect bedding, clothing, restaurant eating utensils, slaughterhouses, barns, and dairies.

Sodium hypochlorite (NaOCl), Clorox®, is used as a household disinfectant and deodorant but is also used to disinfect swimming pools, dairies, food-processing equipment, and kidney dialysis units. It can be used to treat drinking water of questionable quality. Addition of 1/2 teaspoon of household bleach (5.25% NaOCl) to 2 gallons of clear water renders it drinkable after 1/2 hour. The Centers for Disease Control even recommend a 1:10 dilution of bleach as an effective disinfectant against human immunodeficiency virus, the virus that causes acquired immune deficiency syndrome (AIDS).

Chlorine gas (Cl_2) is used to disinfect swimming pool water, sewage, and municipal water supplies. This treatment has successfully eliminated epidemics of waterborne diseases; however, chlorine is inactivated in the presence of some organic materials and, in some cases, may form toxic chlorinated organic compounds. For these reasons, many cities are considering the use of ozone (O_3) rather than chlorine.

Ozone is produced from O_2 by high-voltage electrical discharges. (That fresh smell in the air after an electrical storm is ozone.) Several European cities use ozone to disinfect drinking water. It is a more effective killing agent than chlorine, especially with some viruses; less ozone is required for disinfection; there is no unpleasant residual odor or flavor; and there appear to be fewer toxic by-products. However, ozone is more expensive than chlorine, and maintaining the required concentration in the water is more difficult. Nonetheless, the benefits seem to outweigh the drawbacks, and many U.S. cities may soon follow the example of the European cities and convert to the use of ozone for water treatment.

Reduction is defined as a gain of electrons, gain of hydrogen atoms, or loss of oxygen atoms. A *chlorine atom* is reduced to a *chloride ion* by gaining one electron when it reacts with a metal such as sodium:

$$Cl + \boxed{e^-} \longrightarrow Cl^-$$

Oxidation and reduction are complementary processes. The *oxidation half-reaction* produces an electron that is the reactant for the *reduction half-reaction*. The combination of two half-reactions, one oxidation and one reduction, produces the complete reaction:

a. A strong base is added to the solution.

b. More acetic acid is added to the solution.

9.51 What is $[H_3O^+]$ for a buffer solution that is 0.200 M in acid and 0.500 M in the corresponding salt if the weak acid $K_a = 5.80 \times 10^{-7}$?

9.52 What is the pH of the solution described in Question 9.51?

Oxidation–Reduction Reactions

9.53 Define:

a. oxidation

b. oxidizing agent

9.54 Define:

a. reduction

b. reducing agent

9.55 During an oxidation process in an oxidation–reduction reaction the species oxidized _____ electrons.

9.56 During an oxidation–reduction reaction the species _____ is the oxidizing agent.

9.57 During an oxidation–reduction reaction the species _____ is the reducing agent.

9.58 Metals tend to be good _____ agents.

9.59 In the following reaction, identify the oxidized species, reduced species, oxidizing agent, and reducing agent:

$$Cl_2(aq) + 2KI(aq) \longrightarrow 2KCl(aq) + I_2(aq)$$

9.60 In the following reaction, identify the oxidized species, reduced species, oxidizing agent, and reducing agent:

$$Zn(s) + Cu^{2+}(aq) \longrightarrow Zn^{2+}(aq) + Cu(s)$$

9.61 Explain the relationship between oxidation–reduction and voltaic cells.

9.62 Compare and contrast a battery and electrolysis.

9.63 Describe one application of voltaic cells.

9.64 Describe one application of electrolytic cells.

Critical Thinking Problems

1. Acid rain is a threat to our environment because it can increase the concentration of toxic metal ions, such as Cd^{2+} and Cr^{3+}, in rivers and streams. If cadmium and chromium are present in sediment as $Cd(OH)_2$ and $Cr(OH)_3$, write reactions that demonstrate the effect of acid rain. Use the library to find the properties of cadmium and chromium responsible for their environmental impact.

2. Aluminum carbonate is more soluble in acidic solution, forming aluminum cations. Write a reaction (or series of reactions) that explain this observation.

3. Carbon dioxide reacts with the hydroxide ion to produce the bicarbonate anion. Write the Lewis dot structures for each reactant and product. Label each as a Brønsted acid or base. Explain the reaction using the Brønsted theory. Why would the Arrhenius theory provide an inadequate description of this reaction?

4. Maalox® is an antacid composed of $Mg(OH)_2$ and $Al(OH)_3$. Explain the origin of the trade name Maalox. Write chemical reactions that demonstrate the antacid activity of Maalox.

5. Acid rain has been described as a regional problem, whereas the greenhouse effect is a global problem. Do you agree with this statement? Why or why not?

10 The Nucleus, Radioactivity, and Nuclear Medicine

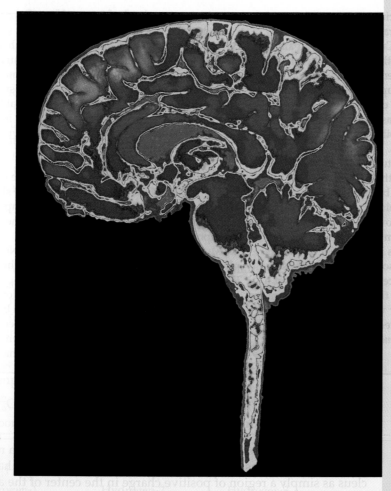

Noninvasive techniques for visualizing internal organs have revolutionized medical diagnosis.

Learning Goals

1 Enumerate the characteristics of alpha, beta, and gamma radiation.

2 Write balanced equations for common nuclear processes.

3 Calculate the amount of radioactive substance remaining after a specified number of half-lives.

4 Describe the various ways in which nuclear energy may be used to generate electricity: fission, fusion, and the breeder reactor.

5 Explain the process of radiocarbon dating.

6 Cite several examples of the use of radioactive isotopes in medicine.

7 Describe the use of ionizing radiation in cancer therapy.

8 Discuss the preparation of radioisotopes for use in diagnostic imaging studies.

9 Explain the difference between natural and artificial radioactivity.

10 Describe the characteristics of radioactive materials that relate to radiation exposure and safety.

11 Be familiar with common techniques for the detection of radioactivity.

12 Know the common units in which radiation intensity is represented: the curie, roentgen, rad, and rem.

Critical Thinking Problems

1. Isotopes used as radioactive tracers have chemical properties that are similar to those of a nonradioactive isotope of the same element. Explain why this is a critical consideration in their use.

2. A chemist proposes a research project to discover a catalyst that will speed up the decay of radioactive isotopes that are waste products of a medical laboratory. Such a discovery would be a potential solution to the problem of nuclear waste disposal. Critique this proposal.

3. A controversial solution to the disposal of nuclear waste involves burial in sealed chambers far below the earth's surface. Describe potential pros and cons of this approach.

4. What type of radioactive decay is favored if the number of protons in the nucleus is much greater than the number of neutrons? Explain.

5. If the proton-to-neutron ratio in question 4 (above) were reversed, what radioactive decay process would be favored? Explain.

6. Radioactive isotopes are often used as "tracers" to follow an atom through a chemical reaction, and the following is an example. Acetic acid reacts with methyl alcohol by eliminating a molecule of water to form methyl acetate. Explain how you would use the radioactive isotope oxygen-18 to show whether the oxygen atom in the water product comes from the —OH of the acid or the —OH of the alcohol.

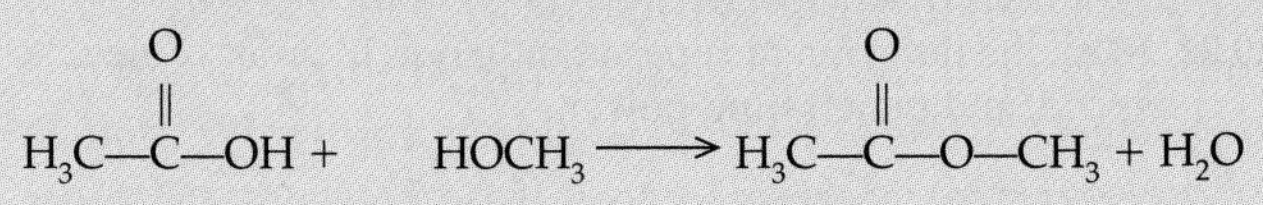

$$
\begin{array}{ccccc}
\text{O} & & & & \text{O} \\
\| & & & & \| \\
\text{H}_3\text{C—C—OH} + & \text{HOCH}_3 & \longrightarrow & \text{H}_3\text{C—C—O—CH}_3 + & \text{H}_2\text{O} \\
\text{Acetic acid} & \text{Methyl alcohol} & & \text{Methyl acetate} &
\end{array}
$$

11 An Introduction to Organic Chemistry: The Saturated Hydrocarbons

The origins of fossil fuels.

Learning Goals

1. Compare and contrast organic and inorganic compounds.

2. Draw structures that represent each of the families of organic compounds.

3. Write the names and draw the structures of the common functional groups.

4. Write condensed and structural formulas for saturated hydrocarbons.

5. Describe the relationship between the structure and physical properties of saturated hydrocarbons.

6. Use the basic rules of the I.U.P.A.C. Nomenclature System to name alkanes and substituted alkanes.

7. Draw constitutional isomers of simple organic compounds.

8. Write the names and draw the structures of simple cycloalkanes.

9. Draw *cis* and *trans* isomers of cycloalkanes.

10. Describe conformations of alkanes.

11. Draw the chair and boat conformations of cyclohexane.

12. Write equations for combustion reactions of alkanes.

13. Write equations for halogenation reactions of alkanes.

295

12

The Unsaturated Hydrocarbons: Alkenes, Alkynes, and Aromatics

Mining the sea for hydrocarbons.

Learning Goals

1 Describe the physical properties of alkenes and alkynes.

2 Draw the structures and write the I.U.P.A.C. names for simple alkenes and alkynes.

3 Write the names and draw the structures of simple geometric isomers of alkenes.

4 Write equations predicting the products of addition reactions of alkenes: hydrogenation, halogenation, hydration, and hydrohalogenation.

5 Apply Markovnikov's rule to predict the major and minor products of the hydration and hydrohalogenation reactions of unsymmetrical alkenes.

6 Write equations representing the oxidation of simple alkenes.

7 Write equations representing the formation of addition polymers of alkenes.

8 Draw the structures and write the names of common aromatic hydrocarbons.

9 Write equations for substitution reactions involving benzene.

10 Describe heterocyclic aromatic compounds and list several biological molecules in which they are found.

323

13

Alcohols, Phenols, Thiols, and Ethers

Scotland is famous for its unblended scotches made in small distilleries.

Learning Goals

1 Rank selected alcohols by relative water solubility, boiling points, or melting points.

2 Write the names and draw the structures for common alcohols.

3 Discuss the biological, medical, or environmental significance of several alcohols.

4 Classify alcohols as primary, secondary, or tertiary.

5 Write equations representing the preparation of alcohols by the hydration of an alkene.

6 Write equations representing the preparation of alcohols by hydrogenation (reduction) of aldehydes or ketones.

7 Write equations showing the dehydration of an alcohol.

8 Write equations representing the oxidation of alcohols.

9 Discuss the role of oxidation and reduction reactions in the chemistry of living systems.

10 Discuss the use of phenols as germicides.

11 Write names and draw structures for common ethers and discuss their use in medicine.

12 Write equations representing the dehydration reaction between two alcohol molecules.

13 Write names and draw structures for simple thiols and discuss their biological significance.

CHEMISTRY CONNECTION

Fetal Alcohol Syndrome

The first months of pregnancy are a time of great joy and anticipation but are not without moments of anxiety. On her first visit to the obstetrician the mother-to-be is tested for previous exposure to a number of infectious diseases that could damage the fetus. She is provided with information about diet, weight gain, and drugs that could harm the baby. Among the drugs that should be avoided are alcoholic beverages.

The use of alcoholic beverages by a pregnant woman can cause *fetal alcohol syndrome (FAS)*. A *syndrome* is a set of symptoms that occur together and are characteristic of a particular disease. In this case, physicians have observed that infants born to women with chronic alcoholism showed a reproducible set of abnormalities including mental retardation, poor growth before and after birth, and facial malformations.

Mothers who report only social drinking may have children with *fetal alcohol effects*, a less severe form of fetal alcohol syndrome. This milder form is characterized by a reduced birth weight, some learning disabilities, and behavioral problems.

How does alcohol consumption cause these varied symptoms? No one is exactly sure, but it is well known that the alcohol consumed by the mother crosses the placenta and enters the bloodstream of the fetus. Within about fifteen minutes the concentration of alcohol in the blood of the fetus is as high as

that of the mother! However, the mother has enzymes to detoxify the alcohol in her blood; the fetus does not. Now consider that alcohol can cause cell division to stop or be radically altered. It is thought that even a single night on the town could be enough to cause FAS by blocking cell division during a critical developmental period.

This raises the question "How much alcohol can a pregnant woman safely drink?" As we have seen, the severity of the symptoms seems to increase with the amount of alcohol consumed by the mother. However, it is virtually impossible to do the scientific studies that would conclusively determine the risk to the fetus caused by different amounts of alcohol. There is some evidence that suggests that there is a risk associated with drinking even one ounce of absolute (100%) alcohol each day. Because of these facts and uncertainties, the American Medical Association and the U.S. Surgeon General recommend that pregnant women completely abstain from alcohol.

In this chapter we will study the alcohols, along with several other families of organic compounds that are important to biological systems. In addition to the structure, properties, and reactions of these compounds, we will consider the biological significance and medical application of these molecules.

Introduction

An aryl group is an aromatic ring with one hydrogen atom removed.

The characteristic functional group of the *alcohols* and *phenols* is the *hydroxyl group (—OH)*. Alcohols have the general structure R—OH, in which R is any alkyl group. Phenols are similar in structure but contain an aryl group in place of the alkyl group. Both can be viewed as substituted water molecules in which one of the hydrogen atoms has been replaced by an alkyl or aryl group.

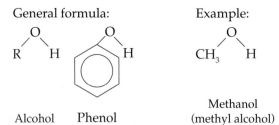

Ethers have two alkyl or aryl groups attached to the oxygen atom and may be thought of as substituted alcohols. The functional group characteristic of an ether is R—O—R. *Thiols* are a family of compounds that contain the sulfhydryl group (—SH). They, too, have a structure similar to that of alcohols.

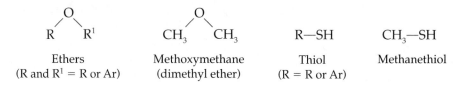

Many important biological molecules, including sugars (carbohydrates), fats (lipids), and proteins, contain hydroxyl and/or thiol groups.

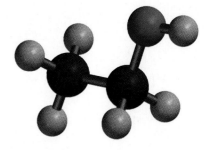

Figure 13.1
Ball-and-stick model of the simple alcohol, ethanol.

D-Gluose, *a sugar*   Lysine vasopressin
                               (partial structure), *a protein*

Monolaurin, *a lipid*

In biological systems the hydroxyl group is often involved in a variety of reactions such as oxidation, reduction, hydration, and dehydration. In glycolysis (a metabolic pathway by which glucose is degraded and energy is harvested in the form of ATP), several steps center on the reactivity of the hydroxyl group. The majority of the consumable alcohol in the world (ethanol) is produced by fermentation reactions carried out by yeasts.

The thiol group is found in the structure of some amino acids and is essential for keeping proteins in the proper three-dimensional shape required for their biological function. Thus these functional groups play a central role in the structure and chemical properties of biological molecules. The thiol group of the amino acid cysteine is highlighted in blue in the structure of lysine vasopressin presented above.

Glycolysis and fermentation are discussed in Chapter 21.

13.1 Alcohols: Structure and Physical Properties

An **alcohol** is an organic compound that contains a **hydroxyl group** (—OH) attached to an alkyl group (Figure 13.1). The R—O—H portion of an alcohol is similar to the structure of water. The oxygen and the two atoms bonded to it lie in the same plane, and the R—O—H bond angle is approximately 104°, which is very similar to the H—O—H bond angle of water.

The hydroxyl groups of alcohols are very polar because the oxygen and hydrogen atoms have significantly different electronegativities. Because the two atoms involved in this polar bond are oxygen and hydrogen, hydrogen bonds can form between alcohol molecules (Figure 13.2).

As a result of this intermolecular hydrogen bonding, alcohols boil at much higher temperatures than hydrocarbons of similar molecular weight. These higher boiling points are caused by the large amount of heat needed to break the hydrogen bonds that attract the alcohol molecules to one another. Compare the boiling points of butane and propanol, which have similar molecular weights:

Learning Goal

1

Electronegativity is discussed in Section 4.1. Hydrogen bonding is described in detail in Section 6.2.

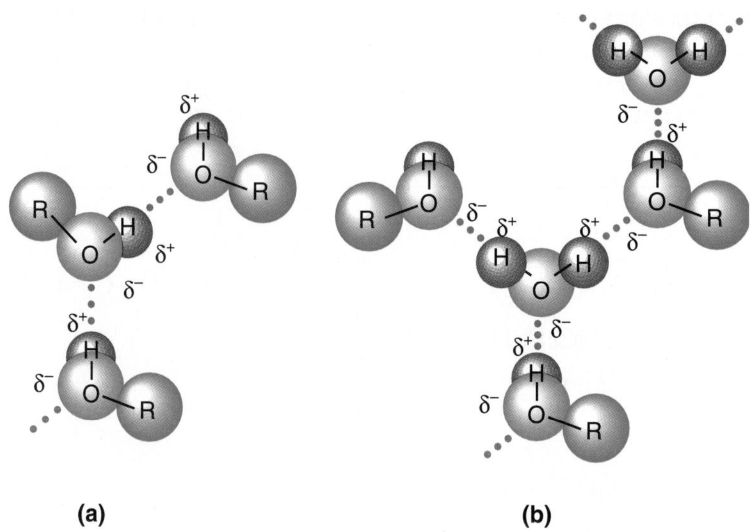

Figure 13.2
(a) Hydrogen bonding between alcohol molecules. (b) Hydrogen bonding between alcohol molecules and water molecules.

(a)   **(b)**

$CH_3CH_2CH_2CH_3$

Butane
M.W. = 58
b.p. = −0.4°C

$CH_3CH_2CH_2OH$

1-Propanol
M.W. = 60
b.p. = 97.2°C

Alcohols with fewer than four or five carbon atoms are very soluble in water, and those with five to eight carbons are moderately soluble in water. This is due to the ability of the alcohol to form intermolecular hydrogen bonds with water molecules (see Figure 13.2b). As the nonpolar, or hydrophobic, portion of an alcohol (the carbon chain) becomes larger relative to the polar, hydrophilic, region (the hydroxyl group), the water solubility of an alcohol decreases. As a result large alcohols are nearly insoluble in water. The term *hydrophobic*, which literally means "water fearing," is used to describe a molecule or a region of a molecule that is nonpolar and, thus, more soluble in nonpolar solvents than in water. Similarly, the term *hydrophilic*, meaning water loving, is used to describe a polar molecule or region of a molecule that is more soluble in the polar solvent water than in a nonpolar solvent.

An increase in the number of hydroxyl groups along a carbon chain will increase the influence of the polar hydroxyl group. It follows, then that diols and triols are more water soluble than alcohols with only a single hydroxyl group.

The presence of polar hydroxyl groups in large biological molecules—for instance, proteins and nucleic acids—allows intramolecular hydrogen bonding that keeps these molecules in the shapes needed for biological function.

Intermolecular hydrogen bonds are attractive forces between two molecules. *Intramolecular* hydrogen bonds are attractive forces between polar groups within the same molecule.

13.2 Alcohols: Nomenclature

I.U.P.A.C. Names

Learning Goal

2

The way to determine the parent compound was described in Section 11.2.

In the I.U.P.A.C. Nomenclature System, alcohols are named according to the following steps:

- Determine the name of the *parent compound*, the longest continuous carbon chain containing the —OH group.
- Replace the *-e* ending of the alkane chain with the *-ol* ending of the alcohol. Following this pattern, an alkane becomes an alkanol. For instance, ethan*e* becomes ethan*ol*, and propan*e* becomes propan*ol*.
- Number the parent chain to give the carbon bearing the hydroxyl group the lowest possible number.

- Name and number all substituents, and add them as prefixes to the "alkanol" name.
- Alcohols containing two hydroxyl groups are named *-diols*. Those bearing three hydroxyl groups are called *-triols*. A number giving the position of each of the hydroxyl groups is needed in these cases.

EXAMPLE 13.1

Using I.U.P.A.C. Nomenclature to Name an Alcohol

Name the following alcohol using I.U.P.A.C. nomenclature.

Learning Goal

2

Solution

$$\underset{OH}{\overset{1 \quad 2 \quad 3 \quad 4 \quad 5 \quad 6 \quad 7}{CH_3CHCH_2CH_2CH_2CHCH_3}}$$
$$\quad\quad\quad OH \quad\quad\quad\quad CH_3$$

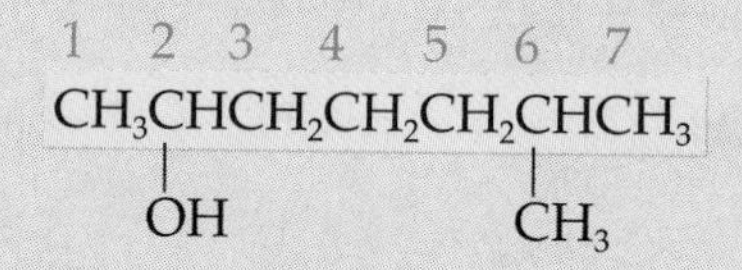

Parent compound: heptane (becomes heptanol)
Position of —OH: carbon-2 (*not* carbon-6)
Substituents: 6-methyl
Name: 6-Methyl-2-heptanol

EXAMPLE 13.2

Using I.U.P.A.C. Nomenclature to Name Alcohols

Name the following cyclic alcohol using I.U.P.A.C. nomenclature.

Learning Goal

2

Solution

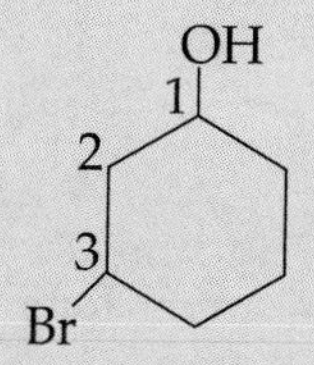

Remember that this line structure represents a cyclic molecule composed of six carbon atoms and associated hydrogen atoms, as follows:

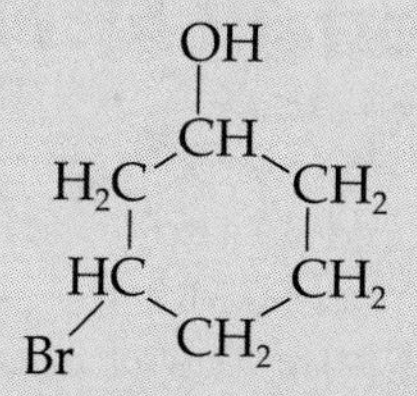

Parent compound: cyclohexane (becomes cyclohexanol)
Position of —OH: carbon-1 (*not* carbon-3)
Substituents: 3-bromo (*not* 5-bromo)
Name: 3-Bromocyclohexanol (it is assumed that the —OH is on carbon-1 in cyclic structures)

Common Names

The common names for alcohols are derived from the alkyl group corresponding to the parent compound. The name of the alkyl group is followed by the word *alcohol*. For some alcohols, such as ethylene glycol and glycerol, historical names are

See Section 11.2 for the names of the common alkyl groups.

used. The following examples provide the I.U.P.A.C. and common names of several alcohols:

$$CH_3CHCH_3$$
$$\overset{|}{OH}$$

$$HOCH_2CH_2OH$$

$$CH_3CH_2OH$$

2-Propanol
(isopropyl
alcohol)

1,2-Ethanediol
(ethylene glycol)

Ethanol
(ethyl alcohol)
(grain alcohol)

Question 13.1

Use the I.U.P.A.C. Nomenclature System to name each of the following compounds.

a. 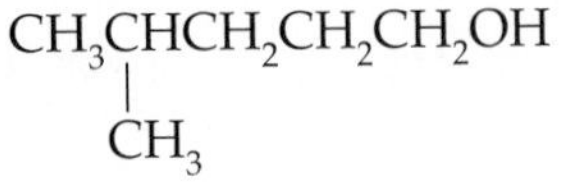 $CH_3CHCH_2CH_2CH_2OH$
   $\overset{|}{CH_3}$

b. 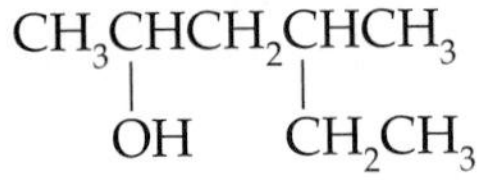 $CH_3CHCH_2CHCH_3$
   $\overset{|}{OH} \quad \overset{|}{CH_2CH_3}$

c. 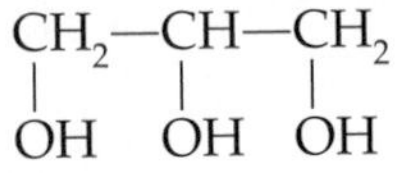 $CH_2\!-\!CH\!-\!CH_2$
   $\overset{|}{OH} \ \overset{|}{OH} \ \overset{|}{OH}$

   (Common name: Glycerol)

d. 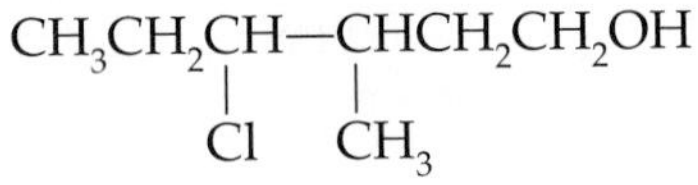 $CH_3CH_2CH\!-\!CHCH_2CH_2OH$
   $\overset{|}{Cl} \quad \overset{|}{CH_3}$

Question 13.2

Give the common name and the I.U.P.A.C. name for each of the following compounds.

a. $CH_3CH_2CH_2CH_2CH_2CH_2CH_2OH$

b. 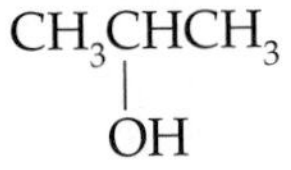 CH_3CHCH_3
   $\overset{|}{OH}$

c. 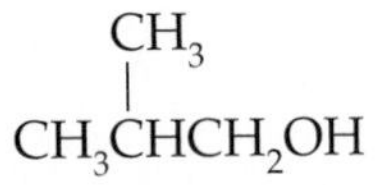 $\quad\quad CH_3$
   $\quad\quad \overset{|}{}$
   CH_3CHCH_2OH

13.3 Medically Important Alcohols

Methanol

Learning Goal

3

Methanol (methyl alcohol), CH_3OH, is a colorless and odorless liquid that is used as a solvent and as the starting material for the synthesis of methanal (formaldehyde). Methanol is often called *wood alcohol* because it can be made by heating wood in the absence of air. Methanol is toxic and can cause blindness and perhaps death if ingested. Methanol may also be used as fuel, especially for "formula" racing cars.

Ethanol

Ethanol (ethyl alcohol), CH_3CH_2OH, is a colorless and odorless liquid and is the alcohol in alcoholic beverages. It is also widely used as a solvent and as a raw material for the preparation of other organic chemicals.

Fermentation reactions are described in detail in Section 21.4 and in "A Human Perspective: Fermentations: The Good, the Bad, and the Ugly."

The ethanol used in alcoholic beverages comes from the **fermentation** of carbohydrates (sugars and starches). The beverage produced depends on the starting material and the fermentation process: scotch (grain), bourbon (corn), burgundy wine (grapes and grape skins), and chablis wine (grapes without red skins). The following equation summarizes the fermentation process:

$$C_6H_{12}O_6 \xrightarrow[\text{enzyme action}]{\text{Several steps involving}} 2CH_3CH_2OH + 2CO_2$$

Sugar (glucose)

Ethanol (ethyl alcohol)

The alcoholic beverages listed have quite different alcohol concentrations. Wines are generally 12–13% alcohol because the yeasts that produce the ethanol are killed by ethanol concentrations of 12–13%. (Wine producers are always trying to find strains of yeast with greater ethanol tolerance.) To produce bourbon or scotch with an alcohol concentration of 40–45% ethanol (80 or 90 proof), the original fermentation products must be distilled.

The sale and use of pure ethanol (100% ethanol) are regulated by the federal government. To prevent illegal use of pure ethanol, it is *denatured* by the addition of a denaturing agent, which makes it unfit to drink but suitable for laboratory applications.

> **Distillation is the separation of compounds in a mixture based on differences in boiling points.**

2-Propanol

2-Propanol (isopropyl alcohol),

$$\underset{\underset{\displaystyle OH}{|}}{CH_3CHCH_3}$$

was commonly called *rubbing alcohol* because patients with high fevers were often given alcohol baths to reduce body temperature. Rapid evaporation of the alcohol results in skin cooling. This practice is no longer commonly used.

It is also used as a disinfectant, an astringent (skin-drying agent), an industrial solvent, and a raw material in the synthesis of organic chemicals. It is colorless, has a very slight odor, and is toxic when ingested.

1,2-Ethanediol

1,2-Ethanediol (ethylene glycol),

$$\underset{\underset{\displaystyle OH}{|}}{CH_2}-\underset{\underset{\displaystyle OH}{|}}{CH_2}$$

is used as automobile antifreeze. When added to water in the radiator, the ethylene glycol solute lowers the freezing point and raises the boiling point of the water. Ethylene glycol has a sweet taste but is extremely poisonous. For this reason, color additives are used in antifreeze to ensure that it is properly identified.

The colligative properties of solutions are discussed in Section 7.6.

1,2,3-Propanetriol

1,2,3-Propanetriol (glycerol),

$$\underset{\underset{\displaystyle OH}{|}}{CH_2}-\underset{\underset{\displaystyle OH}{|}}{CH}-\underset{\underset{\displaystyle OH}{|}}{CH_2}$$

is a viscous, sweet-tasting, nontoxic liquid. It is very soluble in water and is used in cosmetics, pharmaceuticals, and lubricants. Glycerol is obtained as a by-product of the hydrolysis of fats.

13.4 Classification of Alcohols

Alcohols are classified as **primary (1°), secondary (2°),** or **tertiary (3°),** depending on the number of alkyl groups attached to the **carbinol carbon,** the carbon bearing the

Learning Goal

hydroxyl (—OH) group. If no alkyl groups are attached, the alcohol is methyl alcohol; if there is a single alkyl group, the alcohol is a primary alcohol; an alcohol with two alkyl groups bonded to the carbon bearing the hydroxyl group is a secondary alcohol, and if three alkyl groups are attached, the alcohol is a tertiary alcohol.

Methyl alcohol   1° Alcohol   2° Alcohol   3° Alcohol

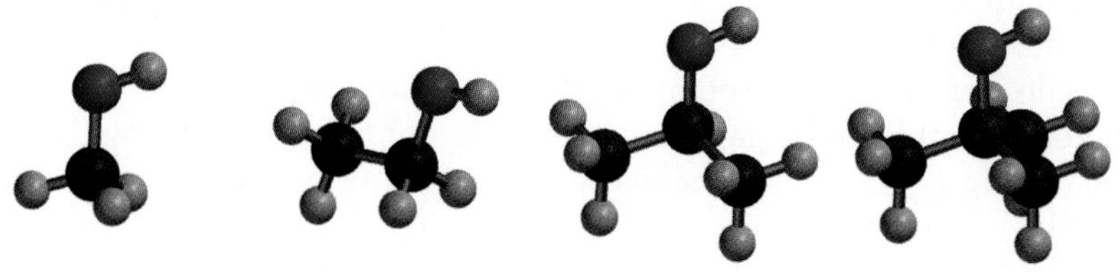

Methanol   Ethanol   2-Propanol   2-Methyl-2-propanol

EXAMPLE 13.3

Learning Goal

4

Classifying Alcohols

Classify each of the following alcohols as primary, secondary, or tertiary.

Solution

In each of the structures shown below, the carbinol carbon is shown in red:

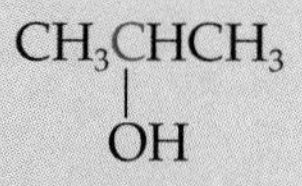

$$CH_3CHCH_3$$
$$\overset{|}{OH}$$

This alcohol, 2-propanol, is a secondary alcohol because there are two alkyl groups attached to the carbinol carbon.

$$\overset{CH_3}{\overset{|}{CH_3CCH_3}}$$
$$\overset{|}{OH}$$

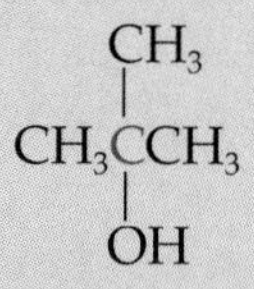

This alcohol, 2-methyl-2-propanol, is a tertiary alcohol because there are three alkyl groups attached to the carbinol carbon.

$$\overset{OH}{\overset{|}{CH_3CH_2CHCH_2}}$$
$$\overset{|}{CH_2CH_3}$$

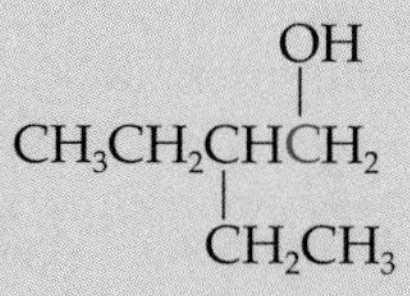

This alcohol, 2-ethyl-1-butanol, is a primary alcohol because there is only one alkyl group attached to the carbinol carbon.

Question 13.3

Classify each of the following alcohols as 1°, 2°, 3°, or aromatic (phenol).
a. $CH_3CH_2CH_2CH_2OH$
b. 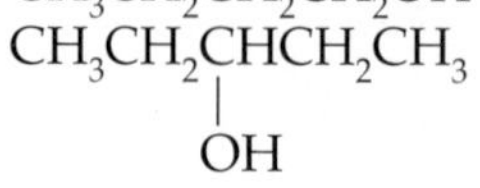 $CH_3CH_2CHCH_2CH_3$
$\qquad\qquad\overset{|}{OH}$
c.

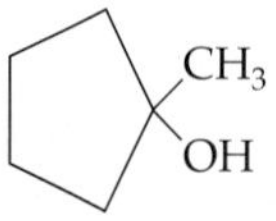

d.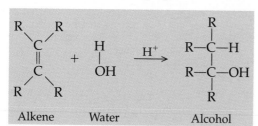

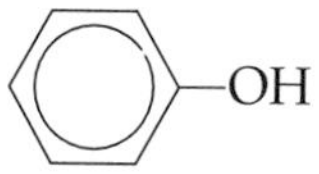

e.

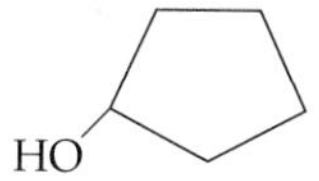

Classify each of the following alcohols as 1°, 2°, or 3°.

a. 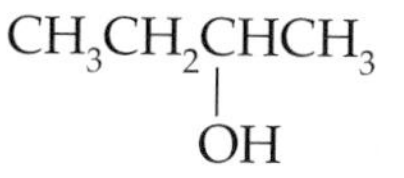 $CH_3CH_2CHCH_3$
       |
       OH

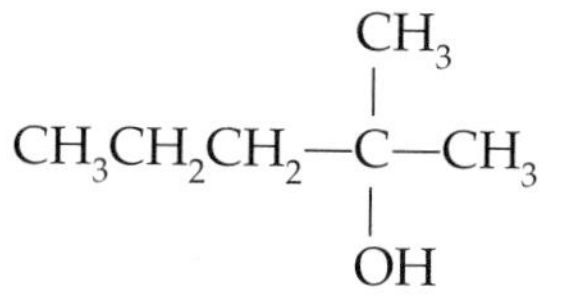

c. $CH_3CH_2—OH$

b.
           CH_3
            |
$CH_3CH_2CH_2—C—CH_3$
            |
           OH

d. 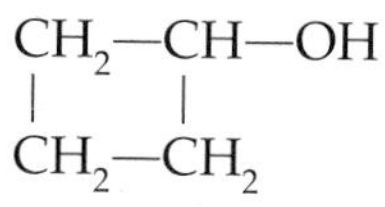 $CH_2—CH—OH$
    | |
   $CH_2—CH_2$

13.5 Reactions Involving Alcohols

Preparation of Alcohols

As we saw in the last chapter, the most important reactions of alkenes are *addition reactions*. Addition of a water molecule to the carbon-carbon double bond of an alkene produces an alcohol. This reaction, called **hydration**, requires a trace of acid (H^+) as a catalyst, as shown in the following equation:

Learning Goal

5

Hydration of alkenes is described in Section 12.4.

| Alkene | Water | Alcohol |

EXAMPLE 13.4

Writing an Equation Representing the Preparation of an Alcohol by the Hydration of an Alkene

Write an equation representing the preparation of cyclohexanol from cyclohexene.

Solution

Begin by writing the structure of cyclohexene. Recall that cyclohexene is a six-carbon cyclic alkene. Now add the water molecule to the equation.

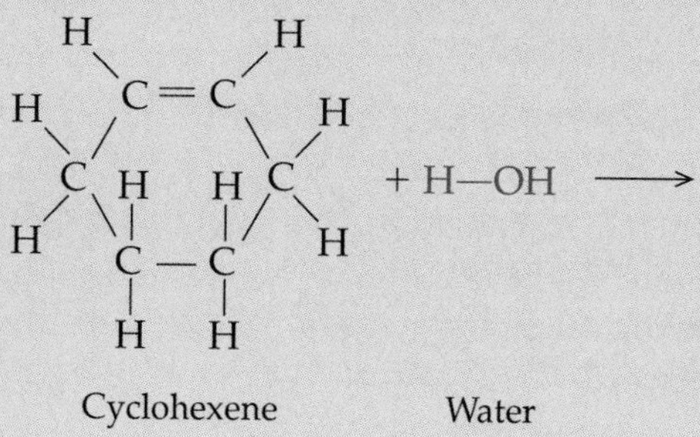

Cyclohexene   Water

You will recognize that the hydration reaction involves the addition of a water molecule to the carbon-carbon double bond. Recall that the reaction requires a trace of acid as a catalyst. Complete the equation by adding the catalyst and product, cyclohexanol.

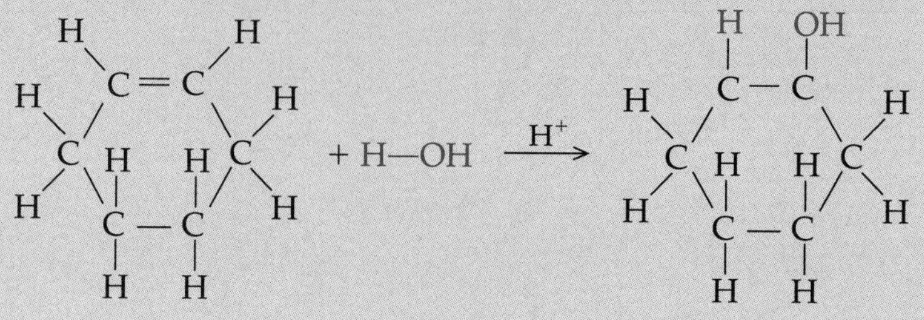

Cyclohexene   Water   Cyclohexanol

Learning Goal

6

Alcohols may also be prepared via the hydrogenation (reduction) of aldehydes and ketones. This reaction, summarized as follows, is discussed in Section 14.4, and is similar to the hydrogenation of alkenes.

In an aldehyde, R^1 and R^2 may be either alkyl groups or H. In ketones, R^1 and R^2 are both alkyl groups.

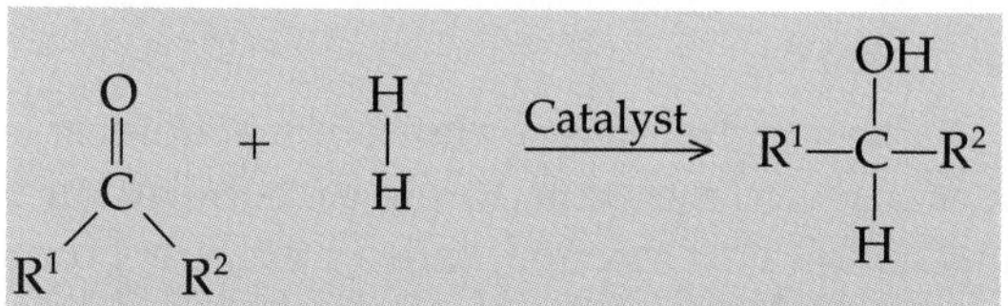

Aldehyde   Hydrogen   Alcohol
or
Ketone

EXAMPLE 13.5

Writing an Equation Representing the Preparation of an Alcohol by the Hydrogenation (Reduction) of an Aldehyde

Write an equation representing the preparation of 1-propanol from propanal.

Solution

Begin by writing the structure of propanal. Propanal is a three-carbon aldehyde. Aldehydes are characterized by the presence of a carbonyl group (—C=O) attached to the end of the carbon chain of the molecule. After you have drawn the structure of propanal, add diatomic hydrogen to the equation.

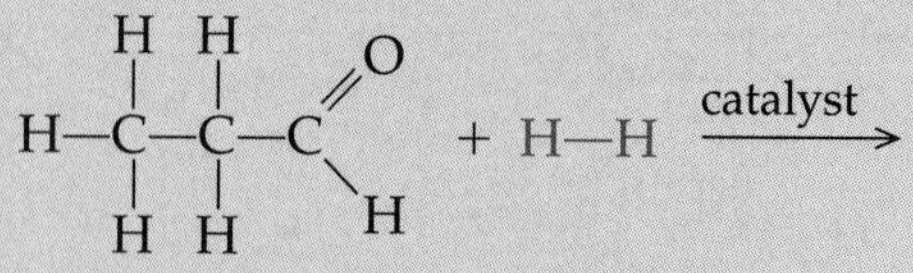

Propanol   Hydrogen

Notice that the general reaction reveals this reaction to be another example of a hydrogenation reaction. As the hydrogens are added to the carbon-oxygen double bond, it is converted to a carbon-oxygen single bond, as the carbonyl group becomes a hydroxyl group.

$$\text{H}-\overset{\overset{\displaystyle\text{H}}{|}}{\underset{\underset{\displaystyle\text{H}}{|}}{\text{C}}}-\overset{\overset{\displaystyle\text{H}}{|}}{\underset{\underset{\displaystyle\text{H}}{|}}{\text{C}}}-\overset{\displaystyle\text{O}}{\underset{\underset{\displaystyle\text{H}}{}}{\text{C}}} + \text{H}-\text{H} \xrightarrow{\text{catalyst}} \text{H}-\overset{\overset{\displaystyle\text{H}}{|}}{\underset{\underset{\displaystyle\text{H}}{|}}{\text{C}}}-\overset{\overset{\displaystyle\text{H}}{|}}{\underset{\underset{\displaystyle\text{H}}{|}}{\text{C}}}-\overset{\overset{\displaystyle\text{H}}{|}}{\underset{\underset{\displaystyle\text{H}}{|}}{\text{C}}}-\text{OH}$$

| Propanal | Hydrogen | 1-Propanol |

Writing an Equation Representing the Preparation of an Alcohol by the Hydrogenation (Reduction) of a Ketone

EXAMPLE 13.6

Write an equation representing the preparation of 2-propanol from 2-propanone.

Learning Goal

Solution

Begin by writing the structure of propanone. Propanone is a three-carbon ketone. Ketones are characterized by the presence of a carbonyl group (—C=O) located anywhere within the carbon chain of the molecule. In the structure of propanone, the carbonyl group must be associated with the center carbon. After you have drawn the structure of propanone, add diatomic hydrogen to the equation.

$$\text{H}-\overset{\overset{\displaystyle\text{H}}{|}}{\underset{\underset{\displaystyle\text{H}}{|}}{\text{C}}}-\overset{\displaystyle\text{O}}{\underset{}{\text{C}}}-\overset{\overset{\displaystyle\text{H}}{|}}{\underset{\underset{\displaystyle\text{H}}{|}}{\text{C}}}-\text{H} + \text{H}-\text{H} \xrightarrow{\text{catalyst}}$$

| Propanone | Hydrogen |

Notice that this reaction is another example of a hydrogenation reaction. As the hydrogens are added to the carbon-oxygen double bond, it is converted to a carbon-oxygen single bond, as the carbonyl group becomes a hydroxyl group.

$$\text{H}-\overset{\overset{\displaystyle\text{H}}{|}}{\underset{\underset{\displaystyle\text{H}}{|}}{\text{C}}}-\overset{\displaystyle\text{O}}{\underset{}{\text{C}}}-\overset{\overset{\displaystyle\text{H}}{|}}{\underset{\underset{\displaystyle\text{H}}{|}}{\text{C}}}-\text{H} + \text{H}-\text{H} \xrightarrow{\text{catalyst}} \text{H}-\overset{\overset{\displaystyle\text{H}}{|}}{\underset{\underset{\displaystyle\text{H}}{|}}{\text{C}}}-\overset{\overset{\displaystyle\text{OH}}{|}}{\underset{\underset{\displaystyle\text{H}}{|}}{\text{C}}}-\overset{\overset{\displaystyle\text{H}}{|}}{\underset{\underset{\displaystyle\text{H}}{|}}{\text{C}}}-\text{H}$$

| Propanone | Hydrogen | 2-Propanol |

Write an equation representing the hydration of cyclopentene. Provide structures and names for the reactants and products.

Question 13.5

Write an equation representing the reduction of 2-butanone. Provide the structures and names for the reactants and products. [*Hint:* 2-butanone is a four-carbon ketone.]

Question 13.6

d. 2 [cyclopentane]—CH$_2$OH $\longrightarrow$?

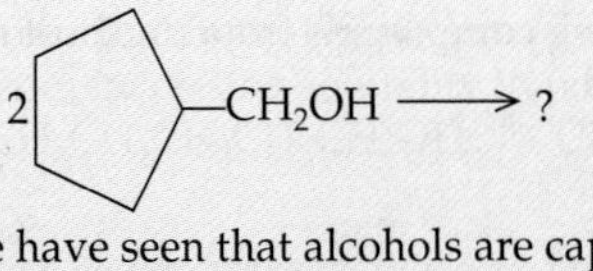

13.56 We have seen that alcohols are capable of hydrogen bonding to each other. Hydrogen bonding is also possible between alcohol molecules and water molecules or between alcohol molecules and ether molecules. Ether molecules *do not* hydrogen bond to each other, however. Explain.

Thiols

13.57 Cystine is an amino acid formed from the oxidation of two cysteine molecules to form a disulfide bond. The molecular formula of cystine is C$_6$H$_{12}$O$_4$N$_2$S$_2$. Draw the structural formula of cystine. (*Hint:* For the structure of cysteine, see Figure 19.11.)

13.58 Explain the way in which British Anti-Lewisite acts as an antidote for mercury poisoning.

13.59 Give the I.U.P.A.C. name for each of the following thiols. (*Hint:* Use the rules for alcohol nomenclature and the suffix -thiol.)

a. CH$_3$CH$_2$CH$_2$—SH

b. CH$_3$CHCH$_2$CH$_3$
$\quad\quad\;$|
$\quad\quad\;$SH

c. $\quad\quad$ CH$_2$CH$_3$
$\quad\quad\quad\;$|
$\quad$CH$_3$—C—CH$_3$
$\quad\quad\quad\;$|
$\quad\quad\quad\;$SH

d. 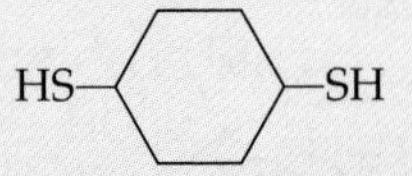 HS—[cyclohexane]—SH

13.60 Give the I.U.P.A.C. name for each of the following thiols. (*Hint:* Use the rules for alcohol nomenclature and the suffix -thiol.)

a. CH$_2$CHCH$_3$
$\quad\;$|$\;\;$|
$\quad\;$SH SH

b. 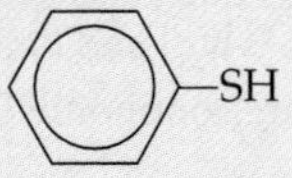 [benzene ring]—SH

c. CH$_3$CHCH$_2$CH$_2$CH$_3$
$\quad\quad\;$|
$\quad\quad\;$SH

d. CH$_3$CH$_2$CH$_2$CH$_2$CH$_2$CH$_2$CH$_2$SH

Critical Thinking Problems

1. You are provided with two solvents: water (H$_2$O) and hexane (CH$_3$CH$_2$CH$_2$CH$_2$CH$_2$CH$_3$). You are also provided with two biological molecules whose structures are shown here:

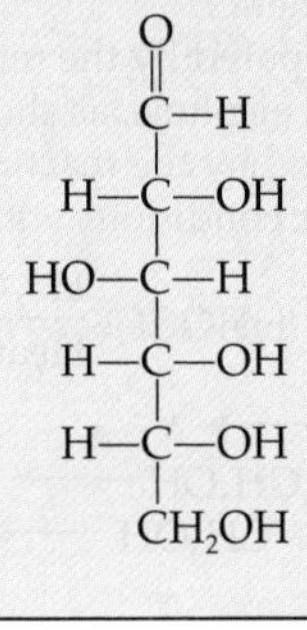

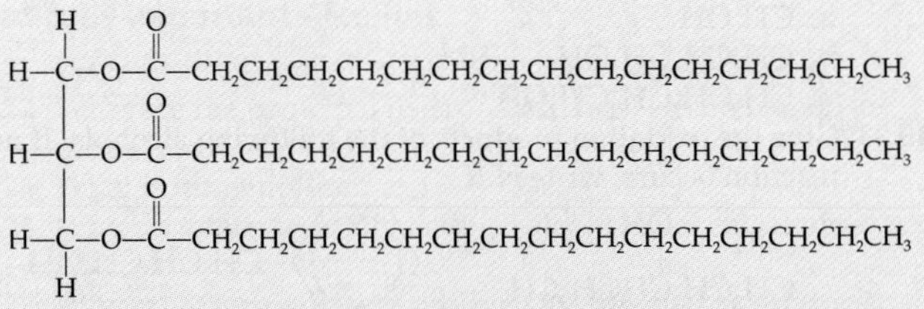

$$\begin{array}{c} O \\ \| \\ C-H \\ | \\ H-C-OH \\ | \\ HO-C-H \\ | \\ H-C-OH \\ | \\ H-C-OH \\ | \\ CH_2OH \end{array}$$

H—C—O—C—CH$_2$CH$_2$CH$_2$CH$_2$CH$_2$CH$_2$CH$_2$CH$_2$CH$_2$CH$_2$CH$_2$CH$_2$CH$_2$CH$_2$CH$_3$
(with O double bonds, glyceryl triester structure shown)

Predict which biological molecule would be more soluble in water and which would be more soluble in hexane. Defend your prediction. Design a careful experiment to test your hypothesis.

Consider the digestion of dietary molecules in the digestive tract. Which of the two biological molecules shown in this problem would be more easily digested under the conditions present in the digestive tract?

2. Cholesterol is an alcohol and a steroid (Chapter 18). Diets that contain large amounts of cholesterol have been linked to heart disease and atherosclerosis, hardening of the arteries. The narrowing of the artery, caused by plaque buildup, is very apparent. Cholesterol is directly involved in this buildup. Describe the various functional groups and principal structural features of the cholesterol molecule. Would you use a polar or nonpolar solvent to dissolve cholesterol? Explain your reasoning.

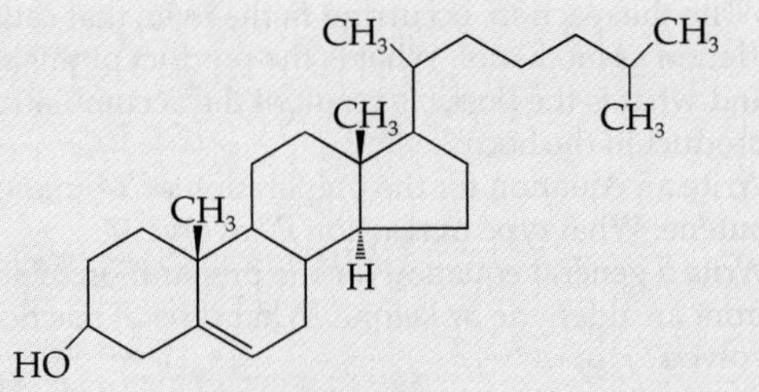

[Cholesterol structure diagram]

Cholesterol

3. An unknown compound A is known to be an alcohol with the molecular formula C$_4$H$_{10}$O. When dehydrated, compound A gave only one alkene product, C$_4$H$_8$, compound B. Compound A could not be oxidized. What are the identities of compound A and compound B?

4. Sulfides are the sulfur analogs of ethers, that is, ethers in which oxygen has been substituted by a sulfur atom. They are named in an analogous manner to the ethers with the term *sulfide* replacing *ether*. For example, CH$_3$—S—CH$_3$ is dimethyl sulfide. Draw the sulfides that correspond to the following ethers and name them:

a. diethyl ether $\quad\quad$ **c.** dibutyl ether

b. methyl propyl ether $\quad$ **d.** ethyl phenyl ether

5. Dimethyl sulfoxide (DMSO) has been used by many sports enthusiasts as a linament for sore joints; it acts as an anti-inflammatory agent and a mild analgesic (pain killer). However, it is no longer recommended for this purpose because it carries toxic impurities into the blood. DMSO is a sulfoxide—it contains the S=O functional group. DMSO is prepared from dimethyl sulfide by mild oxidation, and it has the molecular formula C$_2$H$_6$SO. Draw the structure of DMSO.

14

Aldehydes and Ketones

Vanilla plant blossom.

Learning Goals

1 Draw the structures and discuss the physical properties of aldehydes and ketones.

2 From the structures, write the common and I.U.P.A.C. names of aldehydes and ketones.

3 List several aldehydes and ketones that are of natural, commercial, health, and environmental interest and describe their significance.

4 Write equations for the preparation of aldehydes and ketones by the oxidation of alcohols.

5 Write equations representing the oxidation of carbonyl compounds.

6 Write equations representing the reduction of carbonyl compounds.

7 Write equations for the preparation of hemiacetals, hemiketals, acetals, and ketals.

8 Draw the keto and enol forms of aldehydes and ketones.

9 Write equations showing the aldol condensation.

b. 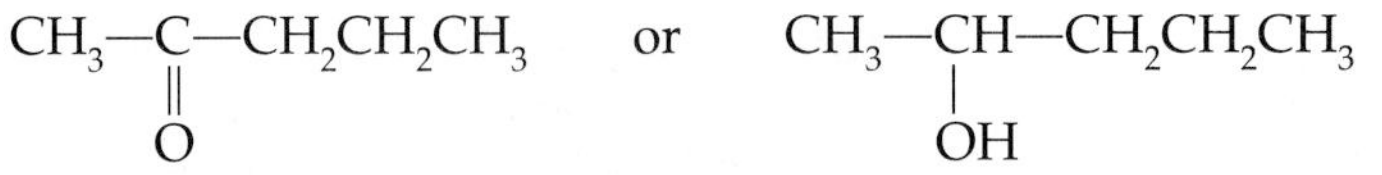 $CH_3-\overset{\overset{\displaystyle O}{\|}}{C}-CH_2CH_2CH_3$ or $CH_3-\overset{\overset{\displaystyle }{|}}{\underset{\underset{\displaystyle OH}{|}}{CH}}-CH_2CH_2CH_3$

c.   or

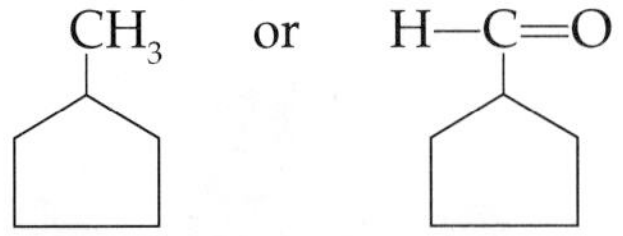

d. 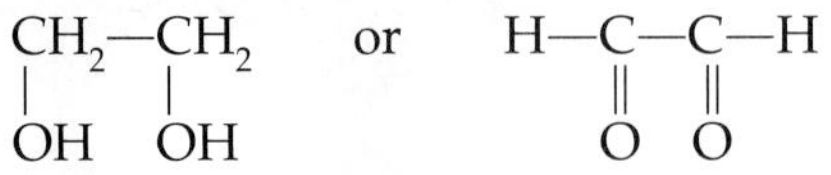 $\underset{\underset{\displaystyle OH}{|}}{CH_2}-\underset{\underset{\displaystyle OH}{|}}{CH_2}$ or $H-\overset{\overset{\displaystyle O}{\|}}{C}-\overset{\overset{\displaystyle O}{\|}}{C}-H$

e. 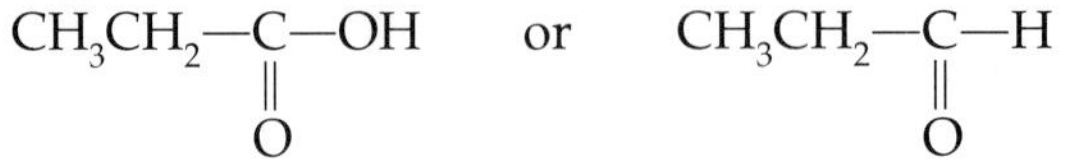 $CH_3CH_2-\overset{\overset{\displaystyle }{}}{\underset{\underset{\displaystyle O}{\|}}{C}}-OH$ or $CH_3CH_2-\overset{\overset{\displaystyle }{}}{\underset{\underset{\displaystyle O}{\|}}{C}}-H$

Question 14.2

Which member in each of the following pairs would have a higher boiling point?

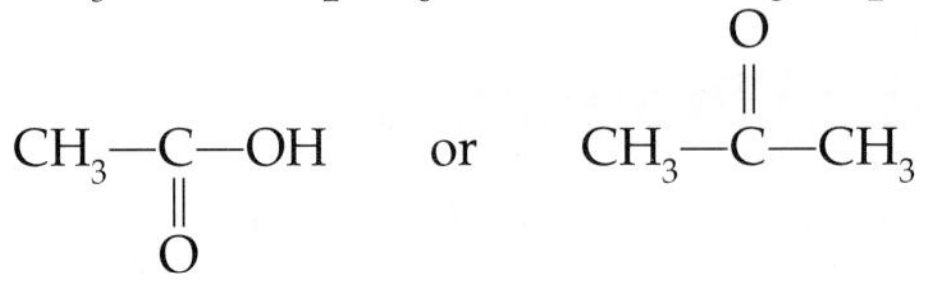

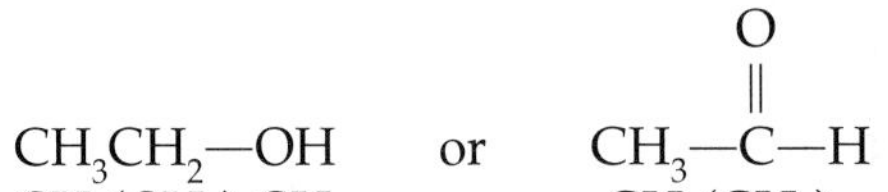

a. 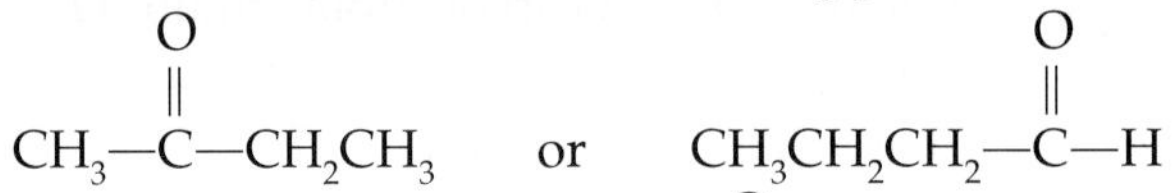 $CH_3-\overset{\overset{\displaystyle O}{\|}}{C}-CH_2CH_3$ or $CH_3CH_2CH_2-\overset{\overset{\displaystyle O}{\|}}{C}-H$

b. $CH_3-\underset{\underset{\displaystyle O}{\|}}{C}-OH$ or $CH_3-\underset{\underset{\displaystyle O}{\|}}{C}-CH_3$

c. CH_3CH_2-OH or $CH_3-\overset{\overset{\displaystyle O}{\|}}{C}-H$
d. 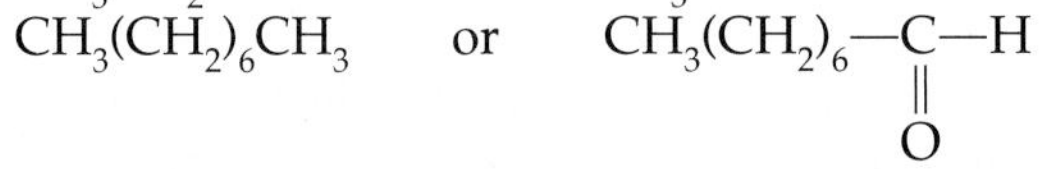 $CH_3(CH_2)_6CH_3$ or $CH_3(CH_2)_6-\underset{\underset{\displaystyle O}{\|}}{C}-H$

14.2 I.U.P.A.C. Nomenclature and Common Names

Naming Aldehydes

Learning Goal

2

In the I.U.P.A.C. system, aldehydes are named according to the following set of rules:

- Determine the parent compound, that is, the longest continuous carbon chain containing the carbonyl group.
- Replace the final -*e* of the parent alkane with -*al*.
- Number the chain beginning with the carbonyl carbon (or aldehyde group) as carbon-1.
- Number and name all substituents as usual.

Several examples are provided here with common names given in parentheses:

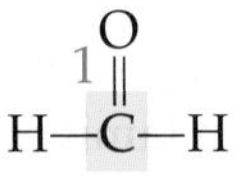

$\underset{\substack{\text{Methanal} \\ \text{(formaldehyde)}}}{H-\overset{\overset{\displaystyle O}{\overset{1}{\|}}}{C}-H}$ $\underset{\substack{\text{Ethanal} \\ \text{(acetaldehyde)}}}{CH_3-\overset{\overset{\displaystyle O}{\overset{1}{\|}}}{\underset{2}{C}}-H}$ $\underset{\substack{\text{Propanal} \\ \text{(propionaldehyde)}}}{CH_3CH_2-\overset{\overset{\displaystyle O}{\overset{1}{\|}}}{\underset{3\quad 2}{C}}-H}$

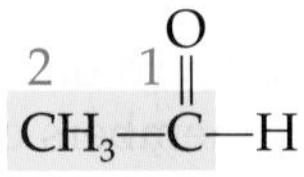

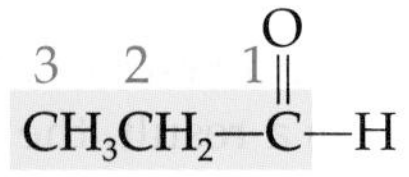

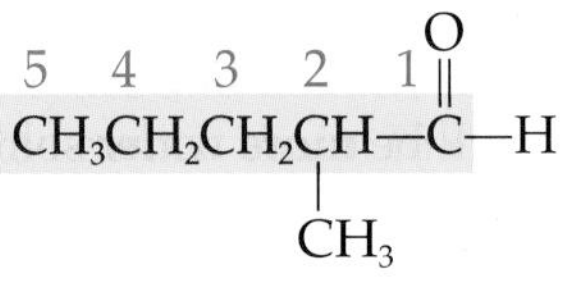

$\underset{\text{2-Methylpentanal}}{\overset{5\quad 4\quad 3\quad 2}{CH_3CH_2CH_2}\underset{\underset{\displaystyle CH_3}{|}}{CH}-\overset{\overset{\displaystyle O}{\overset{1}{\|}}}{C}-H}$

EXAMPLE 14.1

Using the I.U.P.A.C. Nomenclature System to Name Aldehydes

Name the aldehydes represented by the following condensed formulas.

Solution

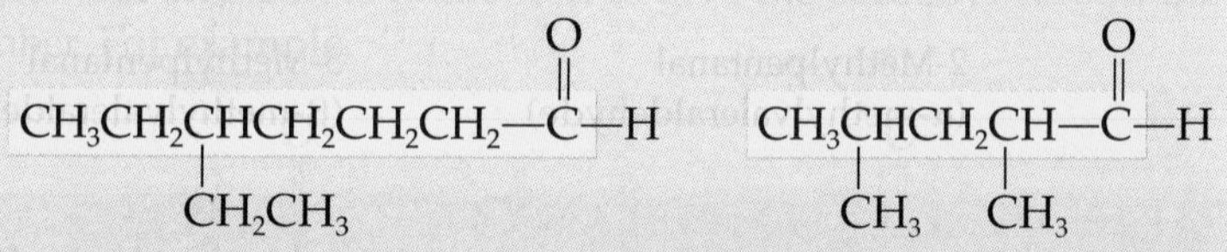

$$CH_3CH_2CHCH_2CH_2CH_2{-}\overset{\overset{\displaystyle O}{\|}}{C}{-}H \qquad CH_3CHCH_2CH{-}\overset{\overset{\displaystyle O}{\|}}{C}{-}H$$

$$\underset{CH_2CH_3}{|} \qquad\qquad\qquad \underset{CH_3}{|}\ \ \ \underset{CH_3}{|}$$

Parent compound:	heptane (becomes heptanal)	pentane (becomes pentanal)
Position of carbonyl group:	carbon-1	carbon-1
Substituents:	5-ethyl	2,4-dimethyl
Name:	5-Ethylheptanal	2,4-Dimethylpentanal

Notice that the position of the carbonyl group is not indicated by a number. By definition, the carbonyl group is located at the end of the carbon chain of an aldehyde. The carbonyl carbon is defined to be carbon-1; thus, it is not necessary to include the position of the carbonyl group in the name of the compound.

The common names of the aldehydes are derived from the same Latin roots as the corresponding carboxylic acids. The common names of the first five aldehydes are presented in Table 14.1.

In the common system of nomenclature, substituted aldehydes are named as derivatives of the straight-chain parent compound (see Table 14.1). Greek letters are used to indicate the position of the substituents. The carbon atom bonded to the carbonyl group is the α-carbon, the next is the β-carbon, and so on.

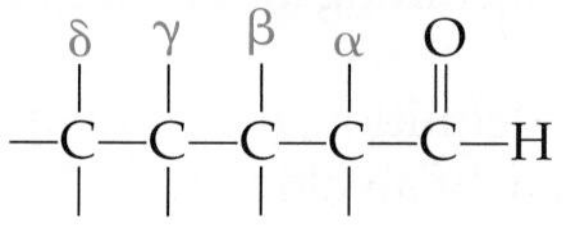

Carboxylic acid nomenclature is described in Section 15.1.

$$\overset{\delta\quad\gamma\quad\beta\quad\alpha\quad O}{\underset{|\ \ \ \ |\ \ \ \ |\ \ \ \ |}{-C-C-C-C-\overset{\|}{C}-H}}$$

Table 14.1	**I.U.P.A.C. and Common Names and Formulas for Several Aldehydes**	
I.U.P.A.C. Name	**Common Name**	**Formula**
Methanal	Formaldehyde	$H{-}\overset{\overset{\displaystyle O}{\|}}{C}{-}H$
Ethanal	Acetaldehyde	$CH_3{-}\overset{\overset{\displaystyle O}{\|}}{C}{-}H$
Propanal	Propionaldehyde	$CH_3CH_2{-}\overset{\overset{\displaystyle O}{\|}}{C}{-}H$
Butanal	Butyraldehyde	$CH_3CH_2CH_2{-}\overset{\overset{\displaystyle O}{\|}}{C}{-}H$
Pentanal	Valeraldehyde	$CH_3CH_2CH_2CH_2{-}\overset{\overset{\displaystyle O}{\|}}{C}{-}H$

14.55 Draw the hemiacetal or hemiketal that results from the reaction of each of the following aldehydes or ketones with ethanol:

a. $CH_3CH_2CH_2-\overset{\overset{\textstyle O}{\|}}{C}-CH_3$

b. 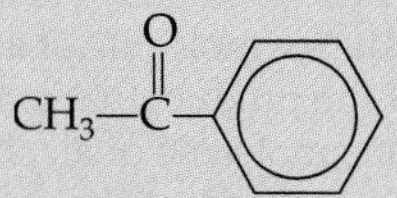 $CH_3-\overset{\overset{\textstyle O}{\|}}{C}-$⬡(benzene ring)

c. cyclopentanone with =O

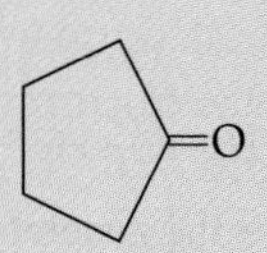

14.56 Identify each of the following compounds as a hemiacetal, hemiketal, acetal, or ketal:

a. 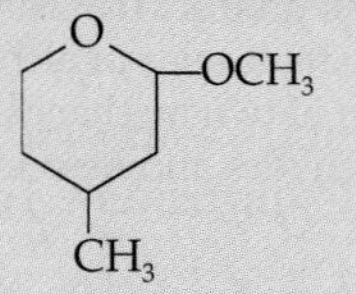 (ring structure with O, OCH$_3$, CH$_3$)

b. 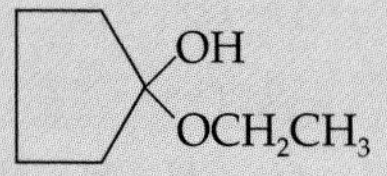 (cyclobutane ring with OH, OCH$_2$CH$_3$)

c. $CH_3\overset{\overset{\textstyle OH}{|}}{\underset{\underset{\textstyle OCH_2CH_3}{|}}{C}}CH_3$

d. 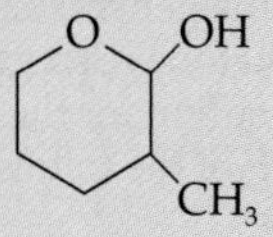 (ring structure with O, OH, CH$_3$)

e. $CH_3\overset{\overset{\textstyle OCH_3}{|}}{\underset{\underset{\textstyle OCH_2CH_3}{|}}{C}}CH_3$

f. $CH_3CH=CH\overset{\overset{\textstyle OCH_3}{|}}{\underset{\underset{\textstyle OH}{|}}{C}}CH_3$

14.57 Complete the following synthesis by supplying the missing reactant(s), reagent(s), or product(s) indicated by the question marks:

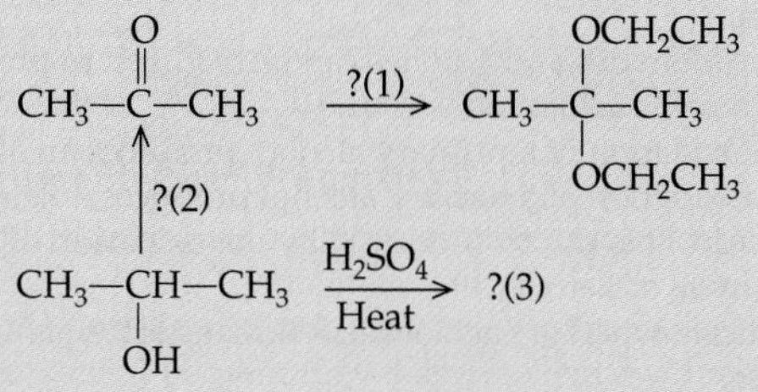

$CH_3-\overset{\overset{\textstyle O}{\|}}{C}-CH_3 \xrightarrow{?(1)} CH_3-\overset{\overset{\textstyle OCH_2CH_3}{|}}{\underset{\underset{\textstyle OCH_2CH_3}{|}}{C}}-CH_3$

$\uparrow ?(2)$

$CH_3-\overset{\underset{\underset{\textstyle OH}{|}}{}}{C}H-CH_3 \xrightarrow[\text{Heat}]{H_2SO_4} ?(3)$

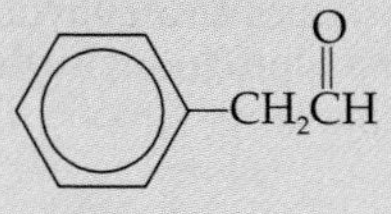

14.58 Which alcohol would you oxidize to produce each of the following compounds?

a. $CH_3\overset{\overset{\textstyle CH_3}{|}}{C}HCH_2\overset{\overset{\textstyle O}{\|}}{C}CH_3$

b. $H\overset{\overset{\textstyle O}{\|}}{C}CH_2CH_2\overset{\overset{\textstyle O}{\|}}{C}H$

c. ⬡$-CH_2\overset{\overset{\textstyle O}{\|}}{C}H$

d. $H\overset{\overset{\textstyle O}{\|}}{C}CH_2\overset{\overset{\textstyle O}{\|}}{C}CH_3$

e. $CH_3\overset{\overset{\textstyle CH_3}{|}}{\underset{\underset{\textstyle CH_3}{|}}{C}}CH_2CH_2\overset{\overset{\textstyle O}{\|}}{C}H$

f. 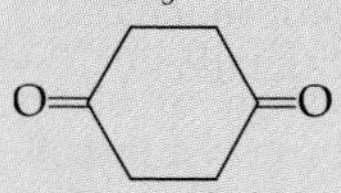 O=⬡=O

1. Review the material on the chemistry of vision and, with respect to the isomers of retinal, discuss the changes in structure that occur as the nerve impulses (that result in vision) are produced. Provide complete structural formulas of the retinal isomers that you discuss.

2. Classify the structure of β-D-fructose as a hemiacetal, hemiketal, acetal, or ketal. Explain your choice.

(structure of β-D-fructose)

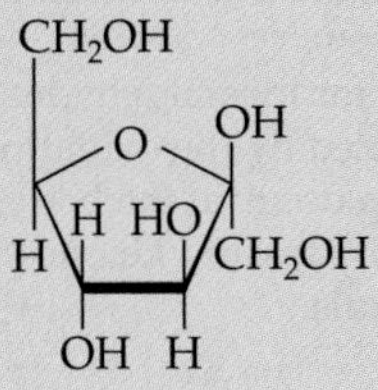

3. Design a synthesis for each of the following compounds, using any inorganic reagent of your choice and any hydrocarbon or alkyl halide of your choice:
   a. Octanal
   b. Cyclohexanone
   c. 2-Phenylethanoic acid

4. When alkenes react with ozone, O_3, the double bond is cleaved, and an aldehyde and/or a ketone is produced. The reaction, called *ozonolysis*, is shown in general as:

$$\overset{}{\underset{}{>}}C=C\overset{}{\underset{}{<}} + O_3 \longrightarrow \overset{}{\underset{}{>}}C=O + O=C\overset{}{\underset{}{<}}$$

   Predict the ozonolysis products that are formed when each of the following alkenes is reacted with ozone:
   a. 1-Butene
   b. 2-Hexene
   c. *cis*-3,6-Dimethyl-3-heptene

5. Lactose is the major sugar found in mammalian milk. It is a disaccharide composed of the monosaccharides glucose and galactose:

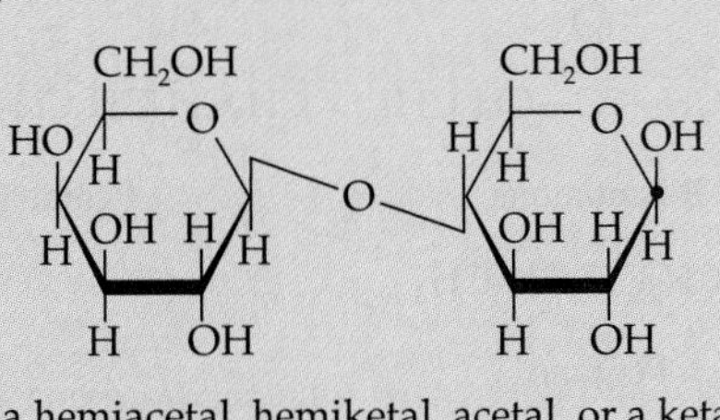

(structure of lactose)

   Is lactose a hemiacetal, hemiketal, acetal, or a ketal? Explain your choice or choices.

6. The following are the keto and enol tautomers of phenol:

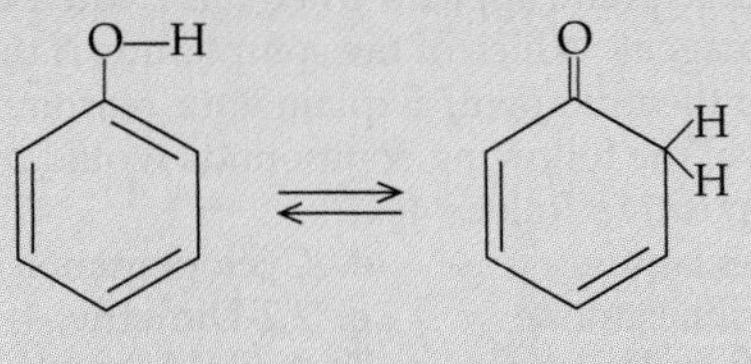

(structures of enol and keto forms of phenol)

Enol form   Keto form
of phenol   of phenol

We have seen that most simple aldehydes and ketones exist mainly in the keto form because it is more stable. Phenol is an exception, existing primarily in the enol form. Propose a hypothesis to explain this.

15

Carboxylic Acids and Carboxylic Acid Derivatives

Natural fruit flavors are complex mixtures of esters and other organic compounds.

Learning Goals

1. Write structures and describe the physical properties of carboxylic acids.

2. Determine the common and I.U.P.A.C. names of carboxylic acids.

3. Describe the biological, medical, or environmental significance of several carboxylic acids.

4. Write equations that show the synthesis of a carboxylic acid.

5. Write equations representing acid–base reactions of carboxylic acids.

6. Write equations representing the preparation of an ester.

7. Write structures and describe the physical properties of esters.

8. Determine the common and I.U.P.A.C. names of esters.

9. Write equations representing the hydrolysis of an ester.

10. Define the term *saponification* and describe how soap works in the emulsification of grease and oil.

11. Determine the common and I.U.P.A.C. names of acid chlorides.

12. Write equations representing the synthesis of acid chlorides.

13. Determine the common and I.U.P.A.C. names of acid anhydrides.

14. Write equations representing the synthesis of acid anhydrides.

15. Discuss the significance of thioesters and phosphoesters in biological systems.

4. Acetyl coenzyme A (acetyl CoA) can serve as a donor of acetate groups in biochemical reactions. One such reaction is the formation of acetylcholine, an important neurotransmitter involved in nerve signal transmission at neuromuscular junctions. The structure of choline is shown below. Draw the structure of acetylcholine.

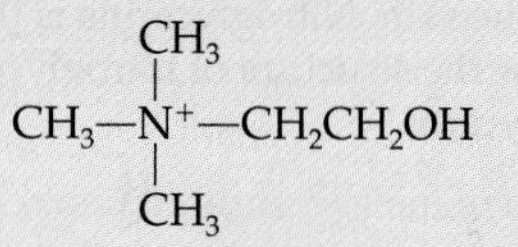

$$CH_3-\overset{\overset{\displaystyle CH_3}{|}}{\underset{\underset{\displaystyle CH_3}{|}}{N^+}}-CH_2CH_2OH$$

Choline

5. Hormones are chemical messengers that are produced in a specialized tissue of the body and travel through the bloodstream to reach receptors on cells of their target tissues. This specific binding to target tissues often stimulates a cascade of enzymatic reactions in the target cells. The work of Earl Sutherland and others led to the realization that there is a *second messenger* within the target cells. Binding of the hormone to the hormone receptor in the cell membrane triggers the enzyme adenyl cyclase to produce adenosine-3′,5′-monophosphate, which is also called *cyclic AMP*, from ATP. The reaction is summarized as follows:

$$ATP \xrightarrow{\text{Mg}^{+2}, \text{ adenyl cyclase}} \text{cyclic AMP} + PP_i + H^+$$

PP_i is the abbreviation for a pyrophosphate group, shown here:

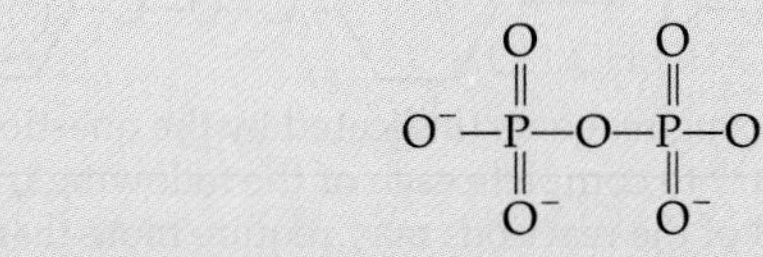

The structure of ATP is shown here with the carbon atoms of the sugar ribose numbered according to the convention used for nucleotides:

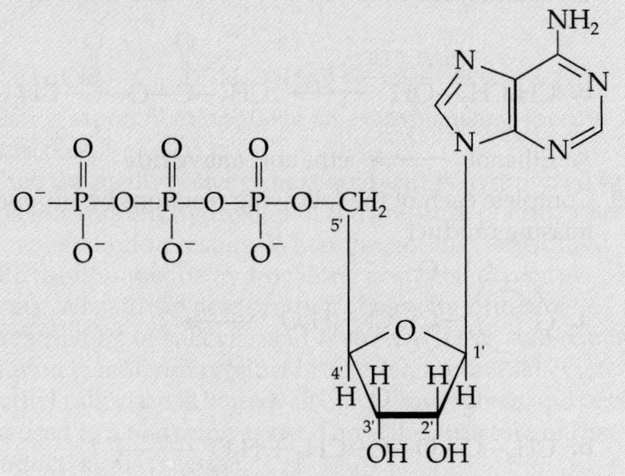

Adenosine-5′-triphosphate

Draw the structure of adenosine-3′,5′-monophosphate.

16

Amines and Amides

Ethnobotanists continue to search for medically active compounds from the rainforest.

Learning Goals

1. Classify amines as primary, secondary, or tertiary.

2. Describe the physical properties of amines.

3. Draw and name simple amines using the Chemical Abstracts, common, and I.U.P.A.C. nomenclature system.

4. Write equations representing the synthesis of amines.

5. Write equations showing the basicity and neutralization of amines.

6. Describe the structure of quaternary ammonium salts and discuss their use as antiseptics and disinfectants.

7. Discuss the biological significance of heterocyclic amines.

8. Describe the physical properties of amides.

9. Draw the structure and write the common and I.U.P.A.C. names of amides.

10. Write equations representing the preparation of amides.

11. Write equations showing the hydrolysis of amides.

12. Draw the general structure of an amino acid.

13. Draw and discuss the structure of a peptide bond.

439

17

Carbohydrates

Would "looking-glass milk" be nutritious?

Learning Goals

1 Explain the difference between complex and simple carbohydrates and know the amounts of each recommended in the daily diet.

2 Apply the systems of classifying and naming monosaccharides according to the functional group and number of carbons in the chain.

3 Determine whether a molecule has a chiral center.

4 Explain stereoisomerism.

5 Identify monosaccharides as either D- or L-.

6 Draw and name the common monosaccharides using structural formulas.

7 Given the linear structure of a monosaccharide, draw the Haworth projection of its α- and β-cyclic forms and vice versa.

8 By inspection of the structure, predict whether a sugar is a reducing or a nonreducing sugar.

9 Discuss the use of the Benedict's reagent to measure the level of glucose in urine.

10 Draw and name the common disaccharides and discuss their significance in biological systems.

11 Describe the difference between galactosemia and lactose intolerance.

12 Discuss the structural, chemical, and biochemical properties of starch, glycogen, and cellulose.

CHEMISTRY CONNECTION

Chemistry Through the Looking Glass

In his children's story *Through the Looking Glass,* Lewis Carroll's heroine Alice wonders whether "looking-glass milk" would be good to drink. As we will see in this chapter, many biological molecules, such as the sugars, exist as two stereoisomers, *enantiomers,* that are mirror images of one another. Because two mirror-image forms occur, it is rather remarkable that in our bodies, and in most of the biological world, only one of the two is found. For instance, the common sugars are members of the D-family, whereas all the common amino acids that make up our proteins are members of the L-family. It is not too surprising, then, that the enzymes in our bodies that break down the sugars and proteins we eat are *stereospecific,* that is, they recognize only one mirror-image isomer. Knowing this, we can make an educated guess that "looking-glass milk" could not be digested by our enzymes and therefore would not be a good source of food for us. It is even possible that it might be toxic to us!

Pharmaceutical chemists are becoming more and more concerned with the stereochemical purity of the drugs that we take. Consider a few examples. In 1960 the drug thalidomide was commonly prescribed in Europe as a sedative. However, during that year, hundreds of women who took thalidomide during pregnancy gave birth to babies with severe birth defects. Thalidomide, it turned out, was a mixture of two enantiomers. One is a sedative; the other is a teratogen, a chemical that causes birth defects.

One of the common side effects of taking antihistamines for colds or allergies is drowsiness. Again, this is the result of the fact that antihistamines are mixtures of enantiomers. One causes drowsiness; the other is a good decongestant.

One enantiomer of the compound carvone is associated with the smell of spearmint; the other produces the aroma of caraway seeds or dill. One mirror-image form of limonene smells like lemons; the other has the aroma of oranges.

The pain reliever ibuprofen is currently sold as a mixture of enantiomers, but one is a much more effective analgesic than the other.

Taste, smell, and the biological effects of drugs in the body all depend on the stereochemical form of compounds and their interactions with cellular enzymes or receptors. As a result, chemists are actively working to devise methods of separating the isomers in pure form. Alternatively, methods of conducting stereospecific syntheses that produce only one stereoisomer are being sought. By preparing pure stereoisomers, the biological activity of a compound can be much more carefully controlled. This will lead to safer medications.

In this chapter we will begin our study of stereochemistry, the spatial arrangement of atoms in molecules, with the carbohydrates. Later, we will examine the stereochemistry of the amino acids that make up our proteins and consider the stereochemical specificity of the metabolic reactions that are essential to life. A more complete treatment of stereochemistry is found in Appendix D: "Stereochemistry and Stereoisomers Revisited."

Introduction

> A kilocalorie is the same as the Calorie referred to in the "count-your-calories" books and on nutrition labels.

See "A Human Perspective: Tooth Decay and Simple Sugars."

Learning Goal

Carbohydrates are produced in plants by photosynthesis (Figure 17.1). Natural carbohydrate sources such as grains and cereals, breads, sugar cane, fruits, milk, and honey are an important source of energy for animals. **Carbohydrates,** especially glucose, are the primary energy source for the brain and nervous system and can be used by many other tissues. When "burned" by cells for energy, each gram of carbohydrate releases approximately four kilocalories of energy.

A healthy diet should contain both complex carbohydrates, such as starches and cellulose, and simple sugars, such as fructose and sucrose. However, the quantity of simple sugars, especially sucrose, should be minimized because large quantities of sucrose in the diet promote obesity and tooth decay.

Complex carbohydrates are better for us than the simple sugars. Starch, found in rice, potatoes, breads, and cereals, is an excellent energy source. In addition, the complex carbohydrates, such as cellulose, provide us with an important supply of dietary fiber.

It is hard to determine exactly what percentage of the daily diet *should* consist of carbohydrates. The *actual* percentage varies widely throughout the world, from 80% in the Far East, where rice is the main component of the diet, to 40–50% in the

Animals release $CO_2 + H_2O$
Plants are consumed by animals
PHOTOSYNTHESIS
$6CO_2 + 6H_2O \rightarrow C_6H_{12}O_6 + 6O_2$
ENERGY METABOLISM
$6O_2 + C_6H_{12}O_6 \rightarrow 6CO_2 + 6H_2O$ + energy
And
BIOSYNTHESIS

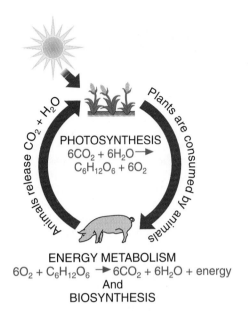

Figure 17.1
Carbohydrates are produced by the process of photosynthesis, which uses the energy of sunlight to produce hexoses from CO_2 and H_2O. The plants use these hexoses to harvest energy and produce ATP to fuel cellular function and to produce macromolecules, including starch, cellulose, fats, nucleic acids, and proteins. Animals depend on plants as a source of organic carbon. The hexoses are metabolized to generate ATP and are used as precursors for the biosynthesis of carbohydrates, fats, proteins, and nucleic acids.

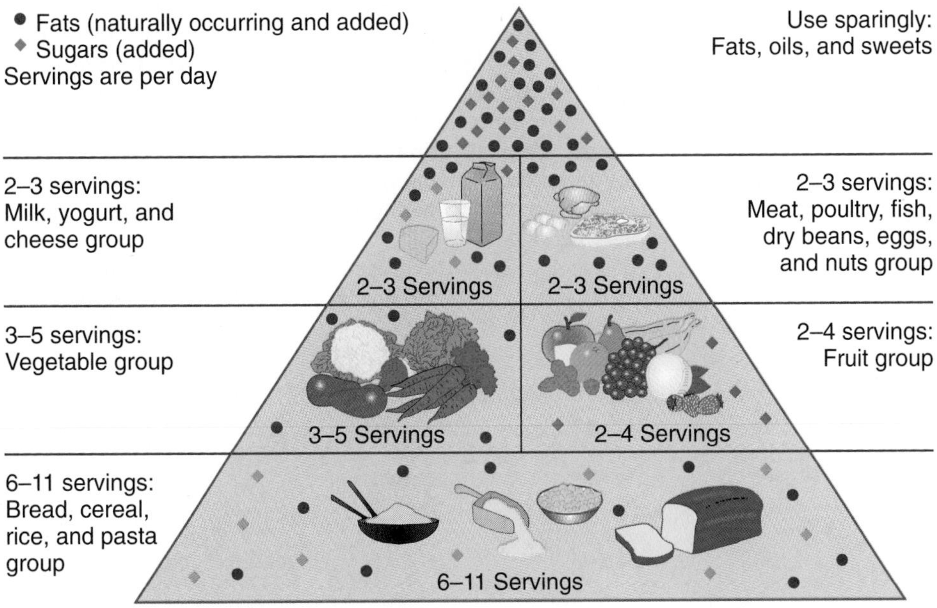

● Fats (naturally occurring and added)
◆ Sugars (added)
Servings are per day
Use sparingly: Fats, oils, and sweets
2–3 servings: Milk, yogurt, and cheese group
2–3 servings: Meat, poultry, fish, dry beans, eggs, and nuts group
2–3 Servings
2–3 Servings
3–5 servings: Vegetable group
2–4 servings: Fruit group
3–5 Servings
2–4 Servings
6–11 servings: Bread, cereal, rice, and pasta group
6–11 Servings

Figure 17.2
The U.S. Department of Agriculture has adopted a food pyramid to explain that carbohydrates in the form of cereals, grains, fruits, and vegetables should make up the majority of our diet. Fats and sweets should be consumed sparingly. Source: U.S. Department of Agriculture.

United States. Currently, it is recommended that about 58% of the calories in the diet should come from carbohydrates and that no more than 10% of the daily caloric intake should be sucrose. The U.S. Department of Agriculture has adopted a food pyramid to show the recommended amounts of various foods in the diet (Figure 17.2). The foods at the bottom of the pyramid, grains (breads, cereals, rice, pasta), and at the next level, fruits and vegetables, should be the most abundant in our diet. These food groups are our major sources of dietary carbohydrates.

Question 17.1

What is the current recommendation for the amount of carbohydrates that should be included in the diet? Of the daily intake of carbohydrates, what percentage should be simple sugar?

Tooth Decay and Simple Sugars

How many times have you heard the lecture from parents or your dentist about brushing your teeth after a sugary snack? Annoying as this lecture might be, it is based on sound scientific data that demonstrate that the cause of tooth decay is plaque and acid formed by the bacterium *Streptococcus mutans* using sucrose as its substrate.

Saliva is teeming with bacteria in concentrations up to one hundred million (10^8) per milliliter of saliva! Within minutes after you brush your teeth, sticky glycoproteins in the saliva adhere to tooth surfaces. Then millions of oral bacteria immediately bind to this surface.

Although all oral bacteria adhere to the tooth surface, only *S. mutans* causes dental caries, or cavities. Why does this bacterium cause cavities when all the others do not? The answer lies in the special enzyme called *glucosyl transferase* that is found on the surface of *S. mutans* cells.

Glucosyl transferase is a very specific enzyme. It can act only on the disaccharide sucrose, which it breaks down into glucose and fructose. As the accompanying diagram shows, the enzyme then adds the glucose to a growing polysaccharide chain called *dextran* that adheres tightly to both the tooth enamel and the bacteria. *Plaque* is made up of huge masses of bacteria, embedded in dextran, adhering to the tooth surface.

This is just the first stage of cavity formation. Note in the accompanying figure that the second sugar released by the cleavage of sucrose is fructose. The bacteria use the fructose in the energy-harvesting pathways of glycolysis and lactic acid fermentation. Production of lactic acid decreases the pH on the tooth surface and begins to dissolve calcium from the tooth enamel.

Why is the acid not washed away from the tooth surface? After all, we produce about 1 L of saliva each day, which should dilute the acid and remove it from the tooth surface. The problem is the dextran plaque; it is not permeable to saliva, and thus plaque keeps the bacteria and the lactic acid localized on the enamel.

What measures can we take to prevent tooth decay? Practice good oral hygiene; brushing after each meal and flossing regularly reduce plaque buildup. Eat a diet rich in calcium; this helps to build strong tooth enamel. Include many complex carbohydrates in the diet; these cannot be used by glucosyl transferase and will not lead to the formation of acid. Further, the complex carbohydrates from fruits and vegetables help prevent decay by mechanically removing plaque from tooth surfaces. Avoid sucrose-containing snacks between meals. Studies have shown that the consumption of a sucrose-rich dessert with a meal, followed by brushing, does not produce many cavities. However, even small amounts of sugar ingested between meals are very cariogenic.

Researchers have developed a vaccine that prevents tooth decay in rats. Such a vaccine may one day be available for humans.

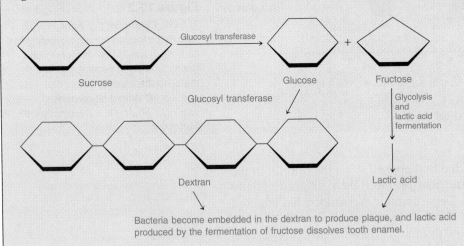

Bacteria become embedded in the dextran to produce plaque, and lactic acid produced by the fermentation of fructose dissolves tooth enamel.

Action of the glucosyl transferase of *Streptococcus mutans*, which is responsible for tooth decay.

Q u e s t i o n 17.2

Distinguish between simple and complex sugars. What are some sources of complex carbohydrates?

17.1 Types of Carbohydrates

Monosaccharides such as glucose and fructose are the simplest carbohydrates because they contain a single (*mono-*) sugar (*saccharide*) unit. **Disaccharides,** including sucrose and lactose, consist of two monosaccharide units joined through bridging oxygen atoms. Such a bond is called a **glycosidic bond. Oligosaccharides** consist of three to ten monosaccharide units joined by glycosidic bonds. The largest and most complex carbohydrates are the **polysaccharides,** which are long, often highly branched, chains of monosaccharides. Starch, glycogen, and cellulose are all examples of polysaccharides.

17.2 Monosaccharides

Monosaccharides are composed of carbon, hydrogen, and oxygen, and most are characterized by the general formula $(CH_2O)_n$, in which n is any integer from 3 to 7. As we will see, this general formula is an oversimplification because several biologically important monosaccharides are modified forms of monosaccharides. For instance, several blood group antigen and bacterial cell wall monosaccharides are substituted with amino groups. Many of the intermediates in carbohydrate metabolism carry phosphate groups. Deoxyribose, the monosaccharide found in DNA, has one fewer oxygen atom than the formula would predict.

The importance of phosphorylated sugars in metabolic reactions is discussed in Sections 15.4 and 21.3.

Nomenclature

Monosaccharides can be named on the basis of the functional groups they contain. A monosaccharide with a ketone (carbonyl) group is a **ketose.** If an aldehyde (carbonyl) group is present, it is called an **aldose.** Because monosaccharides also contain many hydroxyl groups, they are sometimes called *polyhydroxyaldehydes* or *polyhydroxyketones*.

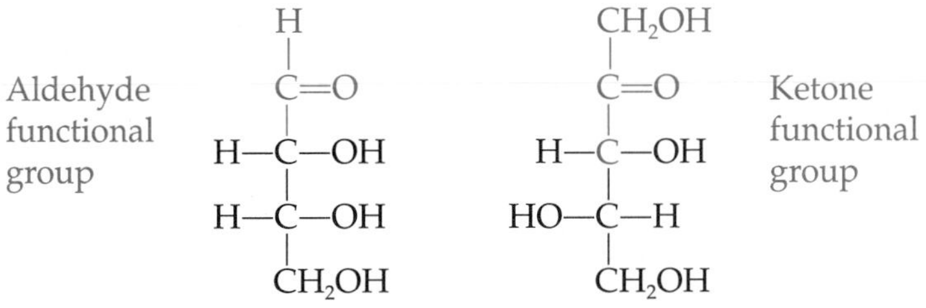

Learning Goal

Aldehyde functional group

```
        H
        |
        C=O
        |
   H—C—OH
        |
   H—C—OH
        |
      CH₂OH
```

An aldose

```
      CH₂OH
        |
        C=O
        |
   H—C—OH
        |
  HO—C—H
        |
      CH₂OH
```

A ketose

Ketone functional group

Another system of nomenclature tells us the number of carbon atoms in the main skeleton. A three-carbon monosaccharide is a *triose,* a four-carbon sugar is a *tetrose,* a five-carbon sugar is a *pentose,* a six-carbon sugar is a *hexose,* and so on. Combining the two naming systems gives even more information about the structure and composition of a sugar. For example, an aldotetrose is a four-carbon sugar that is also an aldehyde.

In addition to these general names, each monosaccharide has a unique name. These names are shown in blue for the following structures. Because the monosaccharides can exist in several different isomeric forms, it is important to provide the complete name. Thus the complete names of the structures shown on the next page are D-glyceraldehyde, D-glucose, and D-fructose. These names tell us that the structure represents one particular sugar and also identifies the sugar as one of two possible isomeric forms (D- or L-).

$$
\begin{array}{ccc}
\text{H} & \text{H} & \text{CH}_2\text{OH} \\
| & | & | \\
\text{C}=\text{O} & \text{C}=\text{O} & \text{C}=\text{O} \\
| & | & | \\
\text{H}-\text{C}-\text{OH} & \text{H}-\text{C}-\text{OH} & \text{HO}-\text{C}-\text{H} \\
| & | & | \\
\text{CH}_2\text{OH} & \text{HO}-\text{C}-\text{H} & \text{H}-\text{C}-\text{OH} \\
 & | & | \\
 & \text{H}-\text{C}-\text{OH} & \text{H}-\text{C}-\text{OH} \\
 & | & | \\
 & \text{H}-\text{C}-\text{OH} & \text{CH}_2\text{OH} \\
 & | & \\
 & \text{CH}_2\text{OH} & \\
\end{array}
$$

Aldose	Aldose	Ketose
Triose	Hexose	Hexose
Aldotriose	Aldohexose	Ketohexose
D-Glyceraldehyde	D-Glucose	D-Fructose

Question 17.3

What is the structural difference between an aldose and a ketose?

Question 17.4

Explain the difference between:

a. A ketohexose and an aldohexose
b. A triose and a pentose

Stereoisomers

Structure

Learning Goal **3**   Learning Goal **4**   Learning Goal **5**

For a more detailed discussion of stereochemistry, see Appendix D, "Stereochemistry and Stereoisomers Revisited."

Build models of these compounds using toothpicks and gumdrops of five different colors to prove this to yourself.

The prefixes D- and L- found in the complete name of a monosaccharide are used to identify one of two possible isomeric forms called **stereoisomers.** By definition, each member of a pair of stereoisomers must have the same molecular formula and the same bonding. How then do isomers of the D-family differ from those of the L-family? D- and L-isomers differ in the spatial arrangements of atoms in the molecule.

Stereochemistry is the study of the different spatial arrangements of atoms. In each member of a pair of stereoisomers, all of the atoms are bonded together with exactly the same bonding pattern; they differ only in the arrangements of their atoms in space. A general example of a pair of stereoisomers is shown in Figure 17.3. In this example the general molecule C-abcd is formed from the bonding of a central carbon to four different groups: a, b, c, and d. This results in two molecules rather than one. Each isomer is bonded together through the exact *same* bonding pattern, yet they are *not* identical. If they were identical, they would be superimposable one upon the other; *they are not.* They are therefore stereoisomers. These two stereoisomers have a mirror-image relationship that is analogous to the mirror-image relationship of the left and right hands (see Figure 17.3b). Two stereoisomers that are nonsuperimposable mirror images of one another are called a pair of **enantiomers.** One is a member of the D-family, and the other is a member of the L-family.

Molecules that are capable of existing in nonsuperimposable mirror image, enantiomeric forms are called **chiral molecules.** The term simply means that as a result of different three-dimensional arrangements of atoms, the molecule can exist in two mirror-image forms. For any pair of nonsuperimposable mirror-image forms (enantiomers), one is always designated D- and the other L-.

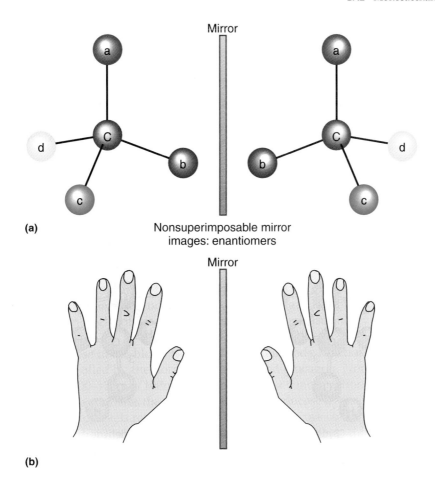

(a)

Nonsuperimposable mirror
images: enantiomers

(b)

Figure 17.3
(a) A pair of enantiomers for the general molecule C-abcd. (b) Mirror-image right and left hands.

A carbon atom that has four different groups bonded to it is called an **asymmetric** (or chiral) **carbon** atom. Any molecule containing a chiral carbon can exist as a pair of enantiomers. Consider the simplest carbohydrate, **glyceraldehyde,** which is shown in Figure 17.4. Note that the second carbon is bonded to four different groups. It is therefore the chiral carbon. As a result, we can draw two enantiomers of glyceraldehyde that are nonsuperimposable mirror images of one another.

> Technically the molecule is chiral, not the carbon atom. We will use the term *chiral carbon* to help us identify the features of the molecule that cause it to be chiral.

Some Important Monosaccharides

D-Glyceraldehyde

The simplest carbohydrate is the three-carbon sugar D-glyceraldehyde (see Figure 17.4). D-Glyceraldehyde is a triose (three-carbon) and an aldose (aldehyde group at the top); therefore it is an aldotriose. First, note that this structure is drawn with the most oxidized carbon, the aldehyde, at the "top." The most oxidized carbon is always placed at the top when the structures of open-chain carbohydrates are written; this carbon is numbered C-1. C-2 of D-glyceraldehyde is covalently bonded to four different groups and thus is chiral. Reflection of D-glyceraldehyde in a mirror gives L-glyceraldehyde.

By convention it is the position of the hydroxyl group on the chiral carbon farthest from the carbonyl group (farthest from the oxidized end) that determines whether a monosaccharide is in the D- or L-configuration. If the —OH group is on the right (your right), the molecule is in its D-configuration. If the —OH group is

Learning Goal
6

EXAMPLE 17.1

Drawing the Structure of a Monosaccharide

Draw the structure for D-glucose.

Solution

Glucose is an aldohexose.

Step 1. Draw six carbons in a straight vertical line; each carbon is separated from the ones above and below it by a bond:

```
1 C
  |
2 C
  |
3 C
  |
4 C
  |
5 C
  |
6 C
```

Step 2. The most highly oxidized carbon is, by convention, drawn as the uppermost carbon (carbon 1). In this case, carbon 1 is an aldehyde carbon:

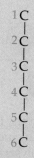

Most oxidized end of carbon chain; aldehyde

Step 3. The atoms are added to the next to the last carbon atom, at the bottom of the chain, to give either the D- or L-configuration as desired. Remember, when the —OH group is to the right, you have D-glucose. When in doubt, compare your structure to D-glyceraldehyde!

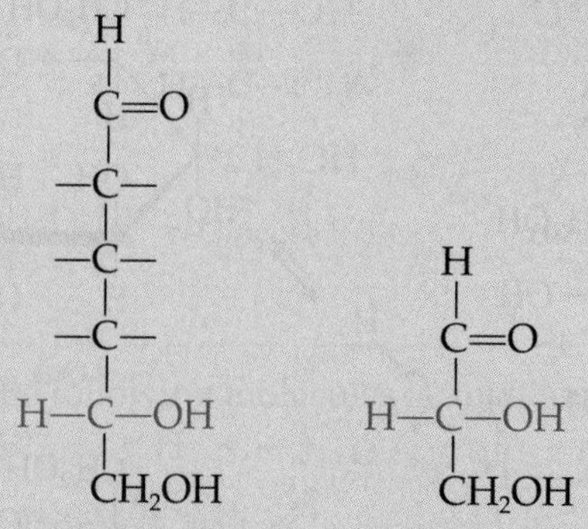

D-Isomer   D-Glyceraldehyde

Compare chiral carbons farthest from the carbonyl group

Step 4. All the remaining atoms are then added to give the desired carbohydrate. For example, one would draw the following structure for D-glucose. Compare this with the structure of L-glucose, the non-superimposable mirror image of D-glucose.

(left margin, partially visible)

Figu
(a) St
glyce
with
oxidi
a mo
orien
attac
from
D-en
the L
(b) A
D- an

The s
L-gly
In fa
gene
amin
D- a
a ne
confi
syste
desc
and

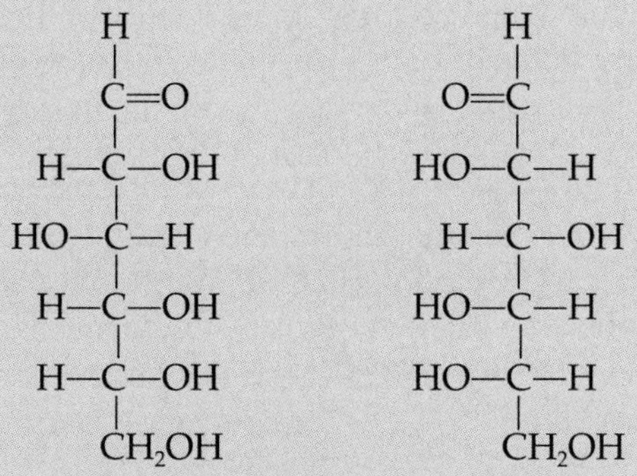

The structures of D-Glucose and L-Glucose are shown in Fischer projections.

D-Glucose   L-Glucose

The positions for the hydrogen atoms and the hydroxyl groups on the remaining carbons must be learned for each sugar. For instance, the complete structures of D-fructose and D-galactose are shown later in this section.

In actuality the open-chain form of glucose is present in very small concentrations in cells. It exists in cyclic form under physiological conditions because the carbonyl group at C-1 of glucose reacts with the hydroxyl group at C-5 to give a six-membered ring. In the discussion of aldehydes, we noted that the reaction between an aldehyde and an alcohol yields a **hemiacetal.** When the aldehyde portion of the glucose molecule reacts with the C-5 hydroxyl group, the product is a cyclic *intramolecular hemiacetal.* For D-glucose, two isomers can be formed in this reaction (see Figure 17.5). These isomers are called α- and β-D-glucose. Two isomers are formed because the cyclization reaction creates a new chiral carbon, in this case C-1. In the α isomers the C-1 hydroxyl group is below the ring, and in the β isomers the C-1 hydroxyl group is above the ring.

In Figure 17.5 a new type of structural formula, called a **Haworth projection,** is presented. Although on first inspection it appears complicated, it is quite simple to derive a Haworth projection from a structural formula, as Example 17.2 shows.

Hemiacetal structure,

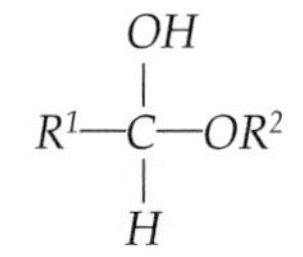

$$\begin{array}{c} OH \\ | \\ R^1\!-\!C\!-\!OR^2 \\ | \\ H \end{array}$$

and formation are described in Section 14.4.

The term *intramolecular* tells us that the reacting carbonyl and hydroxyl groups are part of the same molecule.

EXAMPLE 17.2

Drawing the Haworth Projections of Monosaccharides from the Structural Formula

Draw the Haworth projections of α- and β-D-glucose.

Learning Goal

Solution

1. Before attempting to draw a Haworth projection, look at the first steps of ring formation shown here:

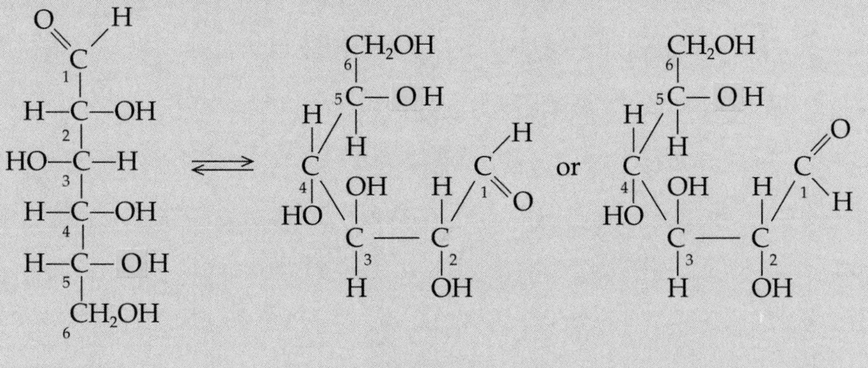

Glucose
(open chain)

Glucose
(intermediates in ring formation)

Try to imagine that you are seeing the molecules on the previous page in three dimensions. Some of the substituent groups on the molecule will be above the ring, and some will be beneath it. The question then becomes: How do you determine which groups to place above the ring and which to place beneath the ring?

2. Look at the two-dimensional structural formula. Note the groups (drawn in blue) to the left of the carbon chain. These are placed above the ring in the Haworth projection.

α-D-Glucose   β-D-Glucose

3. Now note the groups (drawn in red) to the right of the carbon chain. These will be located beneath the carbon ring in the Haworth projection.

α-D-Glucose   β-D-Glucose

4. Thus in the Haworth projection of the cyclic form of any D-sugar the —CH_2OH group is always "up." When the —OH group at C-1 is also "up," *cis* to the —CH_2OH group, the sugar is β-D-glucose. When the —OH group at C-1 is "down," *trans* to the —CH_2OH group, the sugar is α-D-glucose.

Haworth projection
α-D-Glucose

Haworth projection
β-D-Glucose

Question 17.7

Refer to the linear structure of D-galactose. Draw the Haworth projections of α- and β-D-galactose.

Q u e s t i o n   **17.8**

Refer to the linear structure of D-ribose. Draw the Haworth projections of α- and β-D-ribose. Note that D-ribose is a pentose.

Fructose

Fructose, also called levulose and fruit sugar, is the sweetest of all sugars. It is found in large amounts in honey, corn syrup, and sweet fruits. The structure of fructose is similar to that of glucose. When there is a —CH$_2$OH group instead of a —CHO group at carbon-1 and a —C=O group instead of CHOH at carbon-2, the sugar is a ketose. In this case it is D-fructose.

Cyclization of fructose produces α- and β-D-fructose:

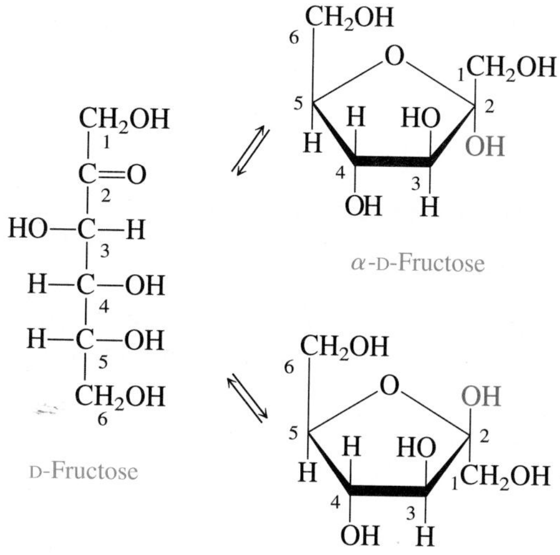

α-D-Fructose

D-Fructose

β-D-Fructose

Fructose is a ketose, or ketone sugar. Recall that the reaction between an alcohol and a ketone yields a **hemiketal.** Thus the reaction between the C-2 keto group and the C-5 hydroxyl group in the fructose molecule produces an *intramolecular hemiketal.* Fructose forms a five-membered ring structure.

Hemiketal formation is described in Section 14.4.

Galactose

Another important hexose is **galactose.** The linear structure of D-galactose and the Haworth projections of α-D-galactose and β-D-galactose are shown here:

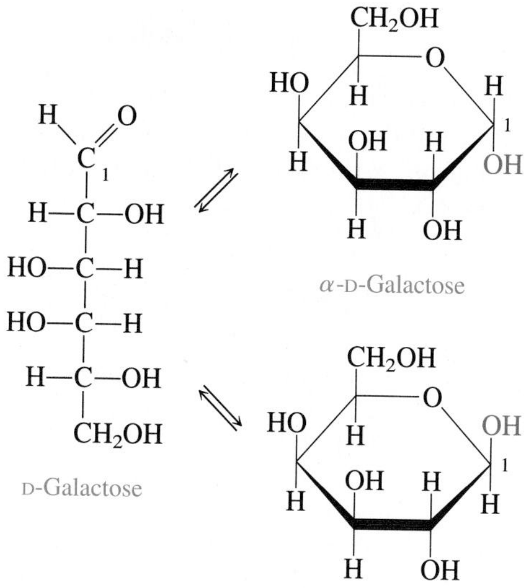

α-D-Galactose

D-Galactose

β-D-Galactose

Galactose is found in biological systems as a component of the disaccharide lactose, or milk sugar. This is the principal sugar found in the milk of all mammals. β-D-Galactose and a modified form, β-D-*N*-acetyl-galactosamine, are also components of the blood group antigens.

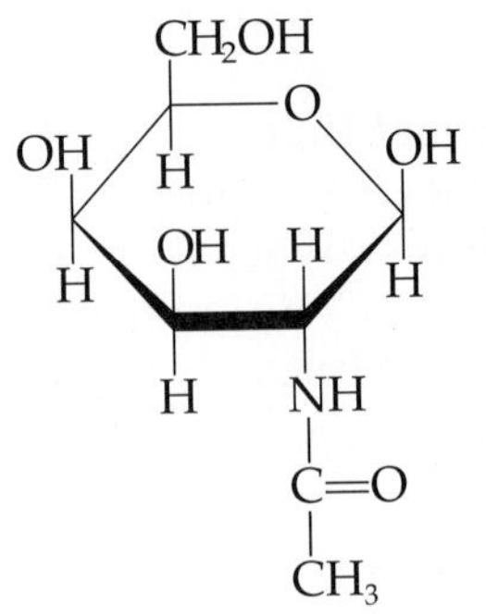

β-D-*N*-Acetylgalactosamine

Ribose and Deoxyribose, Five-Carbon Sugars

Ribose is a component of many biologically important molecules, including RNA, and various coenzymes, a group of compounds required by many of the enzymes that carry out biochemical reactions in the body. The structure of the five-carbon sugar D-ribose is shown in its open-chain form and in the α- and β-cyclic isomer forms.

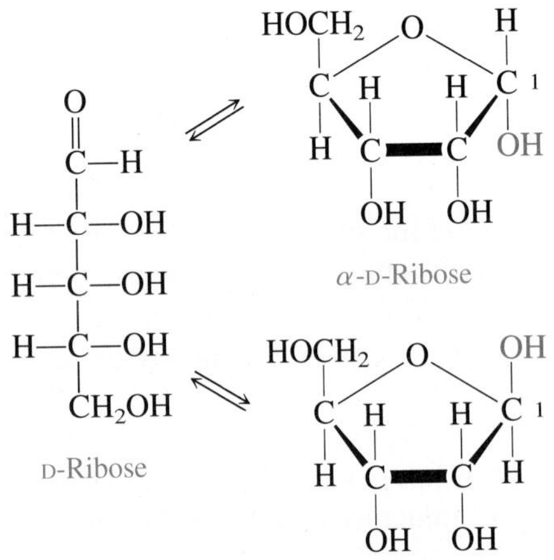

D-Ribose

α-D-Ribose

β-D-Ribose

DNA, the molecule that carries the genetic information of the cell, contains 2-deoxyribose. In this molecule the —OH group at C-2 has been replaced by a hydrogen, hence the designation "2-deoxy," indicating the absence of an oxygen.

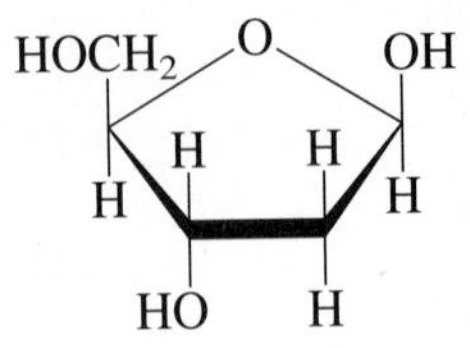

β-D-2-Deoxyribose

Reducing Sugars

Learning Goal 8   **Learning Goal 9**

The aldehyde group of aldoses is readily oxidized by the Benedict's reagent. Recall that the **Benedict's reagent** is a basic buffer solution that contains Cu^{2+} ions. The

Cu^{2+} ions are reduced to Cu^+ ions, which, in basic solution, precipitate as brick-red Cu_2O. The aldehyde group of the aldose is oxidized to a carboxylic acid, which undergoes an acid–base reaction to produce a carboxylate anion.

$$
\begin{array}{c}
\overset{\displaystyle O}{\underset{\displaystyle |}{\underset{\displaystyle H-C-OH}{C}}}\!\!\overset{H}{\diagup} \\
CH_2OH
\end{array}
+ 2Cu^{2+}\,(buffer) + 5OH^- \longrightarrow
\begin{array}{c}
\overset{\displaystyle O \;\; O^-}{C} \\
H-C-OH \\
CH_2OH
\end{array}
+ Cu_2O + 3H_2O
$$

Although ketones generally are not easily oxidized, ketoses are an exception to that rule. Because of the —OH group on the carbon next to the carbonyl group, ketoses can be converted to aldoses, under basic conditions, via an *enediol reaction*:

$$
\underset{\text{D-Fructose}}{
\begin{array}{c}
CH_2OH \\
C=O \\
HO-C-H \\
H-C-OH \\
H-C-OH \\
CH_2OH
\end{array}}
\rightleftharpoons
\underset{\text{Enediol}}{
\begin{array}{c}
HO-C-H \\
C-OH \\
HO-C-H \\
H-C-OH \\
H-C-OH \\
CH_2OH
\end{array}}
\rightleftharpoons
\underset{\text{D-Glucose}}{
\begin{array}{c}
H \\
C=O \\
H-C-OH \\
HO-C-H \\
H-C-OH \\
H-C-OH \\
CH_2OH
\end{array}}
$$

The name of the enediol reaction is derived from the structure of the intermediate through which the ketose is converted to the aldose: It has a double bond (ene), and it has two hydroxyl groups (diol). Because of this enediol reaction, ketoses are also able to react with Benedict's reagent, which is basic. Because the metal ions in the solution are reduced, the sugars are serving as reducing agents and are called **reducing sugars.** All monosaccharides and all the common disaccharides, except sucrose, are reducing sugars.

For many years the Benedict's reagent was used to test for *glucosuria*, the presence of excess glucose in the urine. Individuals suffering from *Type I insulin-dependent diabetes mellitus* do not produce the hormone insulin, which controls the uptake of glucose from the blood. When the blood glucose level rises above 160mg/100mL, the kidney is unable to reabsorb the excess, and glucose is found in the urine. Although the level of blood glucose could be controlled by the injection of insulin, urine glucose levels were monitored to ensure that the amount of insulin injected was correct. The Benedict's reagent was a useful tool because the amount of Cu_2O formed, and hence the degree of color change in the reaction, is directly proportional to the amount of reducing sugar in the urine. A brick-red color indicates a very high concentration of glucose in the urine. Yellow, green, and blue-green solutions indicate decreasing amounts of glucose in the urine, and a blue solution indicates an insignificant concentration. Urine glucose tests have been replaced by blood glucose tests that provide a more accurate indication of how well the diabetic is controlling his or her diet.

See "A Clinical Perspective: Diabetes Mellitus and Ketone Bodies" in Chapter 23.

17.3 Disaccharides

Recall that disaccharides consist of two monosaccharides joined through an "oxygen bridge." In biological systems, monosaccharides exist in the cyclic form and, as we have seen, they are actually hemiacetals or hemiketals. Recall that when a hemiacetal reacts with an alcohol, the product is an *acetal*, and when a hemiketal

Learning Goal

10

Figure 17.6

Glycosidic bond formed between the C-1 hydroxyl group of α-D-glucose and the C-4 hydroxyl group of β-D-glucose. The disaccharide is called β-maltose because the hydroxyl group at the reducing end of the disaccharide has the β-configuration.

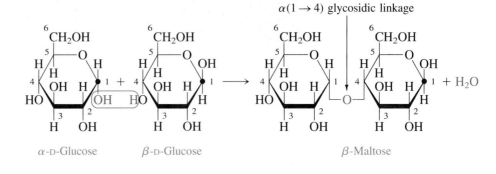

α-D-Glucose β-D-Glucose β-Maltose

Figure 17.7

Comparison of the cyclic forms of glucose and galactose. Note that galactose is identical to glucose except in the position of the C-4 hydroxyl group.

β-D-Glucose β-D-Galactose

Acetals and ketals are described in Section 14.4:

```
    OR              OR
    |               |
R—C—OR          R—C—OR
    |               |
    H               R
  Acetal          Ketal
```

reacts with an alcohol, the product is a *ketal*. In the case of disaccharides, the alcohol comes from a second monosaccharide. The acetals or ketals formed are given the general name *glycosides*, and the carbon-oxygen bonds are called *glycosidic bonds*.

Glycosidic bond formation is nonspecific; that is, it can occur between a hemiacetal or hemiketal and any of the hydroxyl groups on the second monosaccharide. However, in biological systems, we commonly see only particular disaccharides, such as maltose (Figure 17.6), lactose (see Figure 17.8), or sucrose (see Figure 17.9). These specific disaccharides are produced in cells because the reactions are catalyzed by enzymes. Each enzyme catalyzes the synthesis of one specific disaccharide, ensuring that one particular pair of hydroxyl groups on the reacting monosaccharides participates in glycosidic bond formation.

Maltose

If an α-D-glucose and a second glucose are linked, as shown in Figure 17.6, the disaccharide is **maltose,** or malt sugar. This is one of the intermediates in the hydrolysis of starch. Because the C-1 hydroxyl group of α-D-glucose is attached to C-4 of another glucose molecule, the disaccharide is linked by an α (1 → 4) glycosidic bond.

Maltose is a reducing sugar. Any disaccharide that has a hemiacetal hydroxyl group (a free —OH group at C-1) is a reducing sugar. This is because the cyclic structure can open at this position to form a free aldehyde. Disaccharides that do not contain a hemiacetal group on C-1 do not react with the Benedict's reagent and are called **nonreducing sugars.**

Lactose

Learning Goal

11

Milk sugar, or **lactose,** is a disaccharide made up of one molecule of β-D-galactose and one of either α- or β-D-glucose. Galactose differs from glucose only in the configuration of the hydroxyl group at C-4 (Figure 17.7). In the cyclic form of glucose the C-4 hydroxyl group is "down," and in galactose it is "up." In lactose the C-1

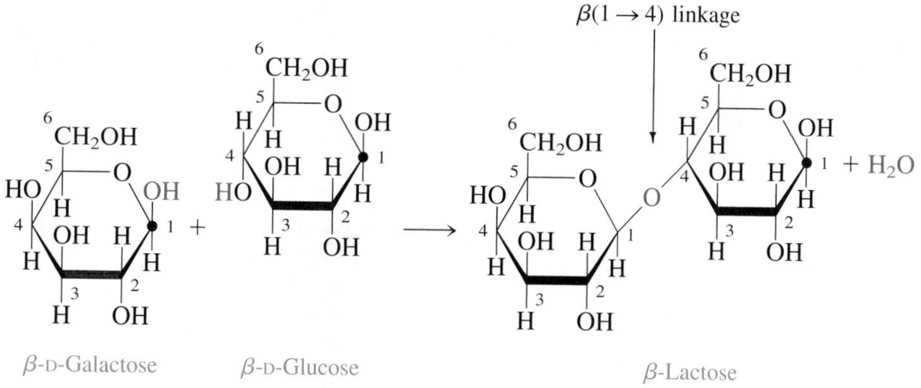

Figure 17.8

Glycosidic bond formed between the C-1 hydroxyl group of β-D-galactose and the C-4 hydroxyl group of β-D-glucose. The disaccharide is called β-lactose because the hydroxyl group at the reducing end of the disaccharide has the β-configuration.

hydroxyl group of β-D-galactose is bonded to the C-4 hydroxyl group of either an α- or β-D-glucose. The bond between the two monosaccharides is therefore a β(1 → 4) glycosidic bond (Figure 17.8).

Lactose is the principal sugar in mammalian milk. To be used by the body as an energy source, lactose must be hydrolyzed to produce glucose and galactose. Note that this is simply the reverse of the reaction shown in Figure 17.8. Glucose liberated by the hydrolysis of lactose is used directly in the energy-harvesting reactions of glycolysis. However, a series of reactions is necessary to convert galactose into a phosphorylated form of glucose that can be used in cellular metabolic reactions. In humans the genetic disease **galactosemia** is caused by the absence of one or more of the enzymes needed for this conversion. A toxic compound formed from galactose accumulates in people who suffer from galactosemia. If the condition is not treated, galactosemia leads to severe mental retardation, cataracts, and early death. However, the effects of this disease can be avoided entirely by providing galactosemic infants with a diet that does not contain galactose. Such a diet, of course, cannot contain lactose and therefore must contain no milk or milk products.

Many adults, and some children, are unable to hydrolyze lactose because they do not make the enzyme *lactase*. This condition, which affects 20% of the population of the United States, is known as **lactose intolerance.** Undigested lactose remains in the intestinal tract and causes cramping and diarrhea that can eventually lead to dehydration. Some of the lactose is metabolized by intestinal bacteria that release organic acids and CO_2 gas into the intestines, causing further discomfort. Lactose intolerance is unpleasant, but its effects can be avoided by a diet that excludes milk and milk products. Alternatively, the enzyme that hydrolyzes lactose is available in tablet form. When ingested with dairy products it breaks down the lactose, preventing symptoms.

Glycolysis is discussed in Chapter 21.

Sucrose

Sucrose is also called table sugar, cane sugar, or beet sugar. Sucrose is an important carbohydrate in plants. It is water soluble and can easily be transported through the circulatory system of the plant. It cannot be synthesized by animals. High concentrations of sucrose produce a high osmotic pressure, which inhibits the growth of microorganisms, so it is used as a preservative. Of course, it is also widely used as a sweetener. In fact, it is estimated that the average American consumes 100–125 pounds of sucrose each year. It has been suggested that sucrose in the diet is undesirable because it represents a source of empty calories; that is, it contains no vitamins or minerals. However, the only negative association that has been scientifically verified is the link between sucrose in the diet and dental caries, or cavities (see "A Human Perspective: Tooth Decay and Simple Sugars.")

Blood Transfusions and the Blood Group Antigens

The first blood transfusions were tried in the seventeenth century, when physicians used animal blood to replace human blood lost by hemorrhages. Unfortunately, many people died as a result of this attempted cure, and transfusions were banned in much of Europe. Transfusions from human donors were somewhat less lethal, but violent reactions often led to the death of the recipient, and by the nineteenth century, transfusions had been abandoned as a medical failure.

In 1904, Dr. Karl Landsteiner performed a series of experiments on the blood of workers in his laboratory. His results explained the mysterious transfusion fatalities, and blood transfusions were reinstated as a life-saving clinical tool. Landsteiner took blood samples from his co-workers. He separated the blood cells from the serum, the liquid component of the blood, and mixed these samples in test tubes. When he mixed serum from one individual with blood cells of another, Landsteiner observed that, in some instances, the serum samples caused clumping, or *agglutination*, of red blood cells (RBC). The agglutination reaction always indicated that the two bloods were incompatible and transfusion could lead to life-threatening reactions. As a result of many such experiments, Landsteiner showed that there are four human blood groups, designated A, B, AB, and O.

We now know that differences among blood groups reflect differences among oligosaccharides attached to the proteins and lipids of the RBC membranes. The oligosaccharides on the RBC surface have a common core, as shown in the accompanying figure, consisting of β-D-N-acetylgalactosamine, galactose, N-acetylneuraminic acid (sialic acid), and L-fucose. It is the terminal monosaccharide of this oligosaccharide that distinguishes the cells of different blood types and governs the compatibility of the blood types.

The A blood group antigen has β-D-N-acetylgalactosamine at its end, whereas the B blood group antigen has α-D-galactose. In type O blood, neither of these sugars is found on the cell surface; only the core oligosaccharide is present. Some of the oligosaccharides on type AB blood cells have a terminal β-D-N-acetylgalactosamine, whereas others have a terminal α-D-galactose.

Why does agglutination occur? The clumping reaction that occurs when incompatible bloods are mixed is an antigen-antibody reaction. Antigens are large molecules, often portions of bacteria or viruses, that stimulate the immune defenses of the body to produce protective antibodies. Antibodies bind to the foreign antigens and help to destroy them.

People with type A blood also have antibodies against type B blood (anti-B antibodies) in the blood serum. If the person with type A blood receives a transfusion of type B blood, the anti-B antibodies bind to the type B blood cells, causing clumping and destruction of those cells that can result in death. Individuals with type B blood also produce anti-A antibodies and therefore cannot receive a transfusion from a type A individual. Those with type AB blood are considered to be *universal recipients* because they have neither anti-A nor anti-B antibodies in their blood. (If they did, they would destroy their own red blood cells!) Thus in emergency situations a patient with type AB blood can receive blood from an individual of any blood type without serious transfusion reactions. Type O blood has no A or B antigens on the RBC but has both anti-A and anti-B antibodies. Because of the presence of both types of antibodies, type O individuals can receive transfusions only from a person who is also type O. On the other hand, the absence of A and B antigens on the red blood cell surface means that type O blood can be safely transfused into patients of any blood type. Hence type O individuals are *universal donors*.

Sucrose is a disaccharide of α-D-glucose joined to β-D-fructose (Figure 17.9). The glycosidic linkage between α-D-glucose and β-D-fructose is quite different from those that we have examined for lactose and maltose. Both of the carbons that were previously part of a hemiacetal or a hemiketal have reacted to form this linkage. Such a bond is called an (α1 → β2) glycosidic linkage, in this case involving the C-1 of glucose and the C-2 of fructose (noted in red in Figure 17.9). As a result of the (α1 → β2) glycosidic bond, the ring structure cannot open up, and no aldehyde or ketone group can be formed. Therefore sucrose will not react with Benedict's reagent and is not a reducing sugar.

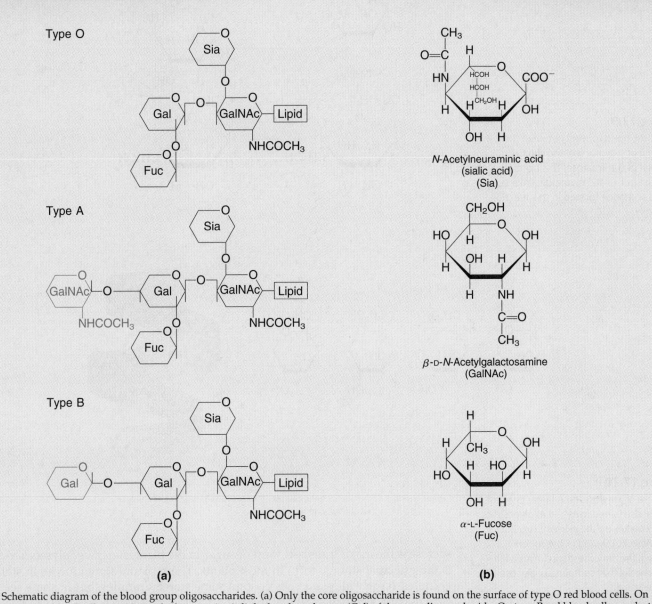

Schematic diagram of the blood group oligosaccharides. (a) Only the core oligosaccharide is found on the surface of type O red blood cells. On type A red blood cells, β-D-N-acetylgalactosamine is linked to the galactose (Gal) of the core oligosaccharide. On type B red blood cells, a galactose molecule is found attached to the galactose of the core oligosaccharide. (b) The structures of some of the unusual monosaccharides found in the blood group oligosaccharides.

17.4 Polysaccharides

Starch

Most carbohydrates that are found in nature are large polymers of glucose. Thus a polysaccharide is a large molecule composed of many monosaccharide units (the monomers) joined in one or more chains.

As seen in Figure 17.1, plants have the ability to use the energy of sunlight to produce monosaccharides, principally glucose, from CO_2 and H_2O. Although

Learning Goal

A polymer (Section 12.4) is a large molecule made up of many small units, the monomers, held together by chemical bonds. See "A Human Perspective: Carboxylic Acid Derivatives of Special Interest" in Chapter 15.

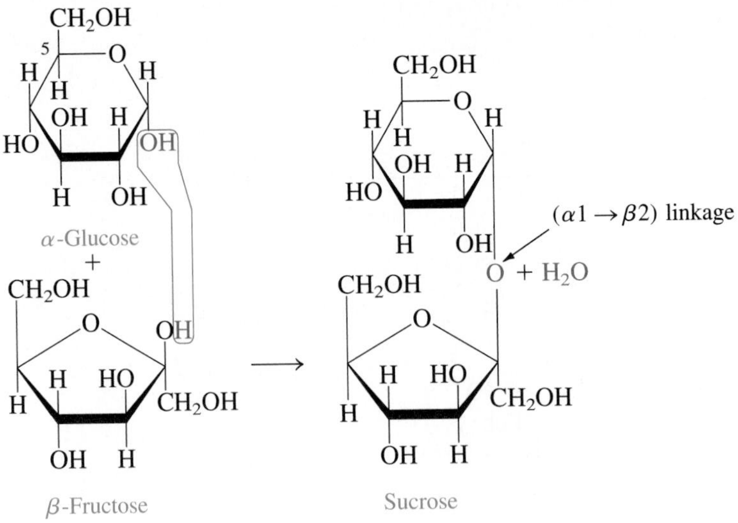

Figure 17.9

Glycosidic bond formed between the C-1 hydroxyl of α-D-glucose and the C-2 hydroxyl of β-D-fructose. This bond is called an (α1 → β2) glycosidic linkage. The disaccharide formed in this reaction is sucrose.

α-Glucose
+
β-Fructose

(α1 → β2) linkage

O + H₂O

Sucrose

α (1 → 4) linkage

(a)

(b)

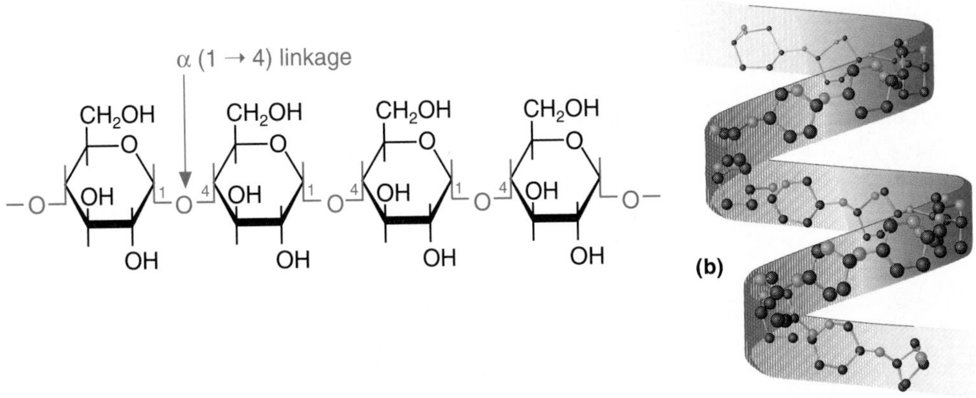

Figure 17.10

Structure of amylose. (a) A linear chain of α-D-glucose joined in α(1 → 4) glycosidic linkage makes up the primary structure of amylose. (b) Owing to hydrogen bonding, the amylose chain forms a left-handed helix that contains six glucose units per turn.

Enzymes are proteins that serve as biological catalysts. They speed up biochemical reactions so that life processes can function. These enzymes are called α(1 → 4) glycosidases because they cleave α(1 → 4) glycosidic bonds.

---

sucrose is the major transport form of sugar in the plant, starch (a polysaccharide) is the principal storage form in most plants. These plants store glucose in starch granules. Nearly all plant cells contain some starch granules, but in some seeds, such as corn, as much as 80% of the cell's dry weight is starch.

Starch is a heterogeneous material composed of the glucose polymers **amylose** and **amylopectin**. Amylose, which accounts for about 80% of the starch of a plant cell, is a linear polymer of α-D-glucose molecules connected by glycosidic bonds between C-1 of one glucose molecule and C-4 of a second glucose. Thus the glucose units in amylose are joined by α(1 → 4) glycosidic bonds. A single chain can contain up to 4000 glucose units. Amylose coils up into a helix that repeats every six glucose units. The structure of amylose is shown in Figure 17.10.

Amylose is degraded by two types of enzymes. They are produced in the pancreas, from which they are secreted into the small intestine, and the salivary glands, from which they are secreted into the saliva. α-*Amylase* cleaves the glycosidic bonds of amylose chains at random along the chain, producing shorter polysaccharide chains. The enzyme β-*amylase* sequentially cleaves the disaccharide maltose from the reducing end of the amylose chain. The maltose is hydrolyzed into glucose by the enzyme *maltase*. The glucose is quickly absorbed by intestinal cells and used by the cells of the body as a source of energy.

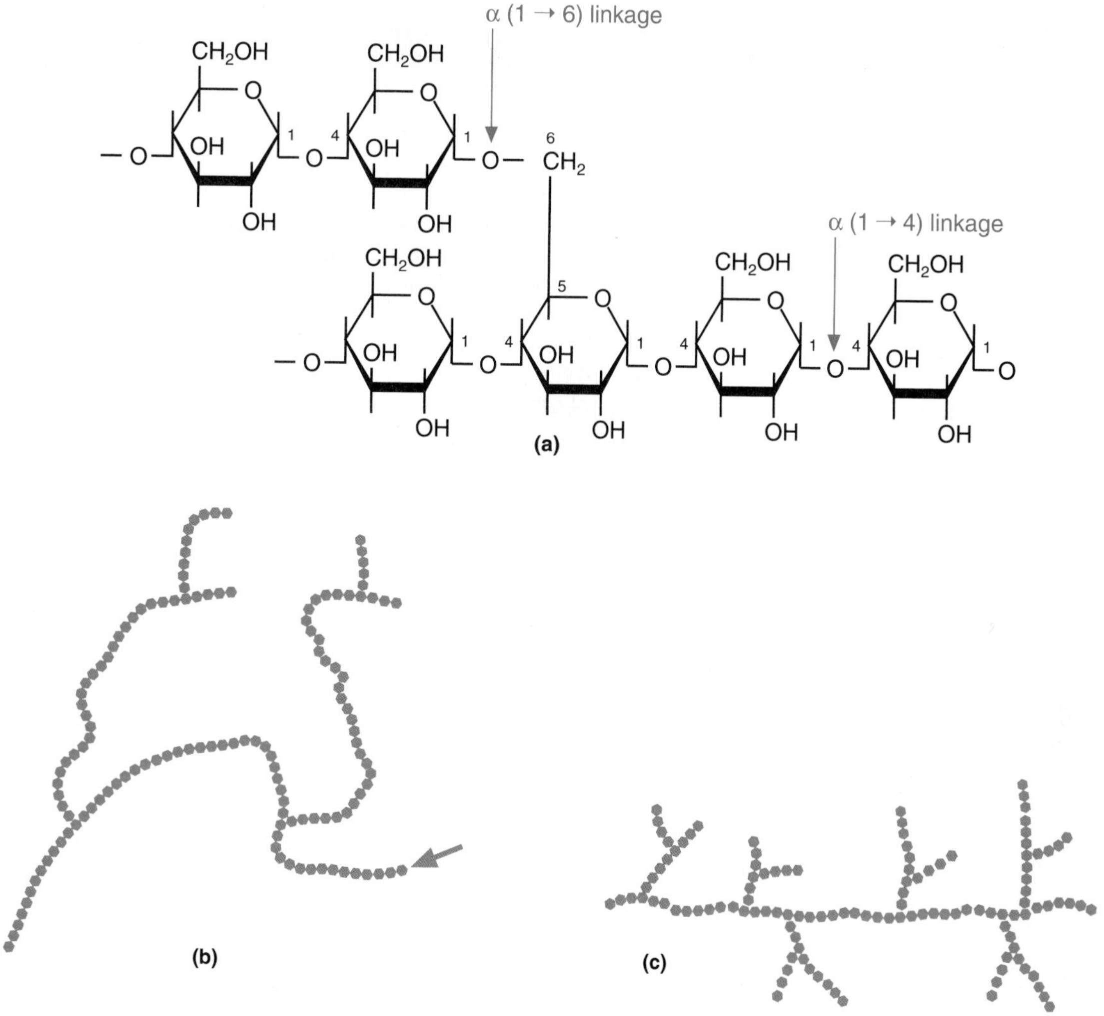

α (1 → 6) linkage

α (1 → 4) linkage

(a)

(b)

(c)

Figure 17.11

Structure of amylopectin and glycogen. (a) Both amylopectin and glycogen consist of chains of α-D-glucose molecules joined in α(1 → 4) glycosidic linkages. Branching from these chains are other chains of the same structure. Branching occurs by formation of α(1 → 6) glycosidic bonds between glucose units. (b) A representation of the branched-chain structure of amylopectin. (c) A representation of the branched-chain structure of glycogen. Glycogen differs from amylopectin only in that the branches are shorter and there are more of them.

Amylopectin is a highly branched amylose in which the branches are attached to the C-6 hydroxyl groups by α(1 → 6) glycosidic bonds (Figure 17.11). The main chains consist of α(1 → 4) glycosidic bonds. Each branch contains 20–25 glucose units, and there are so many branches that the main chain can scarcely be distinguished.

Glycogen

Glycogen is the major glucose storage molecule in animals. The structure of glycogen is similar to that of amylopectin. The "main chain" is linked by α(1 → 4) glycosidic bonds, and it has numerous α(1 → 6) glycosidic bonds, which provide many branch points along the chain. Glycogen differs from amylopectin only by having more and shorter branches. Otherwise, the two molecules are virtually identical. The structure of glycogen is shown in Figure 17.11.

Glycogen is stored in the liver and skeletal muscle. Glycogen synthesis and degradation in the liver are carefully regulated. As we will see in Section 21.7 these two processes are intimately involved in keeping blood glucose levels constant.

Cellulose

The most abundant polysaccharide, indeed the most abundant organic molecule in the world, is **cellulose,** a polymer of β-D-glucose units linked by β(1 → 4) glycosidic

The Bacterial Cell Wall

The major component of bacterial cell walls is a complex polysaccharide known as a *peptidoglycan*. The name tells us that this structure consists of sugar molecules (-glycan) and peptides (peptido-; short polymers of amino acids).

As the accompanying structure shows, the carbohydrate portion of the peptidoglycan is a polymer of alternating units of two modified glucose molecules called *N-acetylglucosamine* and *N-acetylmuramic acid*. These two unusual monosaccharides

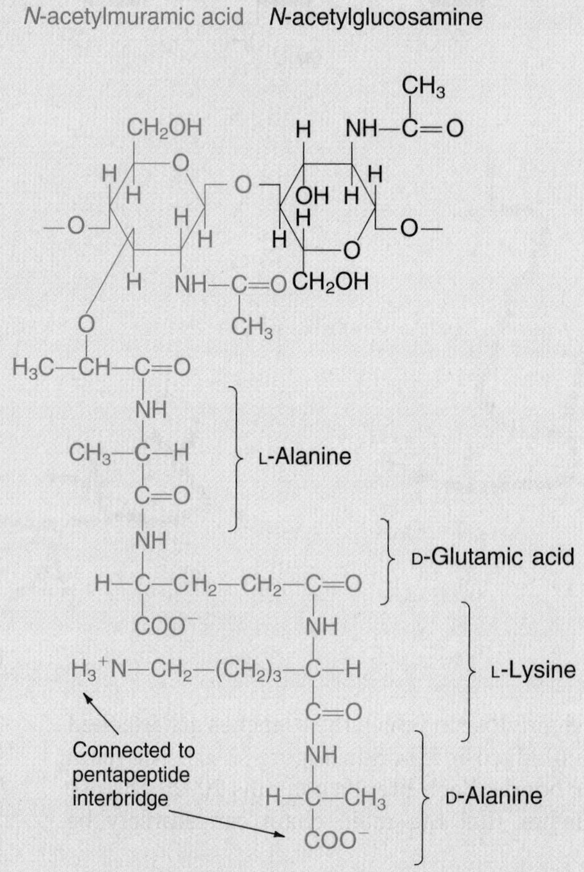

Structures of *N*-acetylglucosamine and *N*-acetylmuramic acid in β(1 → 4) glycosidic linkage. Note the tetrapeptide bridge linked to the *N*-acetylmuramic acid.

bonds (Figure 17.12). A molecule of cellulose typically contains about 3000 glucose units, but the largest known cellulose, produced by the alga *Valonia*, contains 26,000 glucose molecules.

Cellulose is a structural component of the plant cell wall. The unbranched structure of the cellulose polymer and the β(1 → 4) glycosidic linkages allow cellulose molecules to form long, straight chains of parallel cellulose molecules called *fibrils*. These fibrils are quite rigid and are held together tightly by hydrogen bonds; thus it is not surprising that cellulose is a cell wall structural element.

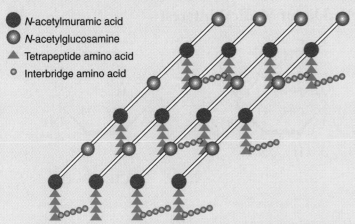

- ● *N*-acetylmuramic acid
- ◉ *N*-acetylglucosamine
- ▲ Tetrapeptide amino acid
- ○ Interbridge amino acid

The three-dimensional structure of one layer of peptidoglycan.

are joined by a β(1 → 4) glycosidic bond. In addition, each *N*-acetylmuramic acid is bonded to a tetrapeptide, a chain of four amino acids.

The structural strength of the cell wall is a result of pentapeptide cross-bridges that link the repeat units to one another (see the figure at the left). Millions of such cross-linkages produce an enormous peptidoglycan molecule, dozens of layers thick, around the bacterium. This thick wall is very rigid. It allows the bacterium to maintain its shape and protects it from bursting if the salt concentration of the environment is too low (hypotonic conditions).

Our bodies are constantly being assaulted by a variety of bacteria, and as you might expect, we have evolved protective mechanisms to minimize the damage. For instance, the enzyme *lysozyme*, found in tears and saliva, catalyzes the hydrolysis of the β(1 → 4) glycosidic bonds of peptidoglycan. As the accompanying figure shows, the enzyme has a deep groove on the surface (the active site) that a six-sugar unit of the cell wall can slip into like a bank card into the slot of an automatic teller machine. Lysozyme then catalyzes bond breakage and destroys the cell wall of the bacterium.

The penicillins are antibiotics that interfere with bacterial cell wall synthesis. The human body has no structures similar to the bacterial cell wall, so treatment with penicillins selectively destroys the bacteria, causing no harm to the patient. In practice, however, it must always be remembered that some individuals may develop an allergy to penicillins.

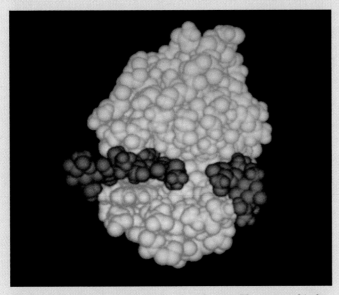

Conformation of lysozyme bound to its substrate. The enzyme binds with a six-sugar portion of the bacterial cell wall and cleaves it. The substrate fits into a deep crevice on the surface of the enzyme.

Penicillin G

The penicillins inhibit the enzyme that catalyzes the formation of the cross-linkage between the tetrapeptides. The antibiotic binds irreversibly to the active site of that enzyme so that it cannot bind to the tetrapeptide tail. Thus no cross-linkage can be made. Without the rigid, highly cross-linked peptidoglycan, the bacterial cells rupture and die.

In contrast to glycogen, amylose, and amylopectin, cellulose *cannot* be digested by humans. The reason is that we cannot synthesize the enzyme *cellulase*, which can hydrolyze the β(1 → 4) glycosidic linkages of the cellulose polymer. Indeed, only a few animals, such as termites, cows, and goats, are able to digest cellulose. These animals have, within their digestive tracts, microorganisms that produce the enzyme cellulase. The sugars released by this microbial digestion can then be absorbed and used by these animals. In humans, cellulose from fruits and vegetables serves as fiber in the diet.

A MEDICAL PERSPECTIVE

Monosaccharide Derivatives and Heteropolysaccharides of Medical Interest

Many of the carbohydrates with important functions in the human body are either derivatives of simple monosaccharides or are complex polymers of monosaccharide derivatives. One type of monosaccharide derivatives, the uronates, is formed when the terminal—CH_2OH group of a monosaccharide is oxidized to a carboxylate group. α-D-Glucuronate is a uronate of glucose:

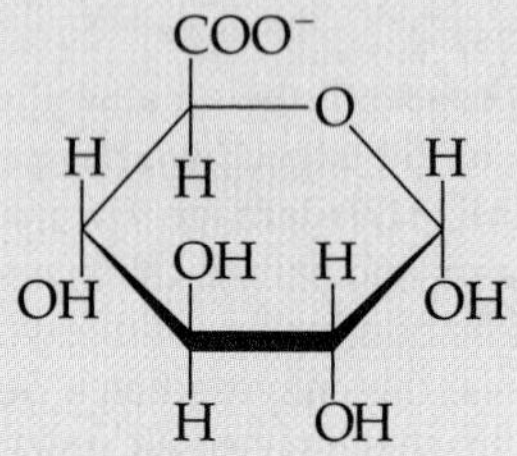

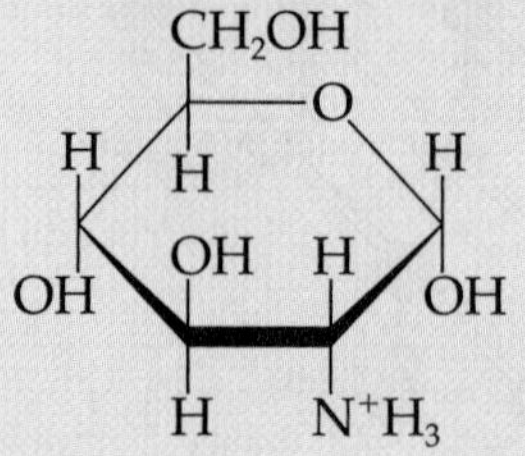

α-D-Glucosamine   α-D-N-Acetylglucosamine

α-D-Glucuronate

In liver cells, α-D-glucuronate is bonded to hydrophobic molecules, such as steroids, to increase their solubility in water. When bonded to the modified sugar, steroids are more readily removed from the body.

Amino sugars are a second important group of monosaccharide derivatives. In amino sugars one of the hydroxyl groups (usually on carbon-2) is replaced by an amino group. Often these are found in complex oligosaccharides that are attached to cellular proteins and lipids. The most common amino sugars, D-glucosamine and D-galactosamine, are often found in the N-acetyl form. N-acetylglucosamine (see "A Clinical Perspective: The Bacterial Cell Wall") is a component of bacterial cell walls and N-acetylgalactosamine is a component of the A, B, O blood group antigens (see "A Human Perspective: Blood Transfusions and the Blood Group Antigens.")

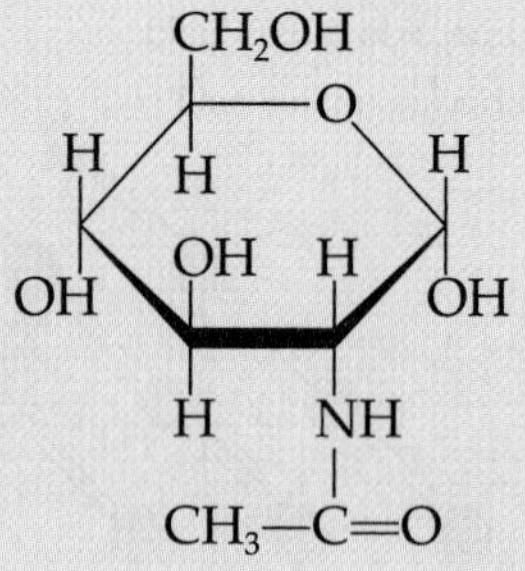

Heteropolysaccharides are long-chain polymers that contain more than one type of monosaccharide, many of which are amino sugars. As a result, they are often referred to as *glycosaminoglycans*, which include chondroitin sulfate, hyaluronic acid, and heparin. Hyaluronic acid is abundant in the fluid of joints and in the vitreous humor of the eye. Chondroitin sulfate is an important component of cartilage; and heparin has anticoagulant function. The structures of the repeat units of these polymers are shown below.

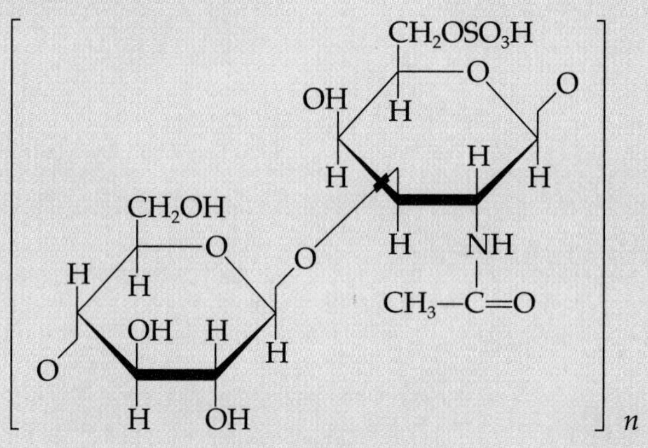

Repeat unit of chondroitin sulfate

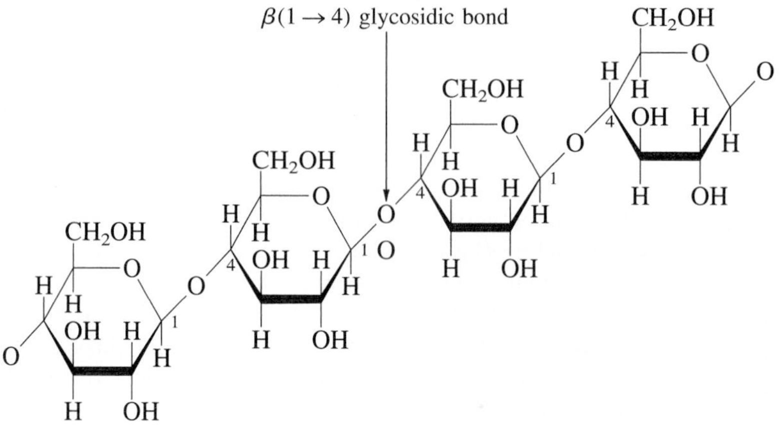

β(1 → 4) glycosidic bond

Figure 17.12
The structure of cellulose.

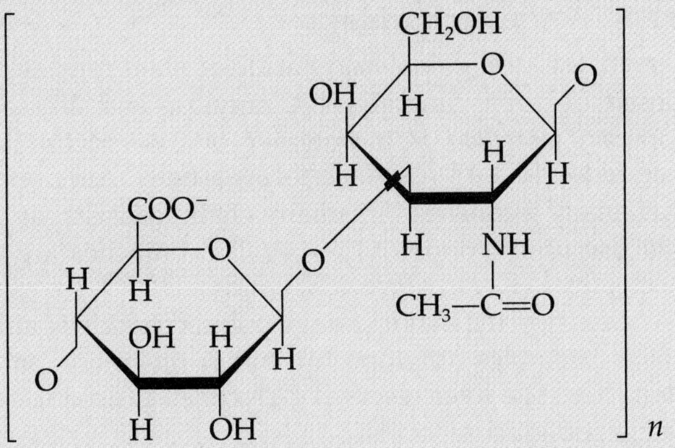

Repeat unit of hyaluronic acid

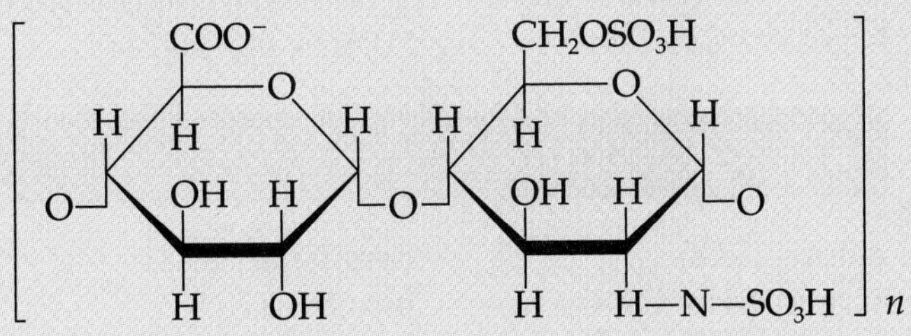

Repeat unit of heparin

Two of these molecules have been studied as potential treatments for osteoarthritis, a painful, degenerative disease of the joints. The amino sugar D-glucosamine is thought to stimulate the production of collagen. Collagen is one of the main components of articular cartilage, which is the shock-absorbing cushion within the joints. With aging, some of the D-glucosamine is lost, leading to a reduced cartilage layer and to the onset and progression of arthritis. It has been suggested that ingestion of D-glucosamine can actually "jump-start" production of cartilage and help repair eroded cartilage in arthritic joints.

It has also been suggested that chondroitin sulfate can protect existing cartilage from premature breakdown. It absorbs large amounts of water, which is thought to facilitate diffusion of nutrients into the cartilage, providing precursors for the synthesis of new cartilage. The increased fluid also acts as a shock absorber.

Studies continue on the effect that D-glucosamine and chondroitin sulfate have on degenerative joint disease. To date the studies are inconclusive because a large placebo effect is observed with sufferers of osteoarthritis. Many people in the control groups of these studies also experience relief of symptoms when they receive treatment with a placebo, such as a sugar pill.

Capsules containing D-glucosamine and chondroitin sulfate are available over-the-counter and many sufferers of osteoarthritis prefer to take this nutritional supplement as an alternative to any nonsteroidal anti-inflammatory drugs (NSAID), such as ibuprofen. Although NSAID can reduce inflammation and pain, long-term use of NSAID can result in stomach ulcers, damage to auditory nerves, and kidney damage.

What chemical reactions are catalyzed by α-amylase and β-amylase?

Question 17.9

What is the function of cellulose in the human diet? How does this relate to the structure of cellulose?

Question 17.10

Summary

Carbohydrates are found in a wide variety of naturally occurring substances and serve as principal energy sources for the body. Dietary carbohydrates include complex carbohydrates, such as starch in potatoes, and simple carbohydrates, such as sucrose.

17.1   Types of Carbohydrates

Carbohydrates are classified as *monosaccharides* (one sugar unit), *disaccharides* (two sugar units), *oligosaccharides* (three to ten sugar units), or *polysaccharides* (many sugar units).

17.2   Monosaccharides

Monosaccharides that have an aldehyde as their most oxidized functional group are *aldoses,* and those having a ketone group as their most oxidized functional group are *ketoses.* They may be classified as *trioses, tetroses, pentoses,* and so forth, depending on the number of carbon atoms in the carbohydrate.

Stereoisomers of monosaccharides exist because of the presence of *chiral* carbon atoms. They are classified as D- or L- depending on the arrangement of the atoms on the chiral carbon farthest from the aldehyde or ketone group. If the —OH on this carbon is to the right, the stereoisomer is of the D-family. If the —OH group is to the left, the stereoisomer is of the L-family.

Important monosaccharides include *glyceraldehyde, glucose, fructose,* and *ribose.* Monosaccharides containing five or six carbon atoms can exist as five-membered or six-membered rings. Formation of a ring produces a new chiral carbon at the original carbonyl carbon, which is designated either α or β depending on the orientation of the groups. The cyclization of an aldose produces an intramolecular *hemiacetal,* and the cyclization of a ketose yields an intramolecular *hemiketal.*

Reducing sugars are oxidized by the *Benedict's reagent.* All monosaccharides and all common disaccharides, except sucrose, are reducing sugars. At one time Benedict's reagent was used to determine the concentration of glucose in urine.

17.3   Disaccharides

Important disaccharides include *lactose* and *sucrose.* Lactose is a disaccharide of β-D-galactose bonded (1 → 4) with D-glucose. In *galactosemia,* defective metabolism of galactose leads to accumulation of a toxic by-product. The ill effects of galactosemia are avoided by exclusion of milk and milk products from the diet of affected infants.

Sucrose is a dimer composed of α-D-glucose bonded (α1 → β2) with β-D-fructose.

17.4   Polysaccharides

Starch, the storage polysaccharide of plant cells, is composed of approximately 80% amylose and 20% amylopectin. *Amylose* is a polymer of α-D-glucose units bonded α(1 → 4). Amylose forms a helix. Amylopectin has many branches. Its main chain consists of α-D-glucose units bonded (1 → 4). The branches are connected by α(1 → 6) glycosidic bonds.

Glycogen, the major storage polysaccharide of animal cells, resembles amylopectin, but it has more, shorter branches. The liver reserve of glycogen is used to regulate blood glucose levels.

Cellulose is a major structural molecule of plants. It is a β(1 → 4) polymer of D-glucose that can contain thousands of glucose monomers. Cellulose cannot be digested by animals because they do not produce an enzyme capable of cleaving the β(1 → 4) glycosidic linkage.

Key Terms

aldose (17.2)	**hemiketal (17.2)**
amylopectin (17.4)	**hexose (17.2)**
amylose (17.4)	**ketose (17.2)**
asymmetric carbon (17.2)	**lactose (17.3)**
Benedict's reagent (17.2)	**lactose intolerance (17.3)**
carbohydrate (Intro)	**maltose (17.3)**
cellulose (17.4)	**monosaccharide (17.1)**
chiral molecule (17.2)	**nonreducing sugar (17.3)**
disaccharide (17.1)	**oligosaccharide (17.1)**
enantiomers (17.2)	**pentose (17.2)**
fructose (17.2)	**polysaccharide (17.1)**
galactose (17.2)	**reducing sugar (17.2)**
galactosemia (17.3)	**ribose (17.2)**
glucose (17.2)	**saccharide (17.1)**
glyceraldehyde (17.2)	**stereochemistry (17.2)**
glycogen (17.4)	**stereoisomers (17.2)**
glycosidic bond (17.1)	**sucrose (17.3)**
Haworth projection (17.2)	**tetrose (17.2)**
hemiacetal (17.2)	**triose (17.2)**

Questions and Problems

Types of Carbohydrates

17.11 What is the difference between a monosaccharide and a disaccharide?

17.12 What is a polysaccharide?

17.13 Read the labels on some of the foods in your kitchen, and see how many products you can find that list one or more carbohydrates among the ingredients in the package. Make a list of these compounds, and attempt to classify them according to parent structure (e.g., monosaccharides, disaccharides, polysaccharides).

17.14 Some disaccharides are often referred to by their common names. What are the chemical names of (a) milk sugar, (b) beet sugar, and (c) cane sugar?

17.15 How many kilocalories of energy are released when 1 g of carbohydrate is "burned" or oxidized?

17.16 List some natural sources of carbohydrates.

17.17 Draw and provide the names of an aldohexose and a ketohexose.

17.18 Draw and provide the name of an aldotriose.

Monosaccharides

17.19 Identify each of the following sugars. Label each as either a hemiacetal or a hemiketal:

a.

c.

b.

17.20 Draw the open-chain form of the sugars in Problem 17.19.

17.21 Draw all of the different possible aldotrioses of molecular formula $C_3H_6O_3$.

17.22 Draw all of the different possible aldotetroses of molecular formula $C_4H_8O_4$.

17.23 Is there any difference between dextrose and D-glucose?

17.24 The structure of D-glucose is shown. Draw its mirror image.

17.25 How are D- and L-glyceraldehyde related?

17.26 Determine whether each of the following is a D- or L-sugar:

a.

b.

c.

d.

17.27 Why does cyclization of D-glucose give two isomers, α- and β-D-glucose?

17.28 Draw the structure of the open chain form of D-fructose, and show how it cyclizes to form α- and β-D-fructose.

17.29 Which of the following would give a positive Benedict's Test?
   a. Sucrose   **c.** β-Maltose
   b. Glycogen   **d.** α-Lactose

17.30 Why was the Benedict's reagent useful for determining the amount of glucose in the urine?

17.31 Describe what is meant by a pair of enantiomers. Draw an example of a pair of enantiomers.

17.32 What is a chiral carbon atom?

17.33 When discussing sugars, what do we mean by an intramolecular hemiacetal?

17.34 When discussing sugars, what do we mean by an intramolecular hemiketal?

Disaccharides

17.35 Maltose is a disaccharide isolated from amylose that consists of two glucose units linked α(1 → 4). Draw the structure of this molecule.

17.36 Sucrose is a disaccharide formed by linking α-D-glucose and β-D-fructose by an (α1 → β2) bond. Draw the structure of this disaccharide. (*Hint:* Refer to Figure 17.9.)

17.37 What is the major biological source of lactose?

17.38 What metabolic defect causes galactosemia?

17.39 What simple treatment prevents most of the ill effects of galactosemia?

17.40 What are the major physiological effects of galactosemia?

17.41 What is lactose intolerance?

17.42 What is the difference between lactose intolerance and galactosemia?

Polysaccharides

17.43 What is the difference between the structure of cellulose and the structure of amylose?

17.44 How does the structure of amylose differ from that of amylopectin and glycogen?

17.45 What is the major physiological purpose of glycogen?

17.46 Where in the body do you find glycogen stored?

17.47 Where are α-amylase and β-amylase produced?

17.48 Where do α-amylase and β-amylase carry out their enzymatic functions?

Critical Thinking Problems

1. The six-member glucose ring structure is not a flat ring. Like cyclohexane, it can exist in the chair conformation. Build models of the chair conformation of α- and β-D-glucose. Draw each of these structures. Which would you predict to be the more stable isomer? Explain your reasoning.

2. The following is the structure of salicin, a bitter-tasting compound found in the bark of willow trees:

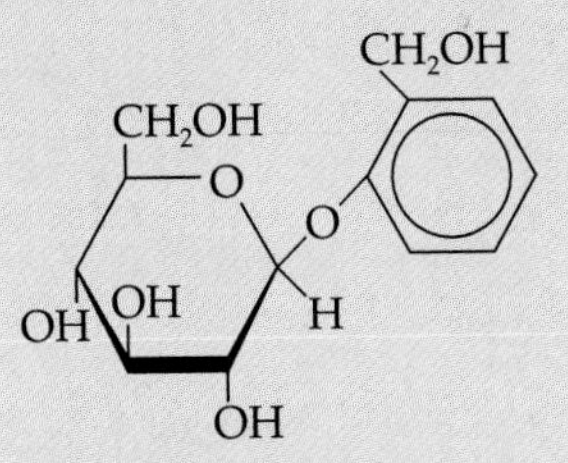

Salicin

The aromatic ring portion of this structure is quite insoluble in water. How would forming a glycosidic bond between the aromatic ring and β-D-glucose alter the solubility? Explain your answer.

3. Ancient peoples used salicin to reduce fevers. Write an equation for the acid-catalyzed hydrolysis of the O-glycosidic bond of salicin. Compare the aromatic product with the structure of acetylsalicylic acid (aspirin). Use this information to develop a hypothesis explaining why ancient peoples used salicin to reduce fevers.

4. Chitin is a modified cellulose in which the C-2 hydroxyl group of each glucose is replaced by

This nitrogen-containing polysaccharide makes up the shells of lobsters, crabs, and the exoskeletons of insects. Draw a portion of a chitin polymer consisting of four monomers.

5. Pectins are polysaccharides obtained from fruits and berries and used to thicken jellies and jams. Pectins are α(1 → 4) linked D-galacturonic acid. D-Galacturonic acid is D-galactose in which the C-6 hydroxyl group has been oxidized to a carboxyl group. Draw a portion of a pectin polymer consisting of four monomers.

6. Peonin is a red pigment found in the petals of peony flowers. Consider the structure of peonin:

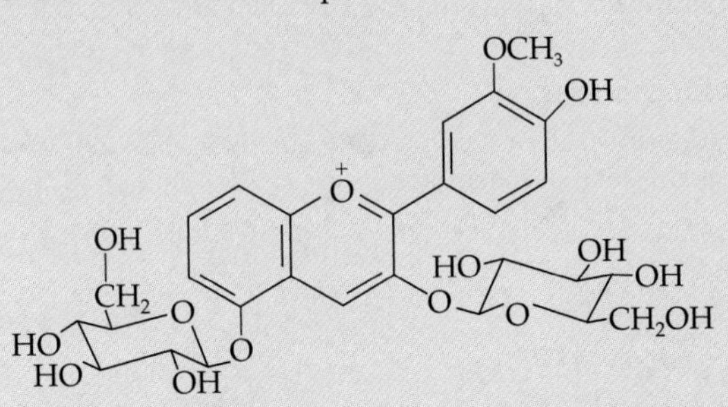

Why do you think peonin is bonded to two hexoses? What monosaccharide(s) would be produced by acid-catalyzed hydrolysis of peonin?

18

Lipids and Their Functions in Biochemical Systems

Honeycombs are constructed from beeswax (myricyl palmitate).

Learning Goals

1. Discuss the physical and chemical properties and biological function of each of the families of lipids.

2. Write the structures of saturated and unsaturated fatty acids.

3. Compare and contrast the structure and properties of saturated and unsaturated fatty acids.

4. Write equations representing the reactions that fatty acids undergo.

5. Describe the functions of prostaglandins.

6. Discuss the mechanism by which aspirin reduces pain.

7. Draw the structure of a phospholipid and discuss its amphipathic nature.

8. Discuss the general classes of sphingolipids and their functions.

9. Draw the structure of the steroid nucleus and discuss the functions of steroid hormones.

10. Describe the function of lipoproteins in triglyceride and cholesterol transport in the body.

11. Draw the structure of the cell membrane and discuss its functions.

12. Discuss passive and facilitated diffusion of materials through a cell membrane.

13. Explain the process of osmosis.

14. Describe the mechanism of action of a Na^+-K^+ ATPase.

495

CHEMISTRY CONNECTION

Life-Saving Lipids

In the intensive-care nursery the premature infant struggles for life. Born three and a half months early, the baby weighs only 1.6 pounds, and the lungs labor to provide enough oxygen to keep the tiny body alive. Premature infants often have respiratory difficulties because they have not yet begun to produce *pulmonary surfactant.*

Pulmonary surfactant is a combination of phospholipids and proteins that reduces surface tension in the alveoli of the lungs. (Alveoli are the small, thin-walled air sacs in the lungs.) This allows efficient gas exchange across the membranes of the alveolar cells; oxygen can more easily diffuse from the air into the tissues and carbon dioxide can easily diffuse from the tissues into the air.

Without pulmonary surfactant, gas exchange in the lungs is very poor. Pulmonary surfactant is not produced until early in the sixth month of pregnancy. Premature babies born before they have begun secretion of natural surfactant suffer from *respiratory distress syndrome (RDS)*, which is caused by the severe difficulty they have obtaining enough oxygen from the air that they breathe.

Until recently, RDS was a major cause of death among premature infants, but now a life-saving treatment is available. A fine aerosol of an artificial surfactant is administered directly into the trachea. The Glaxo-Wellcome Company product EXOSURF® Neonatal™ contains the phospholipid lecithin to reduce surface tension; 1-hexadecanol, which spreads the lecithin; and a polymer called tyloxapol, which disperses both the lecithin and the 1-hexadecanol.

Artificial pulmonary surfactant therapy has dramatically reduced premature infant death caused by RDS and appears to have reduced overall mortality for all babies born weighing less than 700 g (about 1.5 pounds). Advances such as this have come about as a result of research on the makeup of body tissues and secretions in both healthy and diseased individuals. Often, such basic research provides the information needed to develop effective therapies.

In this chapter we will study the chemistry of lipids with a wide variety of structures and biological functions. Among these are the triglycerides that stock our adipose tissue, pain-producing prostaglandins, and steroids that determine our secondary sexual characteristics.

Introduction

Lipids seem to be the most controversial group of biological molecules, particularly in the fields of medicine and nutrition. One concern is the use of anabolic steroids by athletes. Although these hormones increase muscle mass and enhance performance, we are just beginning to understand the damage they cause to the body.

We are concerned about what types of dietary fat we should consume. We hear frequently about the amounts of saturated fats and cholesterol in our diets because a strong correlation has been found between these lipids and heart disease. Large quantities of dietary saturated fats may also predispose an individual to colon, esophageal, stomach, and breast cancers. As a result, we are advised to reduce our intake of cholesterol and saturated fats.

Nonetheless, lipids serve a wide variety of functions essential to living systems and are required in our diet. Standards of fat intake have not been experimentally determined. However, the most recent U.S. Dietary Guidelines recommend that dietary fat not exceed 30% of the daily caloric intake, and no more than 10% should be saturated fats.

18.1 Biological Functions of Lipids

Learning Goal

The term **lipids** actually refers to a collection of organic molecules of varying chemical composition. They are grouped together on the basis of their solubility in nonpolar solvents. Lipids are commonly subdivided into four main groups:

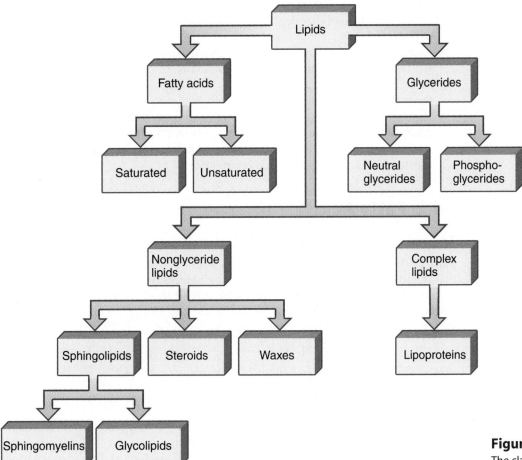

Figure 18.1
The classification of lipids.

1. *Fatty acids* (saturated and unsaturated)
2. *Glycerides* (glycerol-containing lipids)
3. *Nonglyceride lipids* (sphingolipids, steroids, waxes)
4. *Complex lipids* (lipoproteins)

In this chapter we examine the structure, properties, chemical reactions, and biological functions of each of the lipid groups shown in Figure 18.1.

As a result of differences in their structures, lipids serve many different functions in the human body. The following brief list will give you an idea of the importance of lipids in biological processes:

- *Energy source.* Like carbohydrates, lipids are an excellent source of energy for the body. When oxidized, each gram of fat releases 9 kcal of energy, or more than twice the energy released by oxidation of a gram of carbohydrate.
- *Energy storage.* Most of the energy stored in the body is in the form of lipids (triglycerides). Stored in fat cells called *adipocytes,* these fats are a particularly rich source of energy for the body.
- *Cell membrane structural components.* Phosphoglycerides, sphingolipids, and steroids make up the basic structure of all cell membranes. These membranes control the flow of molecules into and out of cells and allow cell-to-cell communication.
- *Hormones.* The steroid hormones are critical chemical messengers that allow tissues of the body to communicate with one another. The prostaglandins exert strong biological effects on both the cells that produce them and other cells of the body.

Lipid-soluble vitamins are discussed in detail in Appendix E.

- *Vitamins.* The lipid-soluble vitamins, A, D, E, and K, play a major role in the regulation of several critical biological processes, including blood clotting and vision.
- *Vitamin absorption.* Dietary fat serves as a carrier of the lipid-soluble vitamins. All are transported into cells of the small intestine in association with fat molecules. Therefore a diet that is too low in fat can result in a deficiency of these four vitamins.
- *Protection.* Fats serve as a shock absorber, or protective layer, for the vital organs. About 4% of the total body fat is reserved for this critical function.
- *Insulation.* Fat stored beneath the skin (subcutaneous fat) serves to insulate the body from extremes of cold temperatures.

18.2 Fatty Acids

Structure and Properties

Learning Goal

Fatty acids are long-chain monocarboxylic acids. As a consequence of their biosynthesis, fatty acids generally contain an *even number* of carbon atoms. The general formula for a **saturated fatty acid** is $CH_3(CH_2)_nCOOH$, in which n in biological systems is an even integer. If $n = 16$, the result is an 18-carbon saturated fatty acid, stearic acid, having the following structural formula:

$$H-\underset{\underset{H}{|}}{\overset{\overset{H}{|}}{C}}-\underset{\underset{H}{|}}{\overset{\overset{H}{|}}{C}}-\underset{\underset{H}{|}}{\overset{\overset{H}{|}}{C}}-\underset{\underset{H}{|}}{\overset{\overset{H}{|}}{C}}-\underset{\underset{H}{|}}{\overset{\overset{H}{|}}{C}}-\underset{\underset{H}{|}}{\overset{\overset{H}{|}}{C}}-\underset{\underset{H}{|}}{\overset{\overset{H}{|}}{C}}-\underset{\underset{H}{|}}{\overset{\overset{H}{|}}{C}}-\underset{\underset{H}{|}}{\overset{\overset{H}{|}}{C}}-\underset{\underset{H}{|}}{\overset{\overset{H}{|}}{C}}-\underset{\underset{H}{|}}{\overset{\overset{H}{|}}{C}}-\underset{\underset{H}{|}}{\overset{\overset{H}{|}}{C}}-\underset{\underset{H}{|}}{\overset{\overset{H}{|}}{C}}-\underset{\underset{H}{|}}{\overset{\overset{H}{|}}{C}}-\underset{\underset{H}{|}}{\overset{\overset{H}{|}}{C}}-\underset{\underset{H}{|}}{\overset{\overset{H}{|}}{C}}-\overset{\overset{O}{||}}{C}-OH$$

The saturated fatty acids may be thought of as derivatives of alkanes, the saturated hydrocarbons described in Chapter 11.

Note that each of the carbons in the chain is bonded to the maximum number of hydrogen atoms. To help remember the structure of a saturated fatty acid, you might think of each carbon in the chain being "saturated" with hydrogen atoms. Examples of common saturated fatty acids are given in Table 18.1. An example of an **unsaturated fatty acid** is the 18-carbon unsaturated fatty acid oleic acid, which has the following structural formula:

The unsaturated fatty acids may be thought of as derivatives of the alkenes, the unsaturated hydrocarbons discussed in Chapter 12.

In the case of unsaturated fatty acids there is at least one carbon-to-carbon double bond. Because of the double bonds, the carbon atoms involved in these bonds are not "saturated" with hydrogen atoms. The double bonds found in almost all naturally occurring unsaturated fatty acids are in the *cis* configuration. In addition, the double bonds are not randomly located in the hydrocarbon chain. Both the placement and the geometric configuration of the double bonds are dictated by the enzymes that catalyze the biosynthesis of unsaturated fatty acids. Examples of common unsaturated fatty acids are also given in Table 18.1.

EXAMPLE 18.1

Writing the Structural Formula of an Unsaturated Fatty Acid

Draw the structural formula for palmitoleic acid

Solution

The I.U.P.A.C. name of palmitoleic acid is *cis*-9-hexadecenoic acid. The name tells us that this is a 16-carbon fatty acid having a carbon-to-carbon double bond between carbons 9 and 10. The name also reveals that this is the *cis* isomer.

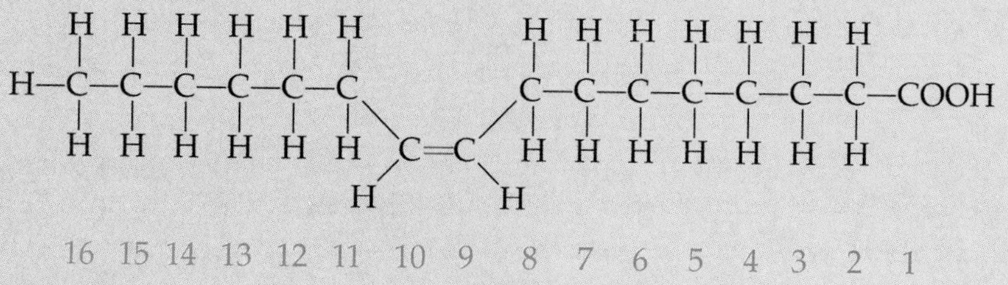

16 15 14 13 12 11 10   9   8   7   6   5   4   3   2   1

Examination of Table 18.1 and Figure 18.2 reveals several interesting and important points about the physical properties of fatty acids.

- The melting points of saturated fatty acids increase with increasing carbon number, as is the case with alkanes. Saturated fatty acids containing ten or more carbons are solids at room temperature.
- The melting point of a saturated fatty acid is greater than that of an unsaturated fatty acid of the same chain length. The reason is that saturated fatty acid chains tend to be fully extended and to stack in a regular structure, thereby causing increased intermolecular attraction. Introduction of a *cis* double bond into the hydrocarbon chain produces a rigid 30° bend. Such

Learning Goal

3

The relationship between alkane chain length and melting point is described in Section 11.2.

Table 18.1   Common Saturated and Unsaturated Fatty Acids

Common Saturated Fatty Acids

Common Name	I.U.P.A.C. Name	Melting Point (°C)	RCOOH Formula	Condensed Formula
Capric	Decanoic	32	$C_9H_{19}COOH$	$CH_3(CH_2)_8COOH$
Lauric	Dodecanoic	44	$C_{11}H_{23}COOH$	$CH_3(CH_2)_{10}COOH$
Myristic	Tetradecanoic	54	$C_{13}H_{27}COOH$	$CH_3(CH_2)_{12}COOH$
Palmitic	Hexadecanoic	63	$C_{15}H_{31}COOH$	$CH_3(CH_2)_{14}COOH$
Stearic	Octadecanoic	70	$C_{17}H_{35}COOH$	$CH_3(CH_2)_{16}COOH$
Arachidic	Eicosanoic	77	$C_{19}H_{39}COOH$	$CH_3(CH_2)_{18}COOH$

Common Unsaturated Fatty Acids

Common Name	I.U.P.A.C. Name	Melting Point (°C)	RCOOH Formula	Number of Double Bonds	Position of Double Bonds
Palmitoleic	*cis*-9-Hexadecenoic	0	$C_{15}H_{29}COOH$	1	9
Oleic	*cis*-9-Octadecenoic	16	$C_{17}H_{33}COOH$	1	9
Linoleic	*cis,cis*-9,12-Octadecadienoic	5	$C_{17}H_{31}COOH$	2	9, 12
Linolenic	All *cis*-9,12,15-Octadecatrienoic	−11	$C_{17}H_{29}COOH$	3	9, 12, 15
Arachidonic	All *cis*-5,8,11,14-Eicosatetraenoic	−50	$C_{19}H_{31}COOH$	4	5, 8, 11, 14

Condensed Formula

Palmitoleic	$CH_3(CH_2)_5CH=CH(CH_2)_7COOH$
Oleic	$CH_3(CH_2)_7CH=CH(CH_2)_7COOH$
Linoleic	$CH_3(CH_2)_4CH=CH-CH_2-CH=CH(CH_2)_7COOH$
Linolenic	$CH_3CH_2CH=CH-CH_2-CH=CH-CH_2-CH=CH(CH_2)_7COOH$
Arachidonic	$CH_3(CH_2)_4CH=CH-CH_2-CH=CH-CH_2-CH=CH-CH_2-CH=CH-(CH_2)_3COOH$

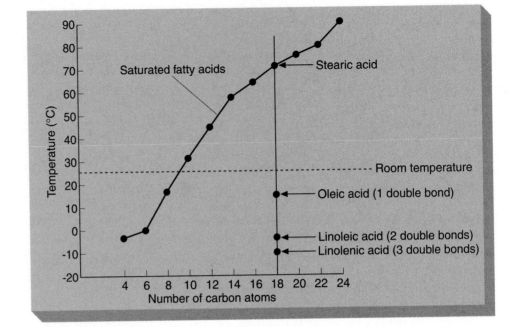

Figure 18.2
The melting points of fatty acids. Melting points of both saturated and unsaturated fatty acids increase as the number of carbon atoms in the chain increases. The melting points of unsaturated fatty acids are lower than those of the corresponding saturated fatty acid with the same number of carbon atoms. Also, as the number of double bonds in the chain increases, the melting points decrease.

The relationship between alkene chain length and melting point is described in Section 12.1.

"kinked" molecules cannot stack in an organized arrangement and thus have lower intermolecular attractions and lower melting points.

- As in the case for saturated fatty acids, the melting points of unsaturated fatty acids increase with increasing hydrocarbon chain length.

EXAMPLE 18.2

Examining the Similarities and Differences between Saturated and Unsaturated Fatty Acids

Construct a table comparing the structure and properties of saturated and unsaturated fatty acids.

Solution

Property	Saturated Fatty Acid	Unsaturated Fatty Acid
Chemical composition	Carbon, hydrogen, oxygen	Carbon, hydrogen, oxygen
Chemical structure	Hydrocarbon chain with a terminal carboxyl group	Hydrocarbon chain with a terminal carboxyl group
Carbon-carbon bonds within the hydrocarbon chain	Only C—C single bonds	At least one C—C double bond
Hydrocarbon chains are characteristic of what group of hydrocarbons	Alkanes	Alkenes
"Shape" of hydrocarbon chain	Linear, fully extended	Bend in carbon chain at site of C—C double bond
Physical state at room temperature	Solid	Liquid
Melting point for two fatty acids of the same hydrocarbon chain length	Higher	Lower
Relationship between melting point and chain length	Longer chain length, higher melting point	Longer chain length, higher melting point

Q u e s t i o n **18.1**

Draw formulas for each of the following fatty acids.
a. Oleic acid
b. Lauric acid
c. Linoleic acid
d. Stearic acid

Q u e s t i o n **18.2**

What is the I.U.P.A.C. name for each of the fatty acids in Question 18.1? (*Hint:* Review the naming of carboxylic acids in Section 15.1 and Table 18.1.)

Chemical Reactions of Fatty Acids

The reactions of fatty acids are identical to those of short-chain carboxylic acids. The major reactions that they undergo include esterification, acid hydrolysis of esters, saponification, and addition at the double bond.

Learning Goal

Esterification

In **esterification,** fatty acids react with alcohols to form esters and water according to the following general reaction:

Esterification is described in Sections 15.1 and 15.2.

$$\underset{\text{Acid}}{R^1-\overset{\overset{\displaystyle O}{\|}}{C}-OH} + \underset{\text{Alcohol}}{HOR^2} \xrightarrow{H^+,\ heat} \underset{\text{Ester}}{R^1-\overset{\overset{\displaystyle O}{\|}}{C}-OR^2} + \underset{\text{Water}}{H-OH}$$

Acid Hydrolysis

Recall that hydrolysis is the reverse of esterification, producing fatty acids from esters:

Acid hydrolysis is discussed in Section 15.2.

$$\underset{\text{Ester}}{R^1-\overset{\overset{\displaystyle O}{\|}}{C}-OR^2} + \underset{\text{Water}}{HO-H} \xrightarrow{H^+,\ heat} \underset{\text{Acid}}{R^1-\overset{\overset{\displaystyle O}{\|}}{C}-OH} + \underset{\text{Alcohol}}{R^2OH}$$

Saponification

Saponification is the base-catalyzed hydrolysis of an ester:

Saponification is described in Section 15.2.

$$\underset{\text{Ester}}{R^1-\overset{\overset{\displaystyle O}{\|}}{C}-OR^2} + \underset{\text{Base}}{NaOH} \longrightarrow \underset{\text{Salt}}{R^1-\overset{\overset{\displaystyle O}{\|}}{C}-O^-Na^+} + \underset{\text{Alcohol}}{R^2OH}$$

The product of this reaction, an ionized salt, is a soap. Because soaps have a long uncharged hydrocarbon tail and a negatively charged terminus (the carboxylate group), they form micelles that dissolve oil and dirt particles. Thus the dirt is emulsified, broken into small particles, and can be rinsed away.

The role of soaps in removal of dirt and grease is described in Section 15.2. Examples of micelles are shown in Figures 15.4 and 23.1.

Problems can arise when "hard" water is used for cleaning because the high concentrations of Ca^{2+} and Mg^{2+} in such water cause fatty acid salts to precipitate. Not only does this interfere with the emulsifying action of the soap, it also leaves a hard scum on the surface of sinks and tubs.

$$2R-\overset{\overset{\displaystyle O}{\|}}{C}-O^- + Ca^{2+} \longrightarrow (R-\overset{\overset{\displaystyle O}{\|}}{C}-O^-)_2Ca^{2+}(s)$$

Hydrogenation is discussed in Section 12.4.

Reaction at the Double Bond (Unsaturated Fatty Acids)

Hydrogenation is an example of an addition reaction. The following is a typical example of the addition of hydrogen to the double bonds of a fatty acid:

$$CH_3(CH_2)_4CH{=}CHCH_2CH{=}CH(CH_2)_7COOH \xrightarrow{\text{2H}_2,\text{ Ni}} CH_3(CH_2)_{16}COOH$$

Linoleic acid   Stearic acid

Hydrogenation is used in the food industry to convert polyunsaturated vegetable oils into saturated solid fats. *Partial hydrogenation* is carried out to add hydrogen to some, but not all, double bonds in polyunsaturated oils. In this way liquid vegetable oils are converted into solid form. Crisco is one example of a hydrogenated vegetable oil.

Margarine is also produced by partial hydrogenation of vegetable oils, such as corn oil or soybean oil. The extent of hydrogenation is carefully controlled so that the solid fat will be spreadable and have the consistency of butter when eaten. If too many double bonds were hydrogenated, the resulting product would have the undesirable consistency of animal fat. Artificial color is added to the product, and it may be mixed with milk to produce a butterlike appearance and flavor.

Hydrogenation of vegetable oils produces a mixture of *cis* and *trans* unsaturated fatty acids. The *trans* unsaturated fatty acids are thought to contribute to atherosclerosis (hardening of the arteries).

EXAMPLE 18.3

Writing Equations Representing the Chemical Reactions of Fatty Acids

Write an equation for each of the following reactions and indicate the I.U.P.A.C. names of each of the organic reactants and products:

a. The esterification of capric acid with propyl alcohol

Solution

$$CH_3(CH_2)_8{-}\overset{\displaystyle O}{\overset{\|}{C}}{-}OH + CH_3CH_2CH_2OH \xrightarrow{\text{H}^+,\text{ heat}}$$

Decanoic acid   Propanol

$$CH_3(CH_2)_8{-}\overset{\displaystyle O}{\overset{\|}{C}}{-}O{-}CH_2CH_2CH_3 + H_2O$$

Propyl decanoate

b. The acid hydrolysis of methyl decanoate

Solution

$$CH_3(CH_2)_8{-}\overset{\displaystyle O}{\overset{\|}{C}}{-}O{-}CH_3 + H_2O \xrightarrow{\text{H}^+,\text{ heat}} CH_3OH + CH_3(CH_2)_8{-}\overset{\displaystyle O}{\overset{\|}{C}}{-}OH$$

Methyl decanoate   Methanol   Decanoic acid

c. The base-catalyzed hydrolysis of ethyl dodecanoate

Solution

$$CH_3(CH_2)_{10}{-}\overset{\displaystyle O}{\overset{\|}{C}}{-}O{-}CH_2CH_3 + NaOH \longrightarrow$$

Ethyl dodecanoate

$$CH_3(CH_2)_{10}{-}\overset{\displaystyle O}{\overset{\|}{C}}{-}O^-Na^+ + CH_3CH_2OH$$

Sodium dodecanoate   Ethanol

d. Hydrogenation of oleic acid

Solution

$$CH_3(CH_2)_7CH{=}CH(CH_2)_7{-}\overset{\overset{\displaystyle O}{\|}}{C}{-}OH \xrightarrow{\text{H}_2,\ \text{Ni}} CH_3(CH_2)_{16}{-}\overset{\overset{\displaystyle O}{\|}}{C}{-}OH$$

cis-9-Octadecenoic acid   Octadecanoic acid

Write the complete equation for each of the following reactions.
a. Esterification of lauric acid and ethanol
b. Reaction of oleic acid with NaOH
c. Hydrogenation of arachidonic acid

Question 18.3

Write the complete equation for each of the following reactions.
a. Esterification of capric acid and 2-pentanol
b. Reaction of lauric acid with KOH
c. Hydrogenation of palmitoleic acid

Question 18.4

Eicosanoids: Prostaglandins, Leukotrienes, and Thromboxanes

Some of the unsaturated fatty acids containing more than one double bond cannot be synthesized by the body. For many years it has been known that linolenic acid and linoleic acid, called the **essential fatty acids,** are necessary for specific biochemical functions and must be supplied in the diet (see Table 18.1). The function of linoleic acid became clear in the 1960s when it was discovered that linoleic acid is required for the biosynthesis of **arachidonic acid,** the precursor of a class of hormonelike molecules known as **eicosanoids.** The name is derived from the Greek word *eikos*, meaning "twenty," because they are all derivatives of twenty-carbon fatty acids. The eicosanoids include three groups of structurally related compounds: the prostaglandins, the leukotrienes, and the thromboxanes.

> A hormone is a chemical signal that is produced by a specialized tissue and is carried by the bloodstream to target tissues. Eicosanoids are referred to as hormonelike because they affect the cells that produce them, as well as other target tissues.

Prostaglandins are extremely potent biological molecules with hormonelike activity. They received the name *prostaglandins* because they were originally isolated from seminal fluid produced in the prostate gland, but more recently they also have been isolated from most animal tissues. Prostaglandins are unsaturated carboxylic acids consisting of a twenty-carbon skeleton that contains a five-carbon ring.

Several general classes of prostaglandins are grouped under the designations A, B, E, and F, among others. The nomenclature of prostaglandins is based on the arrangement of the carbon skeleton and the number and orientation of double bonds, hydroxyl groups, and ketone groups. For example, in the name PGF_2, PG stands for prostaglandin, F indicates a particular group of prostaglandins with a hydroxyl group bonded to carbon-9, and 2 indicates that there are two carbon-carbon double bonds in the compound. The examples in Figure 18.3 illustrate the general structure of prostaglandins and the current nomenclature system.

Prostaglandins are made in most tissues, and exert their biological effects on the cells that produce them and on other cells in the immediate vicinity. The extraordinary range of prostaglandin functions includes

Learning Goal

5

- stimulation of smooth muscle,
- regulation of steroid biosynthesis,
- inhibition of gastric secretion,
- inhibition of hormone-sensitive lipases,
- inhibition of platelet aggregation,

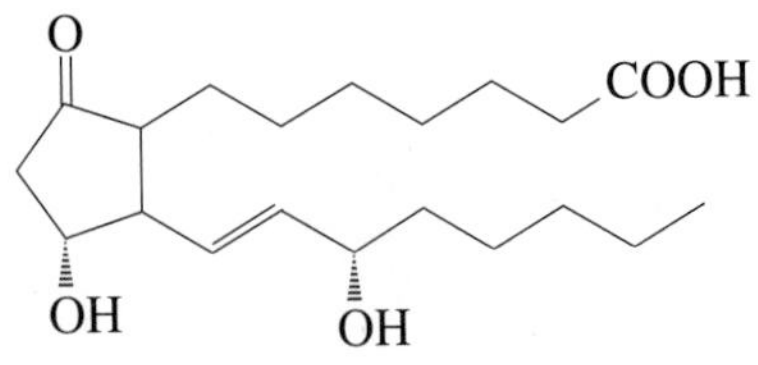

Prostaglandin E$_1$

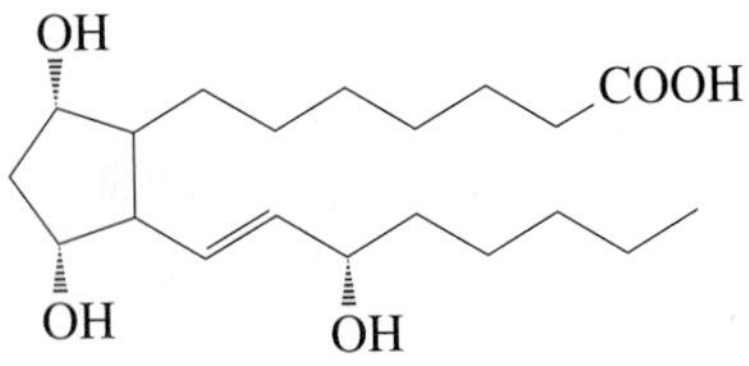

Prostaglandin F$_1$

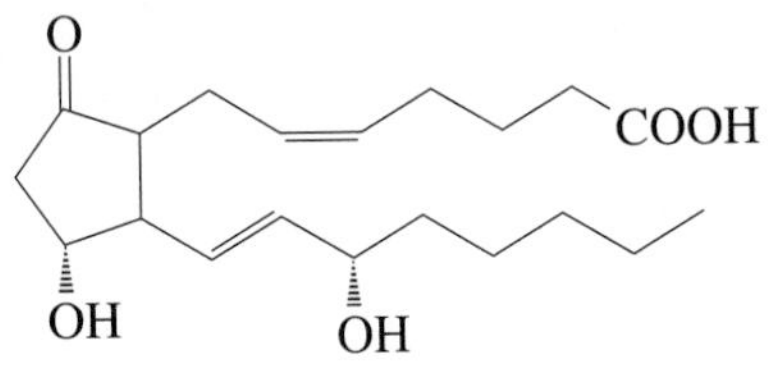

Prostaglandin E$_2$

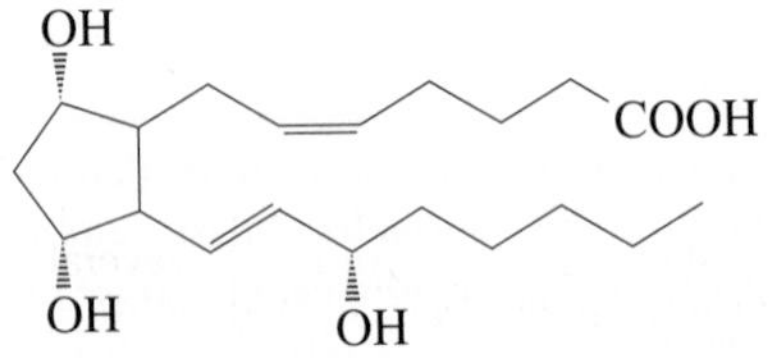

Prostaglandin F$_2$

Figure 18.3

The structures of four prostaglandins.

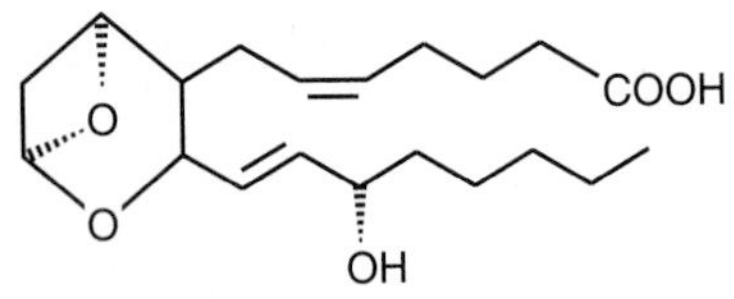

Thromboxane A$_2$

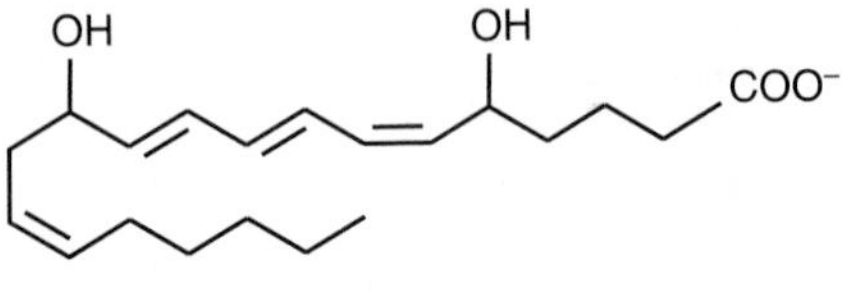

Leukotriene B$_4$

Figure 18.4

The structures of thromboxane A$_2$ and leukotriene B$_4$.

- stimulation of platelet aggregation,
- regulation of nerve transmission,
- sensitization to pain, and
- mediation of the inflammatory response.

Because the prostaglandins and the closely related leukotrienes and thromboxanes affect so many body processes and because they often cause opposing effects in different tissues, it can be difficult to keep track of their many regulatory functions. The following is a brief summary of some of the biological processes that are thought to be regulated by the prostaglandins, leukotrienes, and thromboxanes.

1. **Blood clotting.** Blood clots form when a blood vessel is damaged, yet such clotting along the walls of undamaged vessels could result in heart attack or stroke. *Thromboxane A$_2$* (Figure 18.4) is produced by platelets in the blood and stimulates constriction of the blood vessels and aggregation of the platelets. Conversely, PGI$_2$ (prostacyclin) is produced by the cells lining the blood vessels and has precisely the opposite effect of thromboxane A$_2$. Prostacyclin inhibits platelet aggregation and causes dilation of blood vessels and thus prevents the untimely production of blood clots.

2. **The inflammatory response.** The inflammatory response is another of the body's protective mechanisms. When tissue is damaged by mechanical injury, burns, or invasion by microorganisms, a variety of white blood cells descend on the damaged site to try to minimize the tissue destruction. The result of this response is swelling, redness, fever, and pain. Prostaglandins are thought to promote certain aspects of the inflammatory response, especially pain and fever. Drugs such as aspirin block prostaglandin synthesis and help to relieve the symptoms. We will examine the mechanism of action of these drugs later in this section.

3. **Reproductive system.** PGE$_2$ stimulates smooth muscle contraction, particularly uterine contractions. An increase in the level of prostaglandins has been noted immediately before the onset of labor. PGE$_2$ has also been used to induce second trimester abortions. There is strong evidence that dysmenorrhea (painful menstruation) suffered by many women may be the result of an excess of two prostaglandins. Indeed, drugs, such as ibuprofen, that inhibit prostaglandin synthesis have been approved by the FDA and are found to provide virtually complete relief from these symptoms.

4. **Gastrointestinal tract.** Prostaglandins have been shown to both inhibit the secretion of acid and increase the secretion of a protective mucus layer into the stomach. In this way, prostaglandins help to protect the stomach lining. Consider for a moment the possible side effect that prolonged use of a drug such as aspirin might have on the stomach—ulceration of the stomach lining. Because aspirin inhibits prostaglandin synthesis, it may actually encourage stomach ulcers by inhibiting the formation of the normal protective mucus layer, while simultaneously allowing increased secretion of stomach acid.

5. **Kidneys.** Prostaglandins produced in the kidneys cause the renal blood vessels to dilate. The greater flow of blood though the kidney results in increased water and electrolyte excretion.

6. **Respiratory tract.** Eicosanoids produced by certain white blood cells, the *leukotrienes* (see Figure 18.4), promote the constriction of the bronchi associated with asthma. Other prostaglandins promote bronchodilation.

As this brief survey suggests, the prostaglandins have numerous, often antagonistic effects. Although they do not fit the formal definition of a hormone (a substance produced in a specialized tissue and transported by the circulatory system to target tissues *elsewhere* in the body), the prostaglandins are clearly strong biological regulators with far-reaching effects.

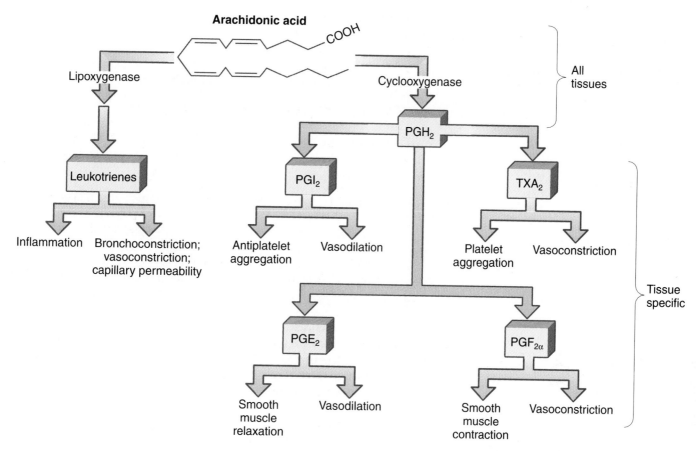

Figure 18.5

A summary of the synthesis of several prostaglandins from arachidonic acid.

As mentioned, prostaglandins stimulate the inflammatory response and, as a result, are partially responsible for the cascade of events that cause pain. Aspirin has long been known to alleviate such pain, and we now know that it does so by inhibiting the synthesis of prostaglandins.

The first two steps of prostaglandin synthesis (Figure 18.5), the release of arachidonic acid from the membrane and its conversion to PGH_2 by the enzyme cyclooxygenase, occur in all tissues that are able to produce prostaglandins. The conversion of PGH_2 into the other biologically active forms is tissue specific and requires the appropriate enzymes, which are found only in certain tissues.

Aspirin works by inhibiting the cyclooxygenase, which catalyzes the first step in the pathway leading from arachidonic acid to PGH_2. The acetyl group of aspirin becomes covalently bound to the enzyme, thereby inactivating it (Figure 18.6). Because the reaction catalyzed by cyclooxygenase occurs in all cells, aspirin effectively inhibits synthesis of all of the prostaglandins.

Learning Goal

18.3 Glycerides

Neutral Glycerides

Glycerides are lipid esters that contain the glycerol molecule and fatty acids. They may be subdivided into two classes: neutral glycerides and phosphoglycerides. Neutral glycerides are nonionic and nonpolar. Phosphoglyceride molecules have a polar region, the phosphoryl group, in addition to the nonpolar fatty acid tails. The structures of each of these types of glycerides are critical to their function.

The esterification of glycerol with a fatty acid produces a **neutral glyceride.** Esterification may occur at one, two, or all three positions, producing **monoglycerides,**

diglycerides, or **triglycerides.** You will also see these referred to as *mono-, di-,* or *tri-acylglycerols.*

EXAMPLE 18.4

Writing an Equation for the Synthesis of a Monoglyceride

Write a general equation for the esterification of glycerol and one fatty acid.

Solution

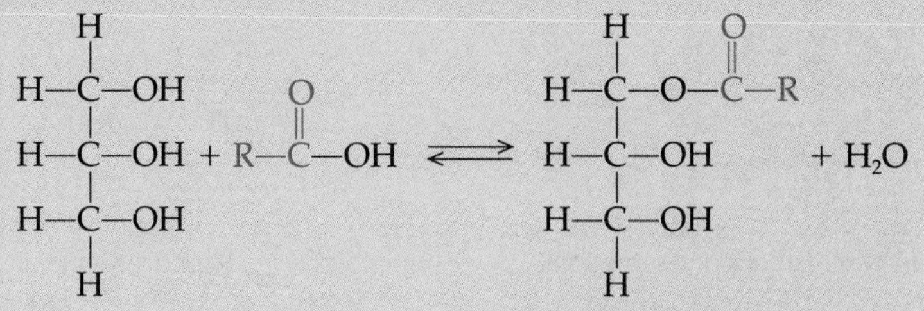

Glycerol Fatty acid Monoglyceride Water

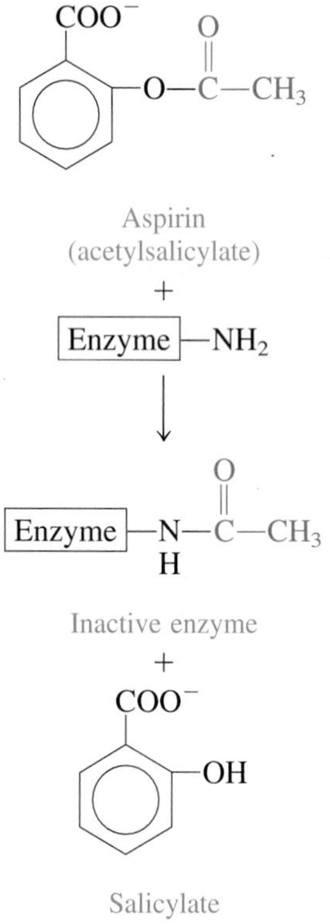

Figure 18.6
Aspirin inhibits the synthesis of prostaglandins by acetylating the enzyme cyclooxygenase. The acetylated enzyme is no longer functional.

Although monoglycerides and diglycerides are present in nature, the most important neutral glycerides are the triglycerides, the major component of fat cells. The triglyceride consists of a glycerol backbone (shown in black) joined to three fatty acid units through ester bonds (shown in red). The formation of a triglyceride is shown in the following reaction:

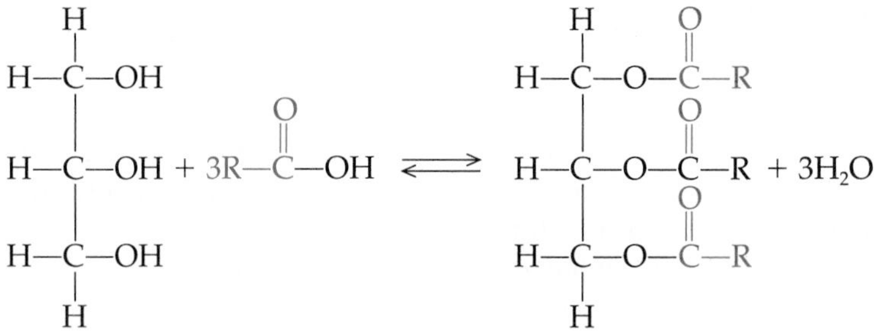

Glycerol Fatty acids Trigylceride Water

Triglycerides (triacylglycerols) are named by using the "backbone" name, *glycerol,* as the suffix. The name(s) of the fatty acyl group(s) are placed before it. The fatty acyl group is named by dropping the ending *-ic acid* and replacing it with the ending *-oyl.* In the examples below, the name of the triglyceride on the left is tristearoylglycerol. The prefix *tri-* tells us that there are three stearic acyl groups attached to the glycerol backbone. The triglyceride on the right is a *mixed triglyceride;* that is, there are three different fatty acyl groups attached to the glycerol backbone. They are listed according to their placement along the glycerol backbone.

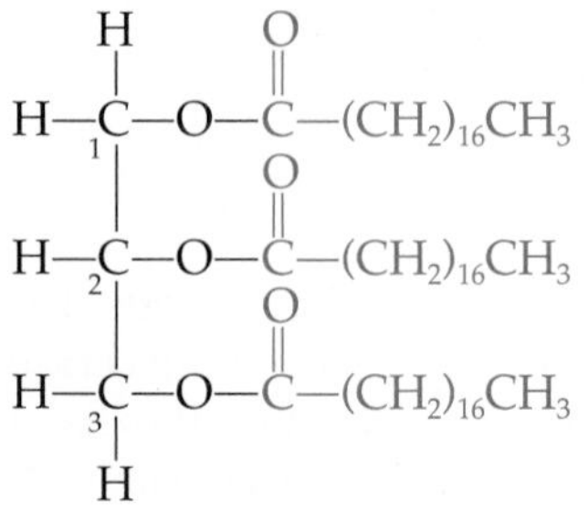

Tristearoylglycerol

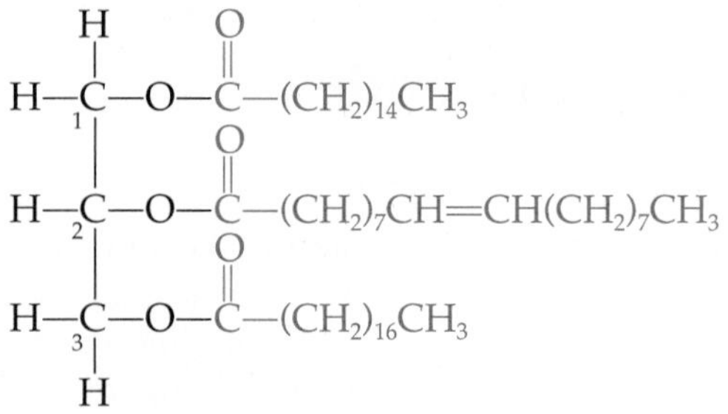

1-Palmitoyl-2-oleoyl-3-stearoylglycerol

Because there are no charges (+ or −) on these molecules, they are called *neutral glycerides*. These long molecules readily stack with one another and constitute the majority of the lipids stored in the body's fat cells.

The principal function of triglycerides in biochemical systems is the storage of energy. If more energy-rich nutrients are consumed than are required for metabolic processes, much of the excess is converted to neutral glycerides and stored as triglycerides in fat cells of *adipose tissue.* When energy is needed, the triglycerides are metabolized by the body, and energy is released. For this reason, exercise, along with moderate reduction in caloric intake, is recommended for overweight individuals. Exercise, an energy-demanding process, increases the rate of metabolism of fats and results in weight loss.

Lipid metabolism is discussed in Chapter 23.

See "A Human Perspective: Losing Those Unwanted Pounds of Adipose Tissue" in Chapter 23.

Phosphoglycerides

Phospholipids are a group of lipids that are phosphate esters. The presence of the phosphoryl group results in a molecule with a polar head (the phosphoryl group) and a nonpolar tail (the alkyl chain of the fatty acid). Because the phosphoryl group ionizes in solution, a charged lipid results.

The most abundant membrane lipids are derived from glycerol-3-phosphate and are known as **phosphoglycerides.** Phosphoglycerides contain acyl groups derived from long-chain fatty acids at C-1 and C-2 of glycerol-3-phosphate. At C-3 the phosphoryl group is joined to glycerol by a phosphoester bond. The simplest phosphoglyceride contains a free phosphoryl group and is known as a **phosphatidate** (Figure 18.7). When the phosphoryl group is attached to another hydrophilic molecule, a more complex phosphoglyceride is formed. For example, *phosphatidylcholine (lecithin)* and *phosphatidylethanolamine (cephalin)* are found in the membranes of most cells (Figure 18.7).

Lecithin possesses a polar "head" and a nonpolar "tail." Thus, it is an *amphipathic* molecule. This structure is similar to that of soap and detergent molecules, discussed earlier. The ionic "head" is hydrophilic and interacts with water molecules, whereas the nonpolar "tail" is hydrophobic and interacts with nonpolar molecules. This amphipathic nature is central to the structure and function of cell membranes.

In addition to being a component of cell membranes, lecithin is the major phospholipid in pulmonary surfactant. It is also found in egg yolks and soybeans and is used as an emulsifying agent in ice cream. An **emulsifying agent** aids in the suspension of triglycerides in water. The amphipathic lecithin serves as a bridge, holding together the highly polar water molecules and the nonpolar triglycerides. Emulsification occurs because the hydrophilic head of lecithin dissolves in water and its hydrophobic tail dissolves in the triglycerides.

Cephalin is similar in general structure to lecithin; the amine group bonded to the phosphoryl group is the only difference.

Learning Goal

7

Phosphoesters are described in Section 15.4.

See the "Chemistry Connection: Life-Saving Lipids," at the beginning of this chapter.

Question 18.5

Using condensed formulas, draw the mono-, di-, and triglycerides that would result from the esterification of glycerol with each of the following acids.
a. Oleic acid
b. Capric acid
c. Palmitic acid
d. Lauric acid

Question 18.6

Name the triglycerides that are produced in the reactions discussed in Question 18.5.

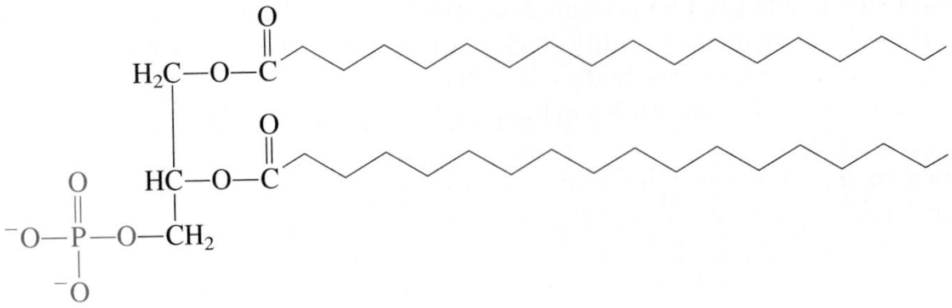

Phosphatidate

(a)

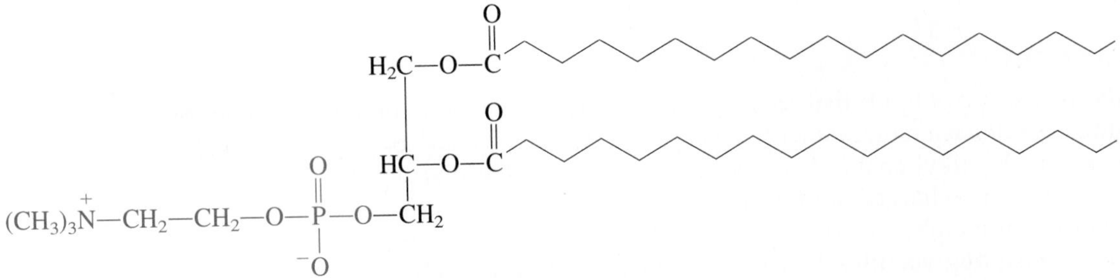

Phosphatidylcholine (lecithin)

(b)

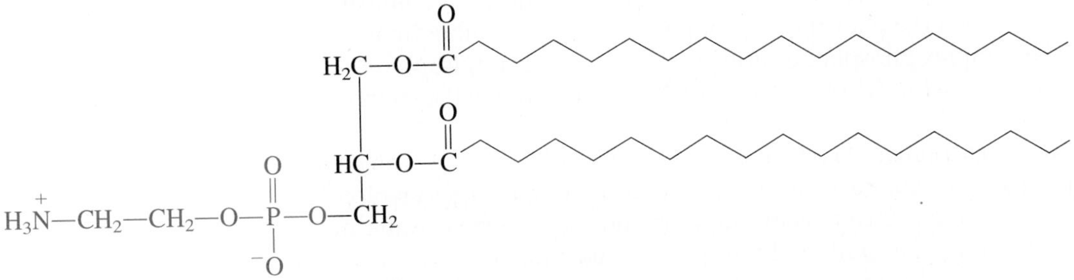

Phosphatidylethanolamine (cephalin)

(c)

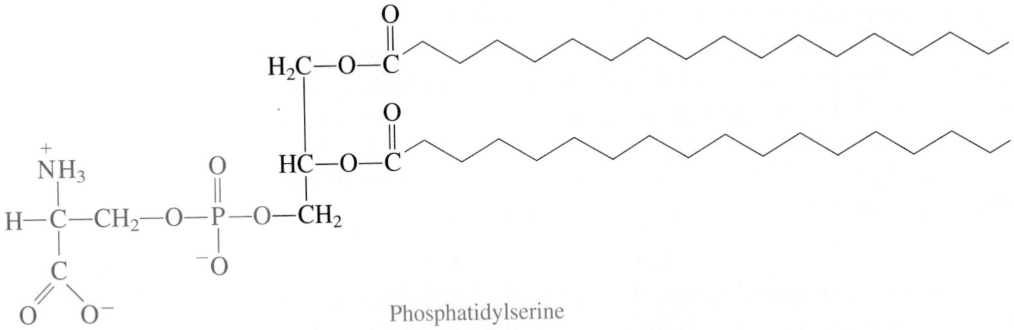

Phosphatidylserine

(d)

Figure 18.7

The structures of (a) phosphatidate and the common membrane phospholipids, (b) phosphatidylcholine (lecithin), (c) phosphatidylethanolamine (cephalin), and (d) phosphatidyl serine.

18.4 Nonglyceride Lipids

Sphingolipids

Sphingolipids are lipids that are not derived from glycerol. Like phospholipids, sphingolipids are amphipathic, having a polar head group and two nonpolar fatty acid tails, and are structural components of cellular membranes. They are derived from sphingosine, a long-chain, nitrogen-containing (amino) alcohol:

Learning Goal
8

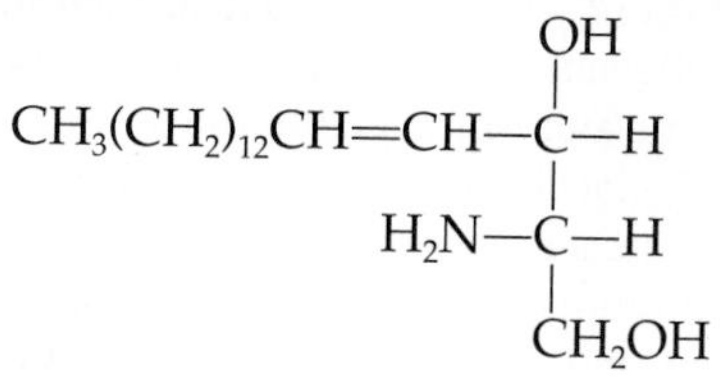

$$CH_3(CH_2)_{12}CH\!=\!CH\!-\!\overset{\overset{\displaystyle OH}{|}}{C}\!-\!H$$
$$H_2N\!-\!\overset{|}{C}\!-\!H$$
$$\overset{|}{C}H_2OH$$

Sphingosine

The sphingolipids include the sphingomyelins and the glycosphingolipids. The **sphingomyelins** are the only class of sphingolipids that are also phospholipids:

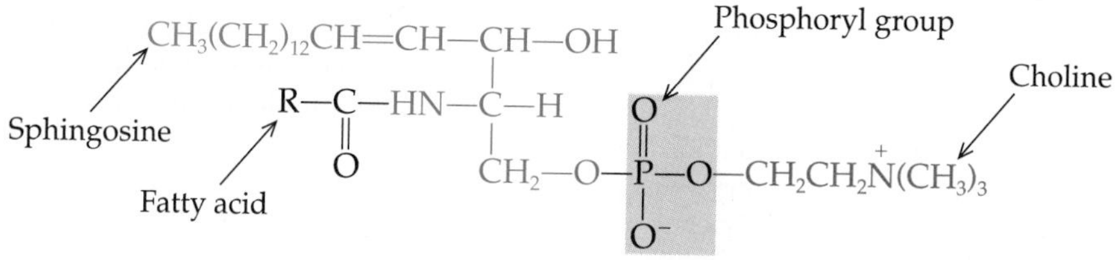

Sphingosine

Fatty acid

$$CH_3(CH_2)_{12}CH\!=\!CH\!-\!CH\!-\!OH$$
$$R\!-\!\overset{}{\underset{\overset{\|}{O}}{C}}\!-\!HN\!-\!\overset{|}{C}\!-\!H$$
$$CH_2\!-\!O\!-\!\overset{\overset{\displaystyle O}{\|}}{\underset{\underset{\displaystyle O^-}{|}}{P}}\!-\!O\!-\!CH_2CH_2\overset{+}{N}(CH_3)_3$$

Phosphoryl group

Choline

Sphingomyelin

Sphingomyelins are located throughout the body, but are particularly important structural lipid components of nerve cell membranes. They are found in abundance in the myelin sheath that surrounds and insulates cells of the central nervous system. In humans, about 25% of the lipids of the myelin sheath are sphingomyelins. Their role is essential to proper cerebral function and nerve transmission.

Glycosphingolipids, or *glycolipids,* include the cerebrosides, sulfatides, and gangliosides and are built on a ceramide backbone structure, which is a fatty acid amide derivative of sphingosine:

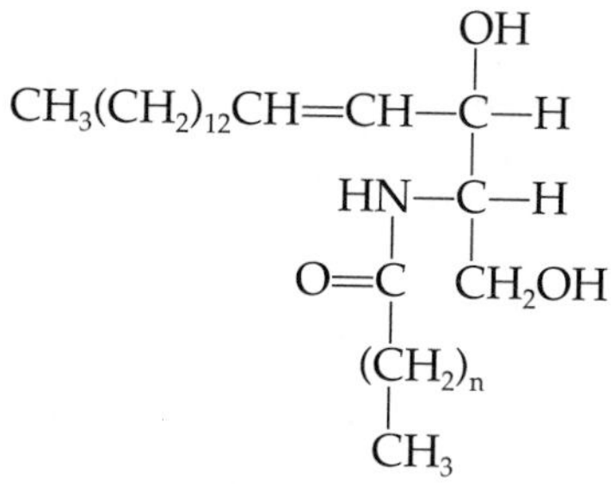

$$CH_3(CH_2)_{12}CH\!=\!CH\!-\!\overset{\overset{\displaystyle OH}{|}}{C}\!-\!H$$
$$HN\!-\!\overset{|}{C}\!-\!H$$
$$O\!=\!\overset{|}{C}\quad CH_2OH$$
$$(\overset{|}{C}H_2)_n$$
$$\overset{|}{C}H_3$$

Ceramide

The *cerebrosides* are characterized by the presence of a single monosaccharide head group. Two common cerebrosides are glucocerebroside, found in the membranes of macrophages (cells that protect the body by ingesting and destroying foreign microorganisms) and galactocerebroside, found almost exclusively in the membranes of brain cells. Glucocerebroside consists of ceramide bonded to the

A CLINICAL PERSPECTIVE

Disorders of Sphingolipid Metabolism

There are a number of human genetic disorders that are caused by a deficiency in one of the enzymes responsible for the breakdown of sphingolipids. In general, the symptoms are caused by the accumulation of abnormally large amounts of these lipids within particular cells. It is interesting to note that three of these diseases, Niemann-Pick disease, Gaucher's disease, and Tay-Sachs disease are found much more frequently among Ashkenazi Jews of Northern European heritage than among other ethnic groups.

Of the four subtypes of Niemann-Pick disease, type A is the most severe. It is inherited as a recessive disorder (i.e., a defective copy of the gene must be inherited from each parent) that results in an absence of the enzyme sphingomyelinase. The absence of this enzyme causes the storage of large amounts of sphingomyelin and cholesterol in the brain, bone marrow, liver, and spleen.

Symptoms may begin when the baby is only a few months old. The parents may notice a delay in motor development and/or problems with feeding. Although the infant may develop some motor skills, they quickly begin to regress as the child loses muscle strength and tone, as well as vision and hearing. The disease progresses rapidly and the child typically dies within the first few years of life.

Tay-Sachs disease is a lipid storage disease caused by an absence of the enzyme hexosaminidase, which functions in ganglioside metabolism. As a result of the enzyme deficiency, the ganglioside, shown in Section 18.4, accumulates in the cells of the brain causing neurological deterioration. Like Niemann-Pick disease, it is an autosomal recessive genetic trait that becomes apparent in the first few months of the life of an infant and rapidly progresses to death within a few years. Symptoms include listlessness, irritability, seizures, paralysis, loss of muscle tone and function, blindness, deafness, and delayed mental and social skills.

Gaucher's disease is an autosomal recessive genetic disorder resulting in a deficiency of the enzyme glucocerebrosidase. In the normal situation, this enzyme breaks down glucocerebroside, which is an intermediate in the synthesis and degradation of complex glycosphingolipids found in cellular membranes. In Gaucher's disease, glucocerebroside builds up in macrophages found in the liver, spleen, and bone marrow. These cells become engorged with excess lipid and displace healthy, normal cells in bone marrow. The symptoms of Gaucher's disease include severe anemia, thrombocytopenia (reduction in the number of platelets), and hepatosplenomegaly (enlargement of the spleen and liver). There can also be skeletal problems including bone deterioration and secondary fractures.

Fabry's disease is an X-linked inherited disorder caused by the deficiency of the enzyme α-galactosidase A. This disease afflicts as many as 50,000 people worldwide. Typically, symptoms, including pain in the fingers and toes and a red rash around the waist, begin to appear when individuals reach their early twenties. A preliminary diagnosis can be confirmed by determining the concentration of the enzyme α-galactosidase A. Patients with Fabry's disease have an increased risk of kidney and heart disease, and a reduced life expectancy. Because this is an X-linked disorder, it is more common among males than females.

hexose glucose; galactocerebroside consists of ceramide joined to the monosaccharide galactose.

Glucocerebroside

Galactocerebroside

Sulfatides are derivatives of galactocerebroside that contain a sulfate group. Notice that they carry a negative charge at physiological pH.

A sulfatide of galactocerebroside

Gangliosides are glycolipids that possess oligosaccharide groups, including one or more molecules of N-acetylneuraminic acid (sialic acid). First isolated from membranes of nerve tissue, gangliosides are found in most tissues of the body.

A ganglioside associated with Tay-Sachs disease

Learning Goal

Lipid digestion is described in Section 23.1.

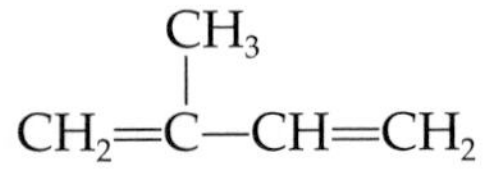

$$CH_2{=}\overset{\overset{\displaystyle CH_3}{|}}{C}{-}CH{=}CH_2$$

Isoprene

The structure and function of the lipid-soluble vitamins are found in Appendix E, "Lipid-Soluble Vitamins."

Steroids

Steroids are a naturally occurring family of organic molecules of biochemical and medical interest. A great deal of controversy has surrounded various steroids. We worry about the amount of cholesterol in the diet and the possible health effects. We are concerned about the use of anabolic steroids by athletes wishing to build muscle mass and improve their performance. However, members of this family of molecules derived from cholesterol have many important functions in the body. The bile salts that aid in the emulsification and digestion of lipids are steroid molecules, as are the sex hormones testosterone and estrone.

The steroids are members of a large, diverse collection of lipids called the *isoprenoids*. All of these compounds are built from one or more five-carbon units called *isoprene*.

Terpene is the general term for lipids that are synthesized from isoprene units. Examples of terpenes include the steroids and bile salts, the lipid-soluble vitamins, chlorophyll, and certain plant hormones.

All steroids contain the steroid nucleus (steroid carbon skeleton) as shown here:

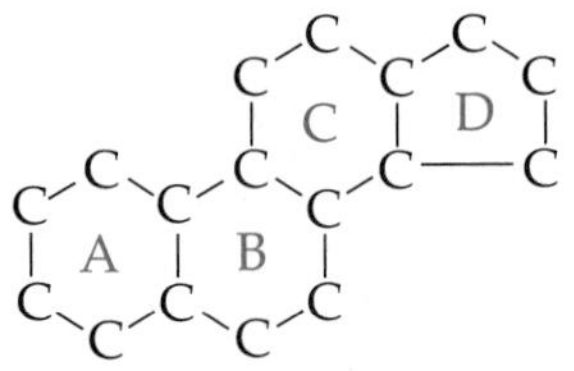

Carbon skeleton of
the steroid nucleus

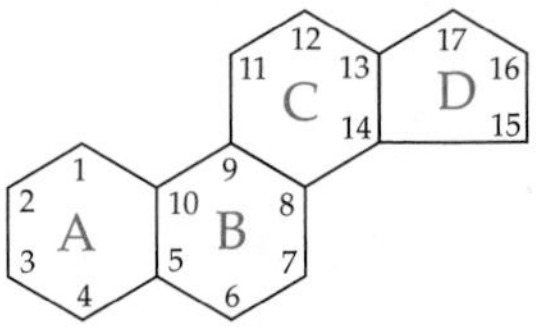

Steroid nucleus

The steroid carbon skeleton consists of four fused rings. Each ring pair has two carbons in common. Thus two fused rings share one or more common bonds as part of their ring backbones. For example, rings A and B, B and C, and C and D are all fused in the preceding structure. Many steroids have methyl groups attached to carbons 10 and 13, as well as an alkyl, alcohol, or ketone group attached to carbon-17.

Cholesterol, a common steroid, is found in the membranes of most animal cells. It is an amphipathic molecule and is readily soluble in the hydrophobic region of membranes. It is involved in regulation of the fluidity of the membrane as a result of the nonpolar fused ring. However, the hydroxyl group is polar and functions like the polar heads of sphingolipids and phospholipids. There is a strong correlation between the concentration of cholesterol found in the blood plasma and heart disease, particularly **atherosclerosis** (hardening of the arteries). Cholesterol, in combination with other substances, contributes to a narrowing of the artery passageway. As narrowing increases, more pressure is necessary to ensure adequate blood flow, and high blood pressure (*hypertension*) develops. Hypertension is also linked to heart disease.

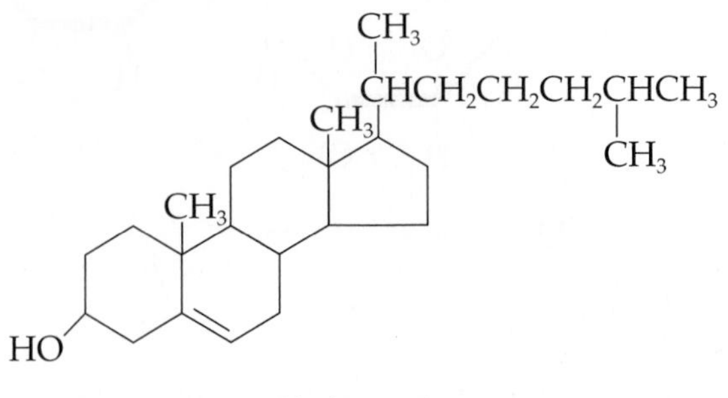

Cholesterol

Egg yolks contain a high concentration of cholesterol, as do many dairy products and animal fats. As a result, it has been recommended that the amounts of these products in the diet be regulated to moderate the dietary intake of cholesterol.

A HUMAN PERSPECTIVE

Anabolic Steroids and Athletics

In the 1988 Summer Olympics, Ben Johnson of Canada ran the fastest 100-meter race in history, 9.79 seconds, and was awarded the Gold Medal. Little more than two days later, Michele Verdier of the International Olympic Committee stood at a press conference and read the following statement: "The urine sample of Ben Johnson, Canada, Athletics, 100 meters, collected Saturday, 24th September 1988, was found to contain metabolites of a banned substance, namely stanozolol, an anabolic steroid." Johnson was disqualified, and Carl Lewis of the United States became the Olympic Gold Medalist in the 100-meter race.

Why do athletes competing in power sports take anabolic steroids? Use of anabolic steroids has a number of desirable effects for the athlete. First, they help build the muscle mass needed to succeed in sprints or weight lifting. They hasten the healing of muscle damage caused by the intense training of the competitive athlete. Finally, anabolic steroids may help the athlete maintain an aggressive attitude, not just during the competition but also throughout training.

If these hormones have such beneficial effects, why not allow all athletes to use them? Unfortunately, the beneficial effects are far outweighed by the negative side effects. These include kidney and liver damage, stroke, impotence and infertility, an increase in cardiovascular disease, and extremely aggressive behavior.

Even though these life-threatening side effects are well known, athletes continue to use anabolic steroids. The temptation must have been too great for Ben Johnson. After a period of suspension from amateur athletics, Johnson again entered competition. In March 1993, testing before a track meet revealed that he had used anabolic steroids to enhance his performance. As a result, he was forever banned from amateur competition.

Stanozolol

Bile salts are amphipathic derivatives of cholesterol that are synthesized in the liver and stored in the gallbladder. The principal bile salts in humans are cholate and chenodeoxycholate.

Bile salts are described in greater detail in Section 23.1.

Cholate

Chenodeoxycholate

Bile salts are emulsifying agents whose polar hydroxyl groups interact with water and whose hydrophobic regions bind to lipids. Following a meal, bile flows from the gallbladder to the duodenum (the uppermost region of the small intestine). Here the bile salts emulsify dietary fats into small droplets that can be more readily digested by lipases (lipid digesting enzymes) also found in the small intestine.

Steroids play a role in the reproductive cycle. In a series of chemical reactions, cholesterol is converted to the steroid *progesterone,* the most important hormone associated with pregnancy. Produced in the ovaries and in the placenta, progesterone is responsible for both the successful initiation and the successful completion of pregnancy. It prepares the lining of the uterus to accept the fertilized egg. Once the egg is attached, progesterone is involved in the development of the fetus and plays a role in the suppression of further ovulation during pregnancy.

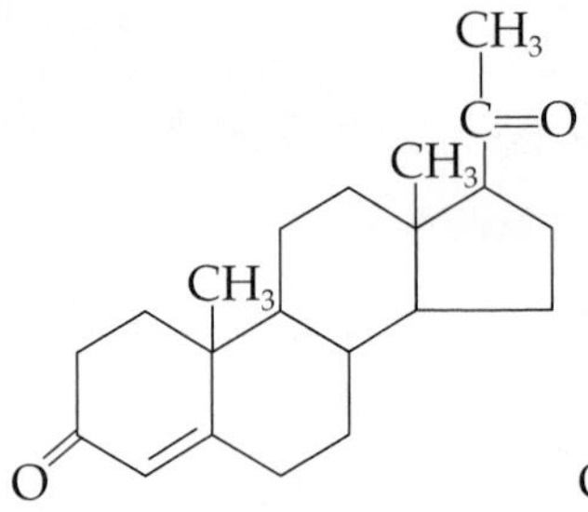

Progesterone

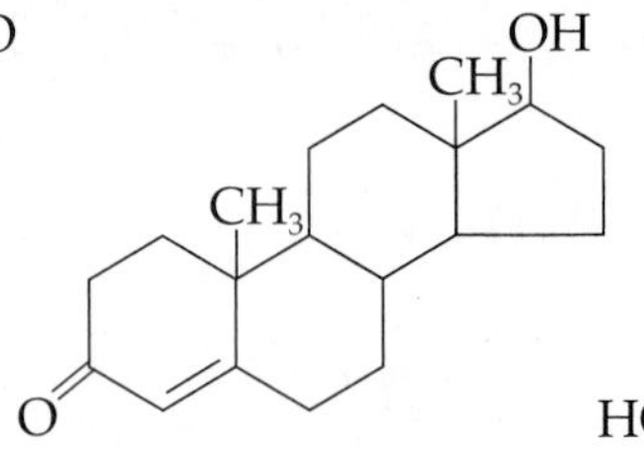

Testosterone

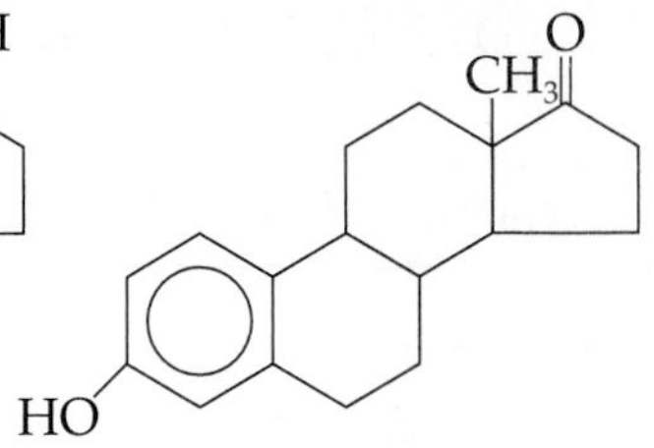

Estrone

19-Norprogesterone

Testosterone, a male sex hormone found in the testes, and *estrone*, a female sex hormone, are both produced by the chemical modification of progesterone. These hormones are involved in the development of male and female sex characteristics.

Many steroids, including progesterone, have played important roles in the development of birth control agents. 19-Norprogesterone was one of the first synthetic birth control agents. It is approximately ten times as effective as progesterone in providing birth control. However, its utility was severely limited because this compound could not be administered orally and had to be taken by injection. A related compound, norlutin (chemical name: 17-α-ethynyl-19-nortestosterone), was found to provide both the strength and the effectiveness of 19-norprogesterone and could be taken orally.

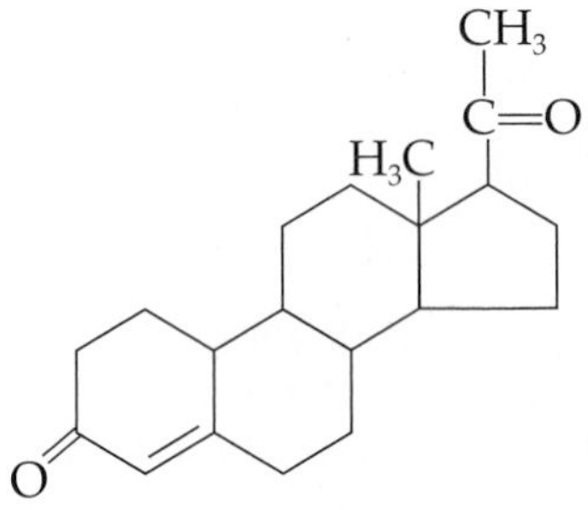

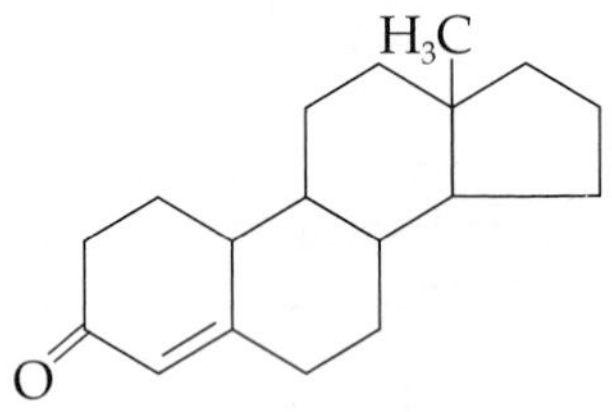

Norlutin

Currently "combination" oral contraceptives are prescribed most frequently. These include a progesterone and an estrogen. These newer products confer better contraceptive protection than either agent administered individually. They are also used to regulate menstruation in patients with heavy menstrual bleeding. First investigated in the late 1950s and approved by the FDA in 1961, there are at least 30 combination pills currently available. In addition, a transdermal patch for the treatment of postmenopausal osteoporosis is being investigated.

All of these compounds act by inducing a false pregnancy, which prevents ovulation. When oral contraception is discontinued, ovulation usually returns within three menstrual cycles. Although there have been problems associated with "the pill," it appears to be an effective and safe method of family planning for much of the population.

Cortisone is also important to the proper regulation of a number of biochemical processes. For example, it is involved in the metabolism of carbohydrates. Cortisone is also used in the treatment of rheumatoid arthritis, asthma, gastrointestinal disorders, many skin conditions, and a variety of other diseases. However, treatment with cortisone is not without risk. Some of the possible side effects of cortisone therapy include fluid retention, sodium retention, and potassium loss that can lead to congestive heart failure. Other side effects include muscle weakness, osteoporosis, gastrointestinal upsets including peptic ulcers, and neurological symptoms, including vertigo, headaches, and convulsions.

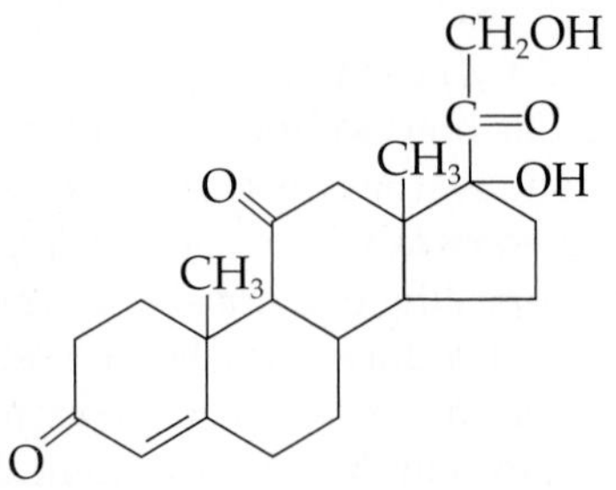

Cortisone

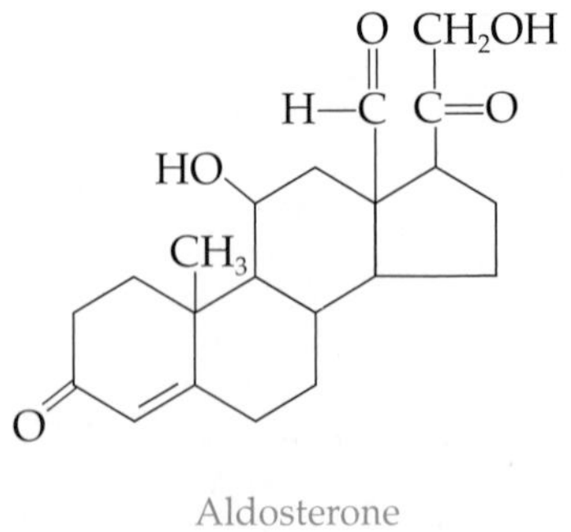

Aldosterone

Aldosterone is a steroid hormone produced by the adrenal cortex and secreted into the bloodstream when blood sodium ion levels are too low. Upon reaching its

Steroids and the Treatment of Heart Disease

The foxglove plant (*Digitalis purpurea*) is an herb that produces one of the most powerful known stimulants of heart muscle. The active ingredients of the foxglove plant (digitalis) are the so-called cardiac glycosides or *cardiotonic steroids*, which include digitoxin, digosin, and gitalin.

Digitalis purpurea, the foxglove plant.

Digitoxin

The structure of digitoxin, one of the cardiotonic steroids produced by the foxglove plant.

These drugs are used clinically in the treatment of congestive heart failure, which results when the heart is not beating with strong, efficient strokes. When the blood is not propelled through the cardiovascular system efficiently, fluid builds up in the lungs and lower extremities (edema). The major symptoms of congestive heart failure are an enlarged heart, weakness, edema, shortness of breath, and fluid accumulation in the lungs.

This condition was originally described in 1785 by a physician, William Withering, who found a peasant woman whose folk medicine was famous as a treatment for chronic heart problems. Her potion contained a mixture of more than twenty herbs, but Dr. Withering, a botanist as well as physician, quickly discovered that foxglove was the active ingredient in the mixture. Withering used *Digitalis purpurea* successfully to treat congestive heart failure and even described some cautions in its use.

The cardiotonic steroids are extremely strong heart stimulants. A dose as low as 1 mg increases the stroke volume of the heart (volume of blood per contraction), increases the strength of the contraction, and reduces the heart rate. When the heart is pumping more efficiently because of stimulation by digitalis, the edema disappears.

Digitalis can be used to control congestive heart failure, but the dose must be carefully determined and monitored because the therapeutic dose is close to the dose that causes toxicity. The symptoms that result from high body levels of cardiotonic steroids include vomiting, blurred vision and lightheadedness, increased water loss, convulsions, and death. Only a physician can determine the initial dose and maintenance schedule for an individual to control congestive heart failure and yet avoid the toxic side effects.

target tissues in the kidney, aldosterone activates a set of reactions that cause sodium ions and water to be returned to the blood. If sodium levels are elevated, aldosterone is not secreted from the adrenal cortex and the sodium ions filtered out of the blood by the kidney will be excreted.

Draw the structure of the steroid nucleus. Note the locations of the A, B, C, and D steroid rings.

Question 18.7

What is meant by the term *fused ring*?

Question 18.8

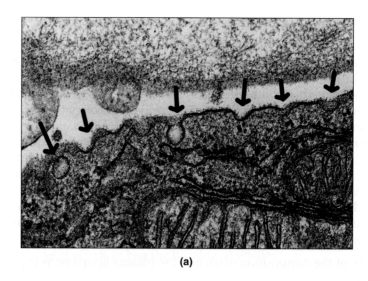

(a)

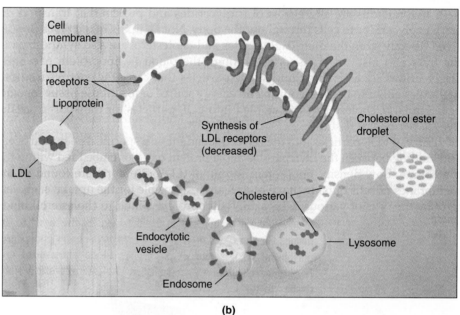

(b)

Figure 18.10

Receptor-mediated endocytosis.
(a) Electron micrographs of the process
of receptor-mediated endocytosis.
(b) Summary of the events of receptor-
mediated endocytosis of LDL.

specifically to the LDL receptor, and the complex is taken into the cell by a process
called *receptor-mediated endocytosis* (Figure 18.10). The membrane begins to invagi-
nate; in other words, it is pulled into the cell at the site of the LDL receptor complexes. This draws the entire LDL particle into the cell. Eventually, the invagination
becomes very deep, and the portion of the membrane surrounding the LDL particles pinches away from the cell membrane and forms a membrane around the LDL
particles. As we will see in Section 18.6, membranes are fluid and readily flow. Thus
they can form a vesicle or endosome containing the LDL particles.

Cellular digestive organelles known as *lysosomes* fuse with the endosomes.
This fusion is accomplished when the membranes of the endosome and the lysosome flow together to create one larger membrane-bound body or vesicle. Hydrolytic enzymes from the lysosome then digest the entire complex to release
cholesterol into the cytoplasm of the cell. There, cholesterol inhibits its own
biosynthesis and activates an enzyme that stores cholesterol in cholesterol ester
droplets. High concentrations of cholesterol inside the cell also inhibit the synthesis of LDL receptors to ensure that the cell will not take up too much cholesterol.
People who have a genetic defect in the gene coding for the LDL receptor do not

take up as much cholesterol. As a result they accumulate LDL cholesterol in the plasma. This excess plasma cholesterol is then deposited on the artery walls, causing atherosclerosis. This disease is called *hypercholesterolemia*.

Liver lipoprotein receptors enable large amounts of cholesterol to be removed from the blood, thus ensuring low concentrations of cholesterol in the blood plasma. Other factors being equal, the person with the most lipoprotein receptors will be the least vulnerable to a high-cholesterol diet and have the least likelihood of developing atherosclerosis.

There is also evidence that high levels of HDL in the blood help reduce the incidence of atherosclerosis, perhaps because HDL carries cholesterol from the peripheral tissues back to the liver. In the liver, some of the cholesterol is used for bile synthesis and secreted into the intestines, from which it is excreted.

A final correlation has been made between diet and atherosclerosis. People whose diet is high in saturated fats tend to have high levels of cholesterol in the blood. Although the relationship between saturated fatty acids and cholesterol metabolism is unclear, it is known that a diet rich in unsaturated fats results in decreased cholesterol levels. In fact, the use of unsaturated fat in the diet results in a decrease in the level of LDL and an increase in the level of HDL. With the positive correlation between heart disease and high cholesterol levels, the current dietary recommendations include a diet that is low in fat and the substitution of unsaturated fats (vegetable oils) for saturated fats (animal fats).

What is the mechanism of uptake of cholesterol from plasma?   *Q u e s t i o n* **18.9**

What is the role of lysosomes in the metabolism of plasma lipoproteins?   *Q u e s t i o n* **18.10**

18.6   The Structure of Biological Membranes

Biological membranes are *lipid bilayers* in which the hydrophobic hydrocarbon tails are packed in the center of the bilayer and the ionic head groups are exposed on the surface to interact with water (Figure 18.11). The hydrocarbon tails of membrane phospholipids provide a thin shell of nonpolar material that prevents mixing of molecules on either side. The nonpolar tails of membrane phospholipids thus provide a barrier between the interior of the cell and its surroundings. The polar heads of lipids are exposed to water, and they are highly solvated. Little exchange, known colloquially as "flip-flop," occurs between lipids on the outer and inner halves of the bilayers (Figure 18.12). The movement of a lipid molecule within one sheath of the bilayer, by contrast, is rapid. A bacterial cell is about 2 μm long, and a lipid molecule diffuses from one end of the cell to the other in a second.

The two layers of the phospholipid bilayer membrane are not identical in composition. For instance, in human red blood cells, approximately 80% of the phospholipids in the outer layer of the membrane are phosphatidylcholine and sphingomyelin; whereas phosphatidylethanolamine and phosphatidylserine make up approximately 80% of the inner layer. In addition, carbohydrate groups are found attached only to those phospholipids found on the outer layer of a membrane. Here they participate in receptor and recognition functions.

Learning Goal

Fluid Mosaic Structure of Biological Membranes

As we have just noted, membranes are not static; they are composed of molecules in motion. The fluidity of biological membranes is determined by the proportions of saturated and unsaturated fatty acid groups in the membrane phospholipids.

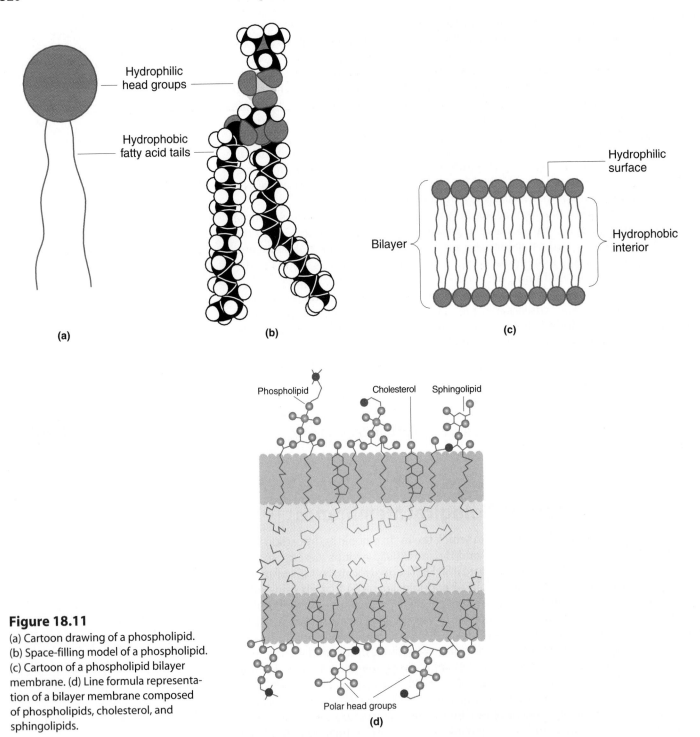

Hydrophilic
head groups

Hydrophobic
fatty acid tails

(a)

(b)

Hydrophilic
surface

Bilayer

Hydrophobic
interior

(c)

Phospholipid   Cholesterol   Sphingolipid

Polar head groups

(d)

Figure 18.11

(a) Cartoon drawing of a phospholipid.
(b) Space-filling model of a phospholipid.
(c) Cartoon of a phospholipid bilayer
membrane. (d) Line formula representation of a bilayer membrane composed
of phospholipids, cholesterol, and
sphingolipids.

About half of the fatty acids that are isolated from membrane lipids from all sources are unsaturated.

The unsaturated fatty acid tails of the phospholipids contribute to membrane fluidity because of the bends introduced into the hydrocarbon chain by the double bonds. Because of these "kinks," the fatty acid tails do not pack together tightly.

We also find that the percentage of unsaturated fatty acid groups in membrane lipids is inversely proportional to the temperature of the environment. Bacteria, for example, have different ratios of saturated and unsaturated fatty acids in their membrane lipids, depending on the temperatures of their surroundings. For instance, the membranes of bacteria that grow in the Arctic Ocean have high levels of unsaturated fatty acids so that their membranes remain fluid even at these frigid temperatures. Conversely, the organisms that live in the hot springs of Yellowstone National Park, with temperatures near the boiling point of water, have membranes with high levels of saturated fatty acids. This flexibility in fatty acid content enables the bacteria to maintain the same membrane fluidity over a temperature range of almost 100°C.

Generally, the body temperatures of mammals are quite constant, and the fatty acid composition of their membrane lipids is therefore usually very uniform. One interesting exception is the reindeer. Much of the year the reindeer must travel through ice and snow. Thus the hooves and lower legs must continue to function at much colder temperatures than the rest of the body. Because of this, the percentage of unsaturation in the membranes varies along the length of the reindeer leg. We find that the proportion of unsaturated fatty acid groups increases closer to the hoof. The lower freezing points and greater fluidity of lipids that contain a high proportion of unsaturated fatty acid groups permit the membranes to function in the low temperatures of ice and snow to which the lower leg is exposed.

Thus membranes are fluid, regardless of the environmental temperature conditions. In fact, it has been estimated that membranes have the consistency of olive oil.

Although the hydrophobic barrier created by the fluid lipid bilayer is an important feature of membranes, the proteins embedded within the lipid bilayer are equally important and are responsible for critical cellular functions. The presence of these membrane proteins was revealed by an electron microscopic technique called *freeze-fracture*. Cells are frozen to very cold temperatures and then fractured with a very fine diamond knife. Some of the cells are fractured between the two layers of the lipid bilayer. When viewed with the electron microscope, the membrane appeared to be a mosaic, studded with proteins. Because of the fluidity of membranes and the appearance of the proteins seen by electron microscopy, our concept of membrane structure is called the **fluid mosaic model** (Figure 18.13).

Some of the observed proteins, called **peripheral membrane proteins,** are bound only to one of the surfaces of the membrane by interactions between ionic head groups of the membrane lipids and ionic amino acids on the surface of the peripheral protein. Other membrane proteins, called **transmembrane proteins,** are embedded within the membrane and extend completely through it, being exposed both inside and outside the cell.

Just as the phospholipid composition of the membrane is asymmetric, so too is the orientation of transmembrane proteins. Each transmembrane protein has hydrophobic regions that associate with the fatty acid tails of membrane phospholipids. Each also has a unique hydrophilic domain that is always found associated with the outer layer of the membrane and is located on the outside of the cell. This region of the protein typically has oligosaccharides covalently attached. Hence these proteins are *glycoproteins*. Similarly, each transmembrane protein has a second hydrophilic domain that is always found associated with the inner layer of the membrane and projects into the cytoplasm of the cell. Typically this region of the transmembrane protein is attached to filaments of the cytoplasmic skeleton.

Because the lipid bilayer is fluid, there can be rapid lateral diffusion of membrane proteins through the lipid bilayer; but membrane proteins, like membrane lipids, do not "flip-flop" across the membrane or turn in the membrane like a revolving door of a department store.

Membranes are dynamic structures, as we may infer from our knowledge about the mobility of membrane proteins and lipids. The mobility of proteins embedded in biological membranes was studied by labeling certain proteins in human and mouse cell membranes with red and green fluorescent dyes. The human and mouse cells were fused; in other words, special techniques were used to cause

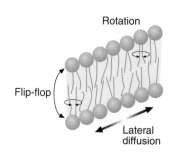

Figure 18.12
Lateral diffusion in a biological membrane is rapid, but "flip-flop" across the membrane almost never occurs.

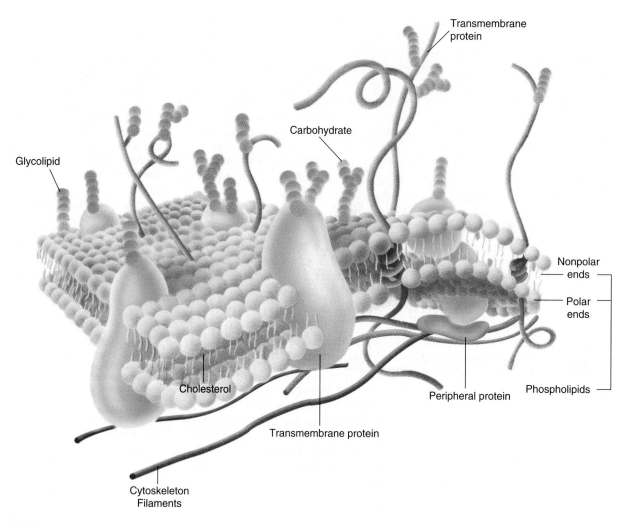

Figure 18.13

The fluid mosaic model of membrane structure.

the membranes of the mouse and human cell to flow together to create a single cell. The new cell was observed through a special ultraviolet or fluorescence microscope. The red and green patches were localized within regions of their original cell membranes when the experiment began. An hour later the color patches were uniformly distributed in the fused cellular membrane (Figure 18.14). This experiment suggests that we can think of the fluid mosaic membrane as an ocean filled with mobile, floating icebergs.

Membrane Transport

Learning Goal

The cell membrane mediates the interaction of the cell with its environment and is responsible for the controlled passage of material into and out of the cell. The external cell membrane controls the entrance of fuel and the exit of waste products. Internal cellular membranes partition metabolites among cell organelles. Most of these transport processes are controlled by transmembrane transport proteins. These transport proteins are the cellular gatekeepers, whose function in membrane transport is analogous to the function of enzymes in carrying out cellular chemical reactions. However, some molecules pass through the membranes unassisted, by the **passive transport** processes of diffusion and osmosis. These are referred to as passive processes because they do not require any energy expenditure by the cell.

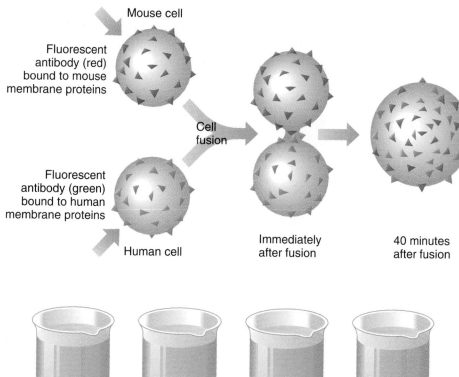

Figure 18.14
Demonstration that membranes are fluid and that proteins move freely in the plane of the lipid bilayer.

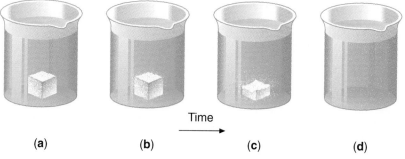

Time

(a) (b) (c) (d)

Figure 18.15
Diffusion results in the *net* movement of sugar and water molecules from the area of high concentration to the area of low concentration. Eventually, the concentrations of sugar and water throughout the beaker will be equal.

Passive Diffusion: The Simplest Form of Membrane Transport

If we put a teaspoon of instant iced tea on the surface of a glass of water, the molecules soon spread throughout the solution. The molecules of both the solute (tea) and the solvent (water) are propelled by random molecular motion. The initially concentrated tea becomes more and more dilute. This process of the *net* movement of a solute with the gradient (from an area of high concentration to an area of low concentration) is called *diffusion* (Figure 18.15).

Diffusion is one means of passive transport across membranes. Let's now suppose that a biological membrane is present and that a substance is found at one concentration outside the membrane and at half that concentration inside the membrane. If the solute can pass through the membrane, diffusion will occur with net transport of material from the region of initial high concentration to the region of initial low concentration, and the substance will equilibrate across the cell membrane (Figure 18.16). After a while, the concentration of the substance will be the same on both sides of the membrane; the system will be at equilibrium, and no more net change will occur.

Of what practical value is the process of diffusion to the cell? Certainly, diffusion is able to distribute metabolites effectively throughout the interior of the cell. But what about the movement of molecules through the membrane? Because of the lipid bilayer structure of the membrane, only a few molecules are able to diffuse freely across a membrane. These include small molecules such as O_2, CO_2, and H_2O. Any large or highly charged molecules or ions are not able to pass through the lipid bilayer directly. Such molecules require an assist from cell membrane proteins. Any membrane that allows the diffusion of some molecules but not others is said to be *selectively permeable*.

Antibiotics That Destroy Membrane Integrity

The "age of antibiotics" began in 1927 when Alexander Fleming discovered, quite by accident, that a product of the mold *Penicillium* can kill susceptible bacteria. We now know that penicillin inhibits bacterial growth by interfering with cell wall synthesis. Since Fleming's time, hundreds of antibiotics, which are microbial products that either kill or inhibit the growth of susceptible bacteria or fungi, have been discovered. The key to antibiotic therapy is to find a "target" in the microbe, a metabolic process or structure that the human does not have. In this way the antibiotic will selectively inhibit the disease-causing organism without harming the patient.

Many antibiotics disrupt cell membranes. The cell membrane is not an ideal target for antibiotic therapy because all cells, human and bacterial, have membranes. Therefore both types of cells are damaged. Because these antibiotics exhibit a wide range of toxic side effects when ingested, they are usually used to combat infections topically (on body surfaces). In this way, damage to the host is minimized but the inhibitory effect on the microbe is maximized.

Polymyxins are antibiotics produced by the bacterium *Bacillus polymyxa*. They are protein derivatives having one end that is hydrophobic because of an attached fatty acid. The opposite end is hydrophilic. Because of these properties, the polymyxins bind to membranes with the hydrophobic end embedded within the membrane, while the hydrophilic end remains outside the cell. As a result, the integrity of the membrane is disrupted, and leakage of cellular constituents occurs, causing cell death.

Amphotericin B

Nystatin

The structures of amphotericin B and nystatin, two antifungal antibiotics.

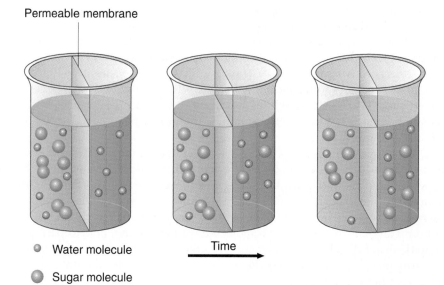

Figure 18.16

Diffusion of a solute through a membrane.

Permeable membrane

Time

● Water molecule

● Sugar molecule

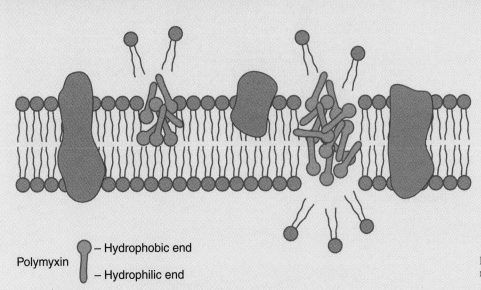

Polymyxin   ▮ – Hydrophobic end

▮ – Hydrophilic end

Polymyxins act like detergents, disrupting membrane integrity and killing the cell.

Although the polymyxins have been found to be useful in treating some urinary tract infections, pneumonias, and infections of burn patients, other antibiotics are now favored because of the toxic effects of the polymyxins on the kidney and central nervous system. Polymyxin B is still used topically and is available as an over-the-counter ointment in combination with two other antibiotics, neomycin and bacitracin.

Two other antibiotics that destroy membranes, amphotericin B and nystatin, are large ring structures that are used in treating serious systemic fungal infections. These antibiotics form complexes with ergosterol in the fungal cell membrane, and they disrupt the membrane permeability and cause leakage of cellular constituents. Neither is useful in treating bacterial infections because most bacteria have no ergosterol in their membranes. Both amphotericin B and nystatin are extremely toxic and cause symptoms that include nausea and vomiting, fever and chills, anemia, and renal failure. It is easy to understand why the use of these drugs is restricted to treatment of life-threatening fungal diseases.

Facilitated Diffusion: Specificity of Molecular Transport

Most molecules are transported across biological membranes by specific protein carriers known as *permeases*. When a solute diffuses through a membrane from an area of high concentration to an area of low concentration by passing through a channel within a permease, the process is known as **facilitated diffusion.** No energy is consumed by facilitated diffusion; thus it is another means of passive transport, and the direction of transport depends upon the concentrations of metabolite on each side of the membrane.

Transport across cell membranes by facilitated diffusion occurs through pores within the permease that have conformations, or shapes, that are complementary to those of the transported molecules. The charge and conformation of the pore define the specificity of the carrier (Figure 18.17). Only molecules that have the correct shape can enter the pore. As a result, the rate of diffusion for any molecule is limited by the number of carrier permease molecules in the membrane that are responsible for the passage of that molecule.

The transport of glucose illustrates the specificity of carrier permease proteins. D-Glucose is transported by the glucose carrier, but its enantiomer, L-glucose, is not. Thus the glucose permease exhibits stereospecificity. In other words, the solute to be brought into the cell must "fit" precisely, like a hand in a glove, into a recognition site within the structure of the permease. In Chapter 20 we will see that

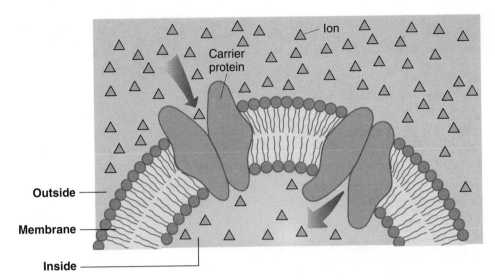

Figure 18.17

Transport by facilitated diffusion occurs through pores in the transport protein whose size and shape are complementary to those of the transported molecule.

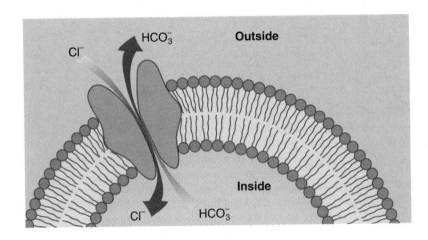

Figure 18.18

Transport of Cl$^-$ and HCO$_3^-$ ions in opposite directions across the red blood cell membrane.

Enzyme specificity is discussed in Section 20.5.

The metabolic effects of insulin are described in Sections 21.7 and 23.6.

enzymes that catalyze the biochemical reactions within cells show this same type of specificity.

The rate of transport of metabolites into the cell has profound effects on the net metabolic rate of many cells. Insulin, a polypeptide hormone synthesized by the islet β-cells of the pancreas, increases the maximum rate of glucose transport by a factor of three to four. The result is that the metabolic activity of the cell is greatly increased.

Red blood cells use the same anion channel to transport Cl$^-$ into the cell in exchange for HCO$_3^-$ ions (Figure 18.18). This two-way transport is known as *antiport*. This transport occurs by facilitated diffusion so that Cl$^-$ flows from a high exterior concentration to a low interior one, and bicarbonate flows from a high interior concentration to a low exterior one.

Osmosis: Passive Movement of a Solvent Across a Membrane

Learning Goal

13

Because a cell membrane is selectively permeable, it is not always possible for solutes to pass through it in response to a concentration gradient. In such cases the solvent diffuses through the membrane. Such membranes, permeable to solvent but not to solute, are specifically called **semipermeable membranes. Osmosis** is the diffusion of a solvent (water in biological systems) through a semipermeable membrane in response to a water concentration gradient.

Suppose that we place a 0.5 *M* glucose solution in a dialysis bag that is composed of a membrane with pores that allow the passage of water molecules but not glucose molecules. Consider what will happen when we place this bag into a beaker of pure water. We have created a gradient in which there is a higher concentration of glucose inside the bag than outside, but the glucose cannot diffuse through the bag to achieve equal concentration on both sides of the membrane.

Now let's think about this situation in another way. We have a higher concentration of water molecules outside the bag (where there is only pure water) than inside the bag (where some of the water molecules are occupied in the hydration of solute particles and are consequently unable to move freely in the system). Because water can diffuse through the membrane, a net diffusion of water will occur through the membrane into the bag. This is the process of osmosis (Figure 18.19).

As you have probably already guessed, this system can never reach equilibrium (equal concentrations inside and outside the bag) because regardless of how much water diffuses into the bag, diluting the glucose solution, the concentration of glucose will always be higher inside the bag (and the accompanying free water concentration will always be lower).

What happens when the bag has taken in as much water as it can, when it has expanded as much as possible? Now the walls of the bag exert a force that will stop the *net* flow of water into the bag. **Osmotic pressure** is the pressure that must be exerted to stop the flow of water across a selectively permeable membrane by osmosis. Stated more precisely, the osmotic pressure of a solution is the net pressure with which water enters it by osmosis from a pure water compartment when the two compartments are separated by a semipermeable membrane.

Osmotic concentration or *osmolarity* is the term used to describe the osmotic strength of a solution. It depends only on the ratio of the number of solute particles to the number of solvent particles. Thus the chemical nature and size of the solute are not important, only the concentration, expressed in molarity. For instance, a 2 *M* solution of glucose (a sugar of molecular weight 180) has the same osmolarity as a 2 *M* solution of albumin (a protein of molecular weight 60,000).

Blood plasma has an osmolarity equivalent to a 0.30 *M* glucose solution or a 0.15 *M* NaCl solution. This latter is true because in solution NaCl dissociates into Na^+ and Cl^- and thus contributes twice the number of solute particles as a molecule that does not ionize. If red blood cells, which have an osmolarity equal to blood plasma, are placed in a 0.30 *M* glucose solution, no net osmosis will occur because the osmolarity and water concentration inside the red blood cell are equal to those of the 0.30 *M* glucose solution. The solutions inside and outside the red blood cell are said to be **isotonic** (*iso* means "same," and *tonic* means "strength") **solutions.** Because the osmolarity is the same inside and outside, the red blood cell will remain the same size (Figure 18.20b).

What happens if we now place the red blood cells into a **hypotonic solution,** in other words, a solution having a lower osmolarity than the cytoplasm of the cell? In this situation there will be a net movement of water into the cell as water diffuses down its concentration gradient. The membrane of the red blood cell does not have the strength to exert a sufficient pressure to stop this flow of water, and the cell will swell and burst (Figure 18.20c). Alternatively, if we place the red blood cells into a **hypertonic solution** (one with a greater osmolarity than the cell), water will pass out of the cells, and they will shrink dramatically in size (Figure 18.20a).

These principles have important applications in the delivery of intravenous (IV) solutions into an individual. Normally, any fluids infused intravenously must have the correct osmolarity; they must be isotonic with the blood cells and the blood plasma. Such infusions are frequently either 5.5% dextrose (glucose) or "normal saline." The first solution is composed of 5.5 g of glucose per 100 mL of solution (0.30 *M*), and the latter of 9.0 g of NaCl per 100 mL of solution (0.15 *M*). In either case they have the same osmotic pressure and osmolarity as the plasma and

Recall that there is an inverse relationship between the osmotic (solute) concentration of a solution and the water concentration of that solution, as discussed in Chapter 7.

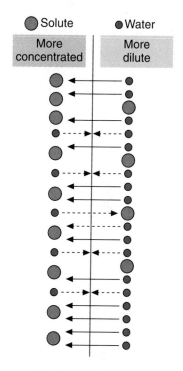

Figure 18.19

Osmosis across a membrane. The solvent, water, diffuses from an area of lower solute concentration to an area of higher solute concentration.

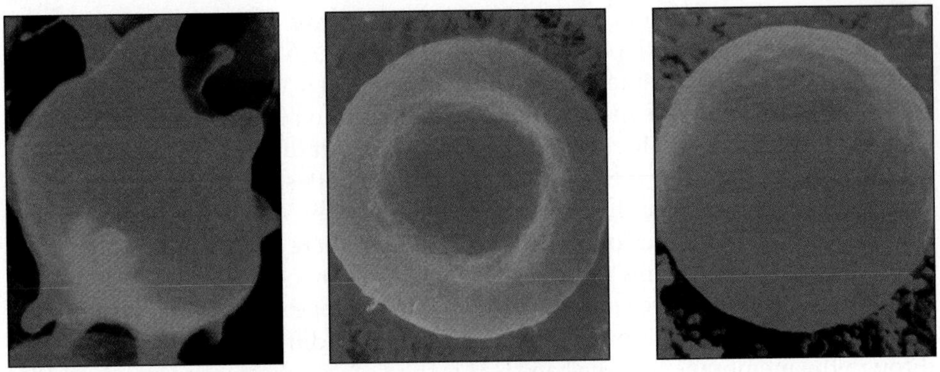

Figure 18.20
Scanning electron micrographs of red
blood cells exposed to (a) hypertonic,
(b) isotonic, and (c) hypotonic solutions.

(a)   (b)   (c)

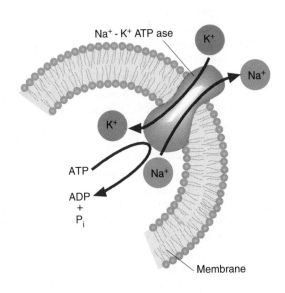

Figure 18.21
Schematic diagram of the operation of
Na⁺-K⁺ ATPase.

blood cells and can therefore be safely administered without upsetting the osmotic
balance between the blood and the blood cells.

Energy Requirements for Transport

Learning Goal

14

Simple diffusion, facilitated diffusion, and osmosis involve the spontaneous flow
of materials from a region of higher concentration to an area of lower concentration (a concentration gradient). To survive, cells often must move substances "up-
hill," against a concentration gradient. This phenomenon, called **active transport,**
requires energy. Many ions and food molecules are imported through the cell
membrane by active transport. The energy used for this process may consume
more than half of the total energy harvested by cellular metabolism.

A good example of active transport is the Na^+-K^+ ATPase, which moves these
ions into and out of the cell against their gradients (Figure 18.21). Cells must maintain a high concentration of Na^+ outside the cell and a high concentration of K^+ inside the cell. This requires a continuous supply of cellular energy in the form of
adenosine triphosphate (ATP). Over one-third of the total ATP produced by the
cell is used to maintain these Na^+ and K^+ concentration gradients across the cell
membrane. Thus, the name *Na^+-K^+ ATPase* refers to the enzymatic activity that hydrolyzes ATP. The hydrolysis of ATP releases the energy needed to move Na^+ and
K^+ ions across the cell membrane. For each ATP molecule hydrolyzed, three Na^+
are moved out of the cell and two K^+ are transported into the cell.

How does membrane transport resemble enzyme catalysis?

Question 18.11

Why is D-glucose transported by the glucose transport protein whereas L-glucose is not?

Question 18.12

Summary

18.1 Biological Functions of Lipids

Lipids are organic molecules characterized by their solubility in nonpolar solvents. Lipids are subdivided into classes based on structural characteristics: *fatty acids*, *glycerides*, nonglycerides, and *complex lipids*. Lipids serve many functions in the body, including energy storage, protection of organs, insulation, and absorption of vitamins. Other lipids are energy sources, hormones, or vitamins. Cells store chemical energy in the form of lipids, and the cell membrane is a lipid bilayer.

18.2 Fatty Acids

Fatty acids are *saturated* and *unsaturated* carboxylic acids containing between twelve and twenty-four carbon atoms. Fatty acids with even numbers of carbon atoms occur most frequently in nature. The reactions of fatty acids are identical to those of carboxylic acids. They include esterification, production by acid hydrolysis of esters, saponification, and addition at the double bond. *Prostaglandins*, thromboxanes, and leukotrienes are derivatives of twenty-carbon fatty acids that have a variety of physiological effects.

18.3 Glycerides

Glycerides are the most abundant lipids. The triesters of glycerol (*triglycerides*) are of greatest importance. Neutral triglycerides are important because of their ability to store energy. The ionic *phospholipids* are important components of all biological membranes.

18.4 Nonglyceride Lipids

Nonglyceride lipids consist of *sphingolipids, steroids,* and *waxes. Sphingomyelin* is a component of the myelin sheath around cells of the central nervous system. The *steroids* are important for many biochemical functions: *Cholesterol* is a membrane component; testosterone, progesterone, and estrone are sex hormones; and cortisone is an anti-inflammatory steroid that is important in the regulation of many biochemical pathways.

18.5 Complex Lipids

Plasma lipoproteins are complex lipids that transport other lipids through the bloodstream. *Chylomicrons* carry dietary triglycerides from the intestine to other tissues. *Very low density lipoproteins* carry triglycerides synthesized in the liver to other tissues for storage. *Low-density lipoproteins* carry cholesterol to peripheral tissues and help regulate blood cholesterol levels. *High-density lipoproteins* transport cholesterol from peripheral tissues to the liver.

18.6 The Structure of Biological Membranes

The *fluid mosaic model* of membrane structure pictures biological membranes that are composed of lipid bilayers in which proteins are embedded. Membrane lipids contain polar head groups and nonpolar hydrocarbon tails. The hydrocarbon tails of phospholipids are derived from saturated and unsaturated long-chain fatty acids containing an even number of carbon atoms. The lipids and proteins diffuse rapidly in the lipid bilayer but seldom cross from one side to the other.

The simplest type of membrane transport is passive diffusion of a substance across the lipid bilayer from the region of highest concentration to that of lowest concentration. Many metabolites are transported across biological membranes by permeases that form pores through the membrane. The conformation of the pore is complementary to that of the substrate to be transported. Cells use energy to transport molecules across the plasma membrane against their concentration gradients, a process known as *active transport.*

The Na^+-K^+ ATPase hydrolyzes one molecule of ATP to provide the driving force for pumping three Na^+ out of the cell in exchange for two K^+.

Key Terms

active transport (18.6)
arachidonic acid (18.2)
atherosclerosis (18.4)
cholesterol (18.4)
chylomicron (18.5)

complex lipid (18.5)
diglyceride (18.3)
eicosanoid (18.2)
emulsifying agent (18.3)
essential fatty acid (18.2)

esterification (18.2)

facilitated diffusion
   (18.6)

fatty acid (18.2)

fluid mosaic model (18.6)

glyceride (18.3)

high-density lipoprotein
   (HDL) (18.5)

hydrogenation (18.2)

hypertonic solution (18.6)

hypotonic solution (18.6)

isotonic solution (18.6)

lipid (18.1)

low-density lipoprotein
   (LDL) (18.5)

monoglyceride (18.3)

neutral glyceride (18.3)

osmosis (18.6)

osmotic pressure (18.6)

passive transport (18.6)

peripheral membrane
   protein (18.6)

phosphatidate (18.3)

phosphoglyceride (18.3)

phospholipid (18.3)

plasma lipoprotein (18.5)

prostaglandin (18.2)

saponification (18.2)

saturated fatty acid (18.2)

semipermeable
   membrane (18.6)

sphingolipid (18.4)

sphingomyelin (18.4)

steroid (18.4)

terpene (18.4)

transmembrane protein
   (18.6)

triglyceride (18.3)

unsaturated fatty acid
   (18.2)

very low density
   lipoprotein (VLDL)
   (18.5)

wax (18.4)

Questions and Problems

Biological Functions of Lipids

18.13 List the four main groups of lipids.

18.14 List the biological functions of lipids.

Fatty Acids: Structure, Properties, and Chemical Reactions

18.15 What is the difference between a saturated and an unsaturated fatty acid?

18.16 Write the structure for a saturated and an unsaturated fatty acid.

18.17 As the length of the hydrocarbon chain of saturated fatty acids increases, what is the effect on the melting points?

18.18 As the number of carbon-carbon double bonds in fatty acids increases, what is the effect on the melting points?

18.19 Draw the structures of each of the following fatty acids:
   a. Decanoic acid
   b. Stearic acid
   c. *trans*-5-Decenoic acid
   d. *cis*-5-Decenoic acid

18.20 What are the common and I.U.P.A.C. names of each of the following fatty acids?
   a. $C_{15}H_{31}COOH$
   b. $C_{11}H_{23}COOH$
   c. $CH_3(CH_2)_5CH=CH(CH_2)_7COOH$
   d. $CH_3(CH_2)_7CH=CH(CH_2)_7COOH$

18.21 Write an equation for each of the following reactions:
   a. Esterification of glycerol with three molecules of myristic acid
   b. Acid hydrolysis of tristearoyl glycerol
   c. Reaction of decanoic acid with KOH
   d. Hydrogenation of linoleic acid

18.22 Write an equation for each of the following reactions:

   a. Esterification of glycerol with three molecules of palmitic acid
   b. Acid hydrolysis of trioleoyl glycerol
   c. Reaction of stearic acid with KOH
   d. Hydrogenation of oleic acid

Fatty Acids: Prostaglandins

18.23 What is the function of the essential fatty acids?

18.24 What molecules are formed from arachidonic acid?

18.25 What is the biochemical basis for the effectiveness of aspirin in decreasing the inflammatory response?

18.26 What is the role of prostaglandins in the inflammatory response?

18.27 List four effects of prostaglandins.

18.28 What are the functions of thromboxane A_2 and leukotrienes?

Glycerides: Neutral Glycerides and Phosphoglycerides

18.29 Draw the structure of the triglyceride molecule formed by esterification at C-1, C-2, and C-3 with hexadecanoic acid, *trans*-9-hexadecenoic acid, and *cis*-9-hexadecenoic acid, respectively.

18.30 Draw one possible structure of a triglyceride that contains the three fatty acids stearic acid, palmitic acid, and oleic acid.

18.31 Draw the structure of the phosphatidate formed between glycerol-3-phosphate that is esterified at C-1 and C-2 with capric and lauric acids, respectively.

18.32 Draw the structure of a lecithin molecule in which the fatty acyl groups are derived from stearic acid.

18.33 What are the structural differences between triglycerides (triacylglycerols) and phospholipids?

18.34 How are the structural differences between triglycerides and phospholipids reflected in their different biological functions?

Nonglyceride Lipids

18.35 What is a sphingolipid?

18.36 What is the function of sphingomyelin?

18.37 What is the role of cholesterol in biological membranes?

18.38 How does cholesterol contribute to atherosclerosis?

18.39 What are the biological functions of progesterone, testosterone, and estrone?

18.40 How has our understanding of the steroid sex hormones contributed to the development of oral contraceptives?

18.41 What is the medical application of cortisone?

18.42 What are the possible side effects of cortisone treatment?

18.43 A wax found in beeswax is myricyl palmitate. What fatty acid and what alcohol are used to form this compound?

18.44 A wax found in the head of sperm whales is cetyl palmitate. What fatty acid and what alcohol are used to form this compound?

18.45 What are isoprenoids?

18.46 What is a terpene?

18.47 List some important biological molecules that are terpenes.

18.48 Draw the five-carbon isoprene unit.

Complex Lipids

18.49 What are the four major types of plasma lipoproteins?

18.50 What is the function of each of the four types of plasma lipoproteins?

18.51 What is the relationship between atherosclerosis and high blood pressure?

18.52 How is LDL taken into cells?

18.53 How does a genetic defect in the LDL receptor contribute to atherosclerosis?

18.54 What is the correlation between saturated fats in the diet and atherosclerosis?

Biological Membranes: Structure

18.55 How will the properties of a biological membrane change if the fatty acid tails of the phospholipids are converted from saturated to unsaturated chains?

18.56 What is the function of unsaturation in the hydrocarbon tails of membrane lipids?

18.57 What is the basic structure of a biological membrane?

18.58 Describe the fluid mosaic model of membrane structure.

18.59 Describe peripheral membrane proteins.

18.60 Describe transmembrane proteins and list some of their functions.

18.61 What is the major effect of cholesterol on the properties of biological membranes?

18.62 Why do the hydrocarbon tails of membrane phospholipids provide a barrier between the inside and outside of the cell?

18.63 What experimental observation shows that proteins diffuse within the lipid bilayers of biological membranes?

18.64 Why don't proteins turn around in biological membranes like revolving doors?

Biological Membranes: Transport

18.65 Explain the difference between simple diffusion across a membrane and facilitated diffusion.

18.66 Explain what would happen to a red blood cell placed in each of the following solutions:

  a. hypotonic **c.** hypertonic

  b. isotonic

18.67 How does active transport differ from facilitated diffusion?

18.68 By what mechanism are Cl^- and HCO_3^- ions transported across the red blood cell membrane?

18.69 What is the meaning of the term *antiport*?

18.70 How does insulin affect the transport of glucose?

18.71 What properties of a transport protein (permease) determine its specificity?

18.72 Why is the function of the Na^+-K^+ ATPase an example of active transport?

18.73 What is the stoichiometry of the Na^+-K^+ ATPase?

18.74 How will the Na^+ and K^+ concentrations of a cell change if the Na^+-K^+ ATPase is inhibited?

18.75 What is meant by the term *active transport?*

Critical Thinking Problems

1. Olestra is a fat substitute that provides no calories, yet has all the properties of a naturally occurring fat. It has a creamy, tongue-pleasing consistency. Unlike other fat substitutes, olestra can withstand heating. Thus, it can be used to prepare foods such as potato chips and crackers. Olestra is a sucrose polyester and is produced by esterification of six, seven, or eight fatty acids to molecules of sucrose. Draw the structure of one such molecule having eight stearic acid acyl groups attached.

2. Liposomes can be made by vigorously mixing phospholipids (like phosphatidylcholine) in water. When the mixture is allowed to settle, spherical vesicles form that are surrounded by a phospholipid bilayer "membrane." Pharmaceutical chemists are trying to develop liposomes as a targeted drug delivery system. By adding the drug of choice to the mixture described above, liposomes form around the solution of drug. Specific proteins can be incorporated into the mixture that will end up within the phospholipid bilayers of the liposomes. These proteins are able to bind to targets on the surface of particular kinds of cells in the body. Explain why injection of liposome encapsulated pharmaceuticals might be a good drug delivery system.

3. "Cholesterol is bad and should be eliminated from the diet." Do you agree or disagree? Defend your answer.

4. Why would a phospholipid such as lecithin be a good emulsifying agent for ice cream?

5. When a plant becomes cold-adapted, the composition of the membranes changes. What changes in fatty acid and cholesterol composition would you predict? Explain your reasoning.

6. In terms of osmosis, explain why it would be preferable for a cell to store 10,000 molecules of glycogen each composed of 10^5 molecules of glucose rather than to store 10^9 individual molecules of glucose.

19 Protein Structure and Function

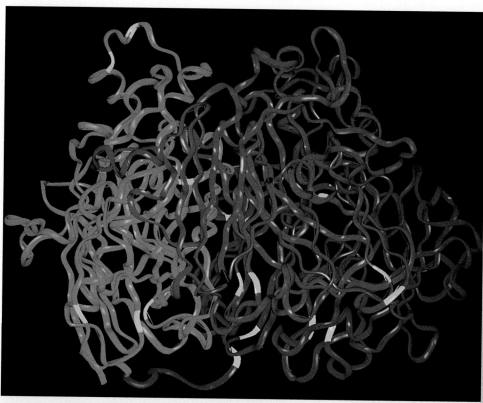

Computer-generated model of the structure of a protein.

Learning Goals

1 List the functions of proteins.

2 Draw the general structure of an amino acid and classify amino acids based on their R groups.

3 Describe the primary structure of proteins and draw the structure of the peptide bond.

4 Draw the structure of small peptides and name them.

5 Describe the types of secondary structure of a protein.

6 Discuss the forces that maintain secondary structure.

7 Describe the structure and functions of fibrous proteins.

8 Describe the tertiary and quaternary structure of a protein.

9 List the R group interactions that maintain protein conformation.

10 List examples of conjugated proteins and prosthetic groups and explain the way in which they function.

11 Discuss the importance of the three-dimensional structure of a protein to its function.

12 Describe the roles of hemoglobin and myoglobin.

13 Describe how extremes of pH and temperature cause denaturation of proteins.

14 Explain the difference between essential and nonessential amino acids.

533

Angiogenesis Inhibitors: Proteins That Inhibit Tumor Growth

Cancer researchers have long known that solid tumors cannot grow larger than the size of a pinhead unless they stimulate the formation of new blood vessels that provide the growing tumor with nutrients and oxygen and remove the waste products of cellular metabolism. Studies of *angiogenesis*, the formation of new blood vessels, in normal tissues have provided new weapons in the arsenal of anti-cancer drugs.

Angiogenesis occurs through a carefully controlled sequence of steps. Consider the process of tissue repair. One of several protein growth factors stimulates the endothelial cells of an existing blood vessel to begin growing, dividing, and migrating into the tissue to be repaired. Threads of new endothelial cells organize themselves into hollow cylinders, or tubules. These tubules become a new network of blood vessels throughout the damaged tissue. These new blood vessels bring the needed nutrients, oxygen, and other factors to the site of damage, allowing the tissue to be repaired and healing to occur.

In addition to the growth factors that stimulate this process, there are several other proteins that inhibit the formation of new blood vessels. In fact, the normal process of angiogenesis is dependent on the appropriate balance of the stimulatory growth factors and the inhibitory proteins.

The normal events of angiogenesis are duplicated at a critical moment in the growth of a tumor. Cells of the tumor secrete one or more of the growth factors known to stimulate angiogenesis. The newly formed blood vessels provide the cells of the growing tumor with everything needed to continue growing and dividing.

Metastasis, the spreading of tumor cells to other sites in the body, also requires angiogenesis. Typically, those tumors having more blood vessels are more likely to metastasize. Clinically, treatment of these tumors has a poorer outcome.

Researchers considered all of this information known about angiogenesis and its impact on tumor formation and metastasis. They developed the hypothesis that proteins that inhibit blood vessel formation might be effective weapons against developing tumors. If this hypothesis turned out to be supported by experimental data, there would be a number of advantages to the use of angiogenesis inhibitors. Because these proteins are normally produced by the human body, they should not have the toxic side-effects caused by so many anti-cancer drugs. In addition, angiogenesis inhibitors can overcome the problem of cancer cell drug resistance. Most cancer cells are prone to mutations and mutant cells resistant to the anti-cancer drugs develop. The angiogenesis inhibitors target normal endothelial cells, which are genetically stable. As a result, drug-resistance is much less likely to occur.

Endostatin is one of the anti-angiogenesis proteins. Discovered in 1997, it was found to be a protein of 20,000 g/mol which is a fragment of the C-terminus of collagen XVIII. Experimentally, endostatin is a potent inhibitor of tumor growth. It binds to the heparin sulfate proteoglycans of the cell surface and interferes with growth factor signaling. As a result, the growth and division of endothelial cells is inhibited and new blood vessels are not formed.

Angiostatin is another anti-angiogenesis protein normally found in the human body. Discovered in 1994, it is a protein fragment of human plasminogen and has a molecular weight of 50,000 g/mol. The role of angiostatin in the human body is to block the growth of diseased tissue by inhibiting the formation of blood vessels. Like endostatin, it is hoped that angiostatin will block the growth of tumors by depriving them of their blood supply.

Currently, there are about twenty angiogenesis inhibitors being tested in clinical trials involving humans. Most are in phase I or II trials which allow scientists to determine a safe dosage and assess the severity of any side effects. Only a small number of people are involved in phase I or II trials. In phase III trials, a large number of patients are divided into two groups. One group receives standard anti-cancer treatment plus a placebo. The other group receives standard treatment and the new drug.

As we await the results of the clinical trials involving the proteins endostatin and angiostatin, scientists explore alternative methods to attack cancer cells. Some of these involve a class of proteins called *antibodies* that can bind specifically to cancer cells and help to inhibit or destroy them.

As we will discover, there are many different classes of proteins that carry out a variety of functions for the body. Endostatin and angiostatin serve as regulatory proteins; the antibodies serve as the body's defense system against infectious diseases. These and many other proteins are the focus of this chapter.

Introduction

In the 1800s, Johannes Mulder came up with the name **protein,** a term derived from a Greek word that means "of first importance." Indeed, proteins are a very important class of food molecules because they provide an organism not only with carbon and hydrogen, but also with nitrogen and sulfur. These latter two elements are unavailable from fats and carbohydrates, the other major classes of food molecules.

In addition to their dietary importance, the proteins are the most abundant macromolecules in the cell, and they carry out most of the work in a cell. Protection of the body from infection, mechanical support and strength, and catalysis of metabolic reactions—all are functions of proteins that are essential to life.

19.1 Cellular Functions of Proteins

Proteins have many biological functions, as the following short list suggests.

Learning Goal

- **Enzymes,** biological catalysts, are proteins. Reactions that would take days or weeks or require extremely high temperatures without enzymes are completed in an instant. Without these remarkable enzymes, life would not be possible.
- **Antibodies** (also called *immunoglobulins*) are specific protein molecules produced by specialized cells of the immune system in response to foreign **antigens.** These foreign invaders include bacteria and viruses that infect the body. Each antibody has regions that precisely fit and bind to a single antigen. It helps to end the infection by binding to the antigen and helping to destroy it or remove it from the body.

In the broadest sense, an antigen is any substance that stimulates an immune response.

- **Transport proteins** carry materials from one place to another in the body. The protein *transferrin* transports iron from the liver to the bone marrow, where it is used to synthesize the heme group for hemoglobin. The proteins *hemoglobin* and *myoglobin* are responsible for transport and storage of oxygen in higher organisms, respectively.
- **Regulatory proteins** control many aspects of cell function, including metabolism and reproduction. We can function only within a limited set of conditions. For life to exist, body temperature, the pH of the blood, and blood glucose levels must be carefully regulated. Many of the hormones that regulate body function, such as *insulin* and *glucagon,* are proteins.
- **Structural proteins** provide mechanical support to large animals and provide them with their outer coverings. Our hair and fingernails are largely composed of the protein *keratin.* Other proteins provide mechanical strength for our bones, tendons, and skin. Without such support, large, multicellular organisms like ourselves could not exist.
- **Movement proteins** are necessary for all forms of movement. Our muscles, including that most important muscle, the heart, contract and expand through the interaction of actin and myosin proteins. Sperm can swim because they have long flagella made up of proteins.
- **Nutrient proteins** serve as sources of amino acids for embryos or infants. Egg albumin and casein in milk are examples of nutrient storage proteins.

19.2 The α-Amino Acids

The proteins of the body are made up of some combination of twenty different subunits called **α-amino acids.** The general structure of an α-amino acid is shown in Figure 19.1. We find that nineteen of the twenty amino acids that are commonly isolated from proteins have this same general structure; they are primary amines on the α-carbon. The remaining amino acid, proline, is a secondary amine.

Learning Goal

The α-carbon in the general structure is attached to a carboxylate group (a carboxyl group that has lost a proton, —COO⁻) and a protonated amino group (an amino group that has gained a proton, —NH₃⁺). In aqueous solution of approximately pH 7, conditions required for life functions, amino acids in which the carboxylate group is protonated (—COOH) and the amino group is unprotonated (—NH₂) do not exist. Under these conditions, the carboxyl group ionizes, and the

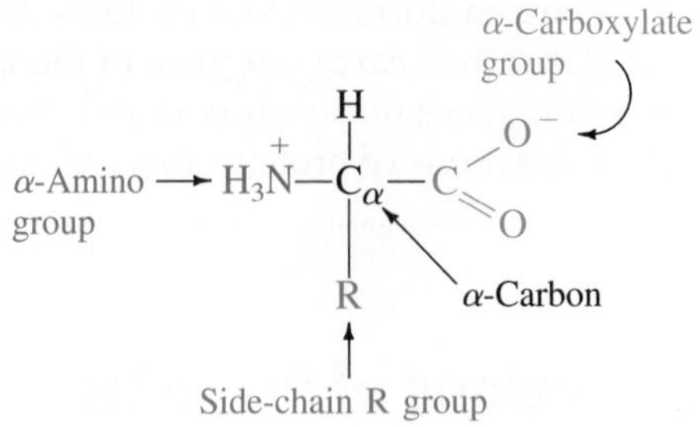

Figure 19.1

General structure of an α-amino acid. All amino acids isolated from proteins, with the exception of proline, have this general structure.

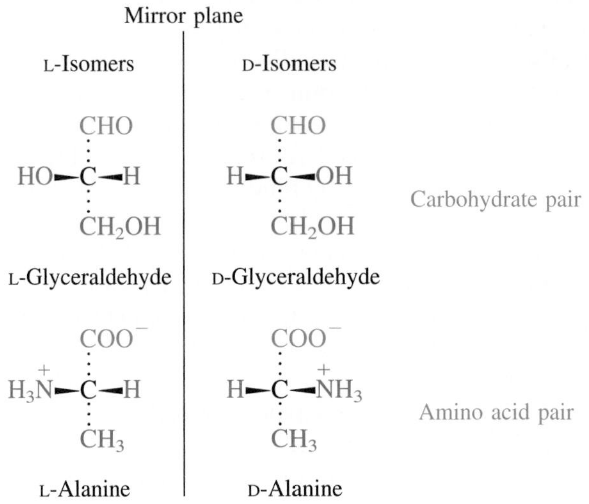

Figure 19.2

Structure of D- and L-glyceraldehyde and their relationship to D- and L-alanine. (The student should build models of these compounds, from which it will be immediately apparent that the members of each pair are nonsuperimposable mirror images.)

basic amino group picks up the proton that is released. As a result, amino acids in water exist as dipolar ions called *zwitterions*.

The α-carbon of each amino acid is also bonded to a hydrogen atom and a side chain, or R group. In a protein, the R groups interact with one another through a variety of weak attractive forces. These interactions participate in folding the protein chain into a precise three-dimensional shape that determines its ultimate function. They also serve to maintain that three-dimensional conformation.

The α-carbon is attached to four different groups in all amino acids except glycine. The α-carbon of α-amino acids is therefore chiral. That is, an α-amino acid isolated from a protein cannot be superimposed on its mirror image. Glycine has two hydrogen atoms attached to the α-carbon and is the only amino acid commonly found in proteins that is not chiral.

Stereochemistry is discussed in Section 17.2 and in Appendix D, "Stereochemistry and Stereoisomers Revisited."

The configuration of α-amino acids isolated from proteins is L-. This is based on comparison of amino acids with D-glyceraldehyde (Figure 19.2). The configuration of α-amino acids isolated from proteins is opposite to that of D-glyceraldehyde; that is, the orientation of the four groups around L-alanine resembles the orientation of the four substituents around the chiral carbon of L-glyceraldehyde.

Because all of the amino acids have a carboxyl group and an amino group, all differences between amino acids depend upon their side-chain R groups. The amino acids are grouped in Figure 19.3 according to the polarity of their side chains.

The hydrophobic interaction between nonpolar R groups is one of the forces that helps maintain the proper three-dimensional shape of a protein.

The side chains of some amino acids are nonpolar. They prefer contact with one another over contact with water and are said to be **hydrophobic** ("water-fearing") **amino acids.** They are generally found buried in the interior of proteins, where they can associate with one another and remain isolated from water. Nine amino acids fall into this category: alanine, valine, leucine, isoleucine, proline,

(a)

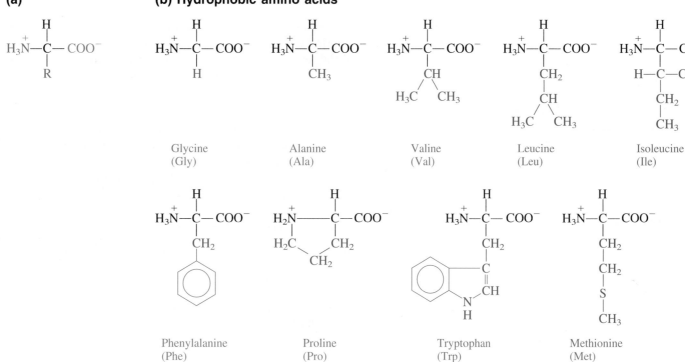

(b) Hydrophobic amino acids

Glycine
(Gly)

Alanine
(Ala)

Valine
(Val)

Leucine
(Leu)

Isoleucine
(Ile)

Phenylalanine
(Phe)

Proline
(Pro)

Tryptophan
(Trp)

Methionine
(Met)

(c) Polar, neutral amino acids

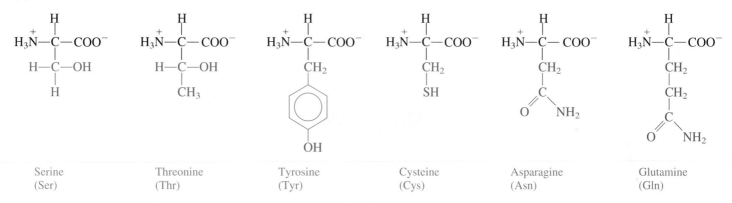

Serine
(Ser)

Threonine
(Thr)

Tyrosine
(Tyr)

Cysteine
(Cys)

Asparagine
(Asn)

Glutamine
(Gln)

(d) Negatively charged amino acids　　**(e) Positively charged amino acids**

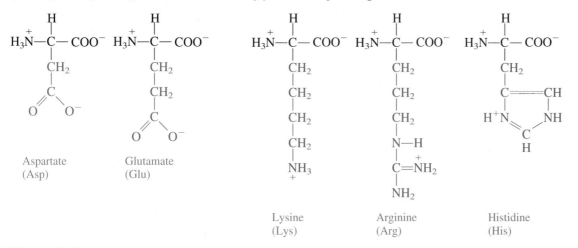

Aspartate
(Asp)

Glutamate
(Glu)

Lysine
(Lys)

Arginine
(Arg)

Histidine
(His)

Figure 19.3

Structures of the amino acids. (a) The general structure of an amino acid. Structures of (b) the hydrophobic, (c) polar, neutral, (d) negatively charged, and (e) positively charged amino acids.

Proline (Pro)

Hydrogen bonding (Section 6.2) is another weak interaction that helps maintain the proper three-dimensional structure of a protein. The positively and negatively charged amino acids within a protein can interact with one another to form ionic bridges. This is yet another attractive force that helps to keep the protein chain folded in a precise way.

glycine, methionine, phenylalanine, and tryptophan. The R group of proline is unique; it is actually bonded to the α-amino group, forming a secondary amine.

The side chains of the remaining amino acids are polar. Because they are attracted to polar water molecules, they are said to be **hydrophilic** ("water-loving") **amino acids.** The hydrophilic side chains are often found on the surfaces of proteins. The polar amino acids can be subdivided into three classes.

- *Polar, neutral amino acids* have R groups that have a high affinity for water but that are not ionic at pH 7. Serine, threonine, tyrosine, cysteine, asparagine, and glutamine fall into this category. Most of these amino acids associate with one another by hydrogen bonding; but cysteine molecules form disulfide bonds with one another, as we will discuss in Section 19.6.
- *Negatively charged amino acids* have ionized carboxyl groups in their side chains. At pH 7 these amino acids have a net charge of −1. Aspartate and glutamate are the two amino acids in this category. They are acidic amino acids because ionization of the carboxylic acid releases a proton.
- *Positively charged amino acids.* At pH 7, lysine, arginine, and histidine have a net positive charge because their side chains contain positive groups. These amino groups are basic because the side chain reacts with water, picking up a proton and releasing a hydroxide anion.

The names of the amino acids can be abbreviated by a three letter code. These abbreviations are shown in Table 19.1.

Table 19.1	Names and Three-Letter Abbreviations of the α-Amino Acids
Amino Acid	**Three-Letter Abbreviation**
Alanine	ala
Arginine	arg
Asparagine	asn
Aspartate	asp
Cysteine	cys
Glutamine	gln
Glutamic acid	glu
Glycine	gly
Histidine	his
Isoleucine	ile
Leucine	leu
Lysine	lys
Methionine	met
Phenylalanine	phe
Proline	pro
Serine	ser
Threonine	thr
Tryptophan	trp
Tyrosine	tyr
Valine	val

Write the three-letter abbreviation and draw the structure of each of the following amino acids.

a. Glycine
b. Proline
c. Threonine
d. Aspartate
e. Lysine

Indicate whether each of the amino acids listed in Question 19.1 is polar, nonpolar, basic, or acidic.

19.3 The Peptide Bond

Proteins are linear polymers of L-α-amino acids. The carboxyl group of one amino acid is linked to the amino group of another amino acid. The amide bond formed in the reaction is called a **peptide bond** (Figure 19.4).

Learning Goal

3

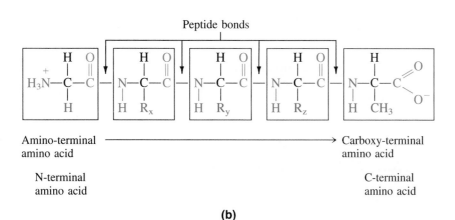

The molecule formed by condensing two amino acids is called a *dipeptide*. The amino acid with a free α-N^+H_3 group is known as the amino terminal, or simply

Figure 19.4
(a) Condensation of two α-amino acids to give a dipeptide. The two amino acids shown are glycine and alanine.
(b) Structure of a pentapeptide. Amino acid residues are enclosed in boxes. Glycine is the amino-terminal amino acid, and alanine is the carboxy-terminal amino acid.

To understand why the N-terminal amino acid is placed first and the C-terminal amino acid is placed last, we need to look at the process of protein synthesis. As we will see in Section 24.6, the N-terminal amino acid is the first amino acid of the protein. It forms a peptide bond involving its carboxyl group and the amino group of the second amino acid in the protein. Thus a free amino group literally projects from the "left" end of the protein. Similarly, the C-terminal amino acid is the last amino acid added to the protein during protein synthesis. Because the peptide bond is formed between the amino group of this amino acid and the carboxyl group of the previous amino acid, a free carboxyl group projects from the "right" end of the protein chain.

the **N-terminal amino acid,** and the amino acid with a free —COO⁻ group is known as the carboxyl, or **C-terminal amino acid.** Structures of proteins are conventionally written with their N-terminal amino acid on the left.

The number of amino acids in small peptides is indicated by the prefixes *di-* (two units), *tri-* (three units), *tetra-* (four units), and so forth. Peptides are named as derivatives of the C-terminal amino acid, which receives its entire name. For all other amino acids the ending *-ine* is changed to *-yl*. Thus the dipeptide alanyl-glycine has glycine as its C-terminal amino acid, as indicated by its full name, *glycine:*

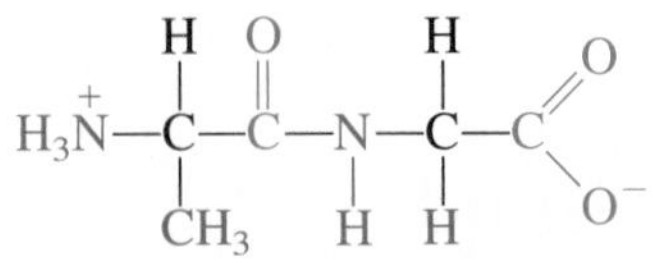

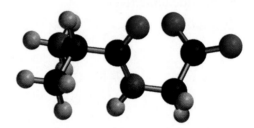

Alanyl-glycine
(ala-gly)

Alanyl-glycine

The dipeptide formed from alanine and glycine that has alanine as its C-terminal amino acid is glycyl-alanine:

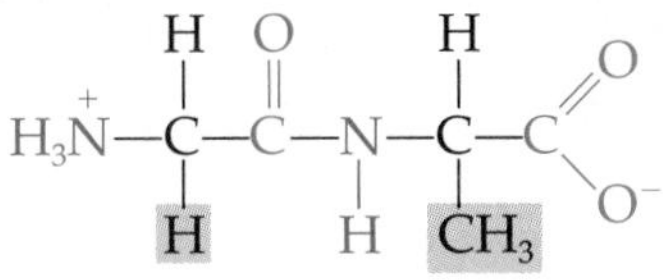

Glycyl-alanine
(gly-ala)

These two dipeptides have the same amino acid composition, but different amino acid sequences.

The structures of small peptides can easily be drawn with practice if certain rules are followed. First note that the backbone of the peptide contains the repeating sequence

N—C—C—N—C—C—N—C—C

1 2   1 2   1 2

in which N is the α-amino group, carbon-1 is the α-carbon, and carbon-2 is the carboxyl group. Carbon-1 is always bonded to a hydrogen atom and to the R group side chain that is unique to each amino acid. Continue drawing as outlined in Example 19.1.

EXAMPLE 19.1

Learning Goal

Writing the Structure of a Tripeptide

Draw the structure of the tripeptide alanyl-glycyl-valine.

Solution

Step 1. Write the backbone for a tripeptide. It will contain three sets of three atoms, or nine atoms in all. Remember that the N-terminal amino acid is written to the left.

$$N—C—C \qquad N—C—C \qquad N—C—C$$

Set 1   Set 2   Set 3

Step 2. Add oxygens to the carboxyl carbons and hydrogens to the amino nitrogens:

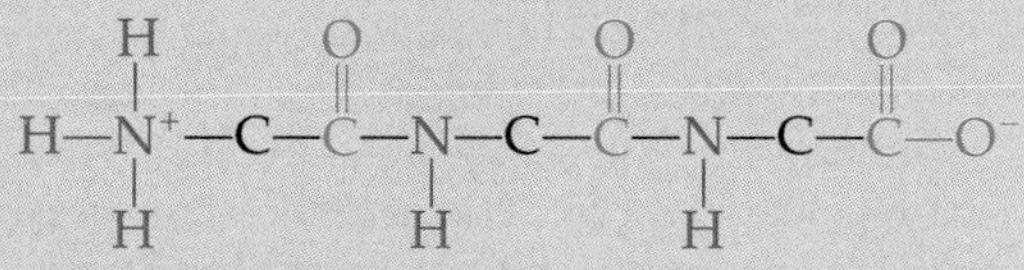

Step 3. Add hydrogens to the α-carbons:

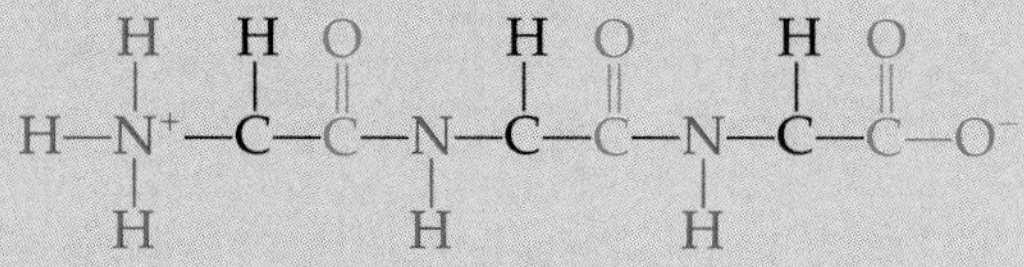

Step 4. Add the side chains. In this example (ala-gly-val) they are, from left to right, —CH₃, H, and —CH(CH₃)₂:

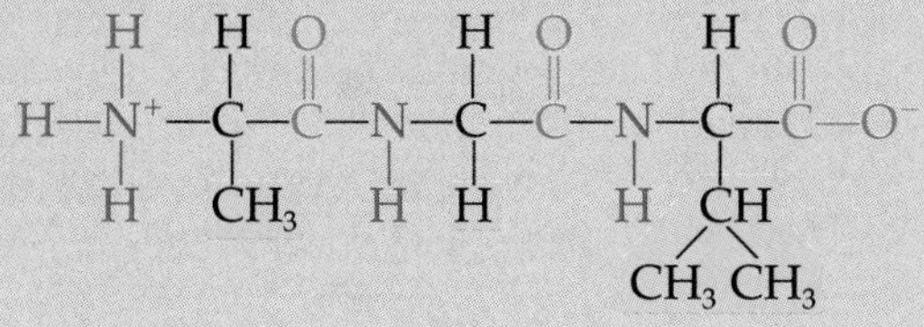

Write the structure of each of the following peptides at pH 7.

a. Alanyl-phenylalanine
b. Lysyl-alanine
c. Phenylalanyl-tyrosyl-leucine

Q u e s t i o n 19.3

Write the structure of each of the following peptides at pH 7.

a. Glycyl-valyl-serine
b. Threonyl-cysteine
c. Isoleucyl-methionyl-aspartate

Q u e s t i o n 19.4

Although you might expect free rotation about the peptide bond, this is not the case. Because the lone pair of electrons of the nitrogen atom interacts with the carbon and oxygen of the carbonyl group, the molecule exhibits resonance. This gives the peptide bond a partially double bond character:

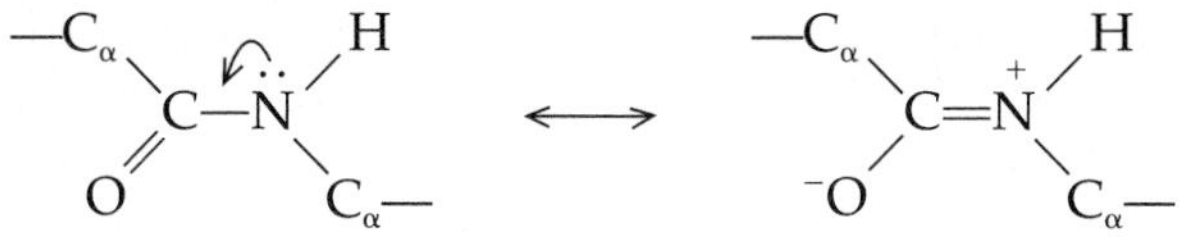

The Opium Poppy and Peptide Synthesis in the Brain

The seeds of the oriental poppy contain morphine. *Morphine* is a narcotic that has a variety of effects on the body and the brain, including drowsiness, euphoria, mental confusion, and chronic constipation. Although morphine was first isolated in 1805, not until the 1850s and the advent of the hypodermic was it effectively used as a painkiller. During the American Civil War, morphine was used extensively to relieve the pain of wounds and amputations. It was at this time that the addictive properties were noticed. By the end of the Civil War, over 100,000 soldiers were addicted to morphine.

As a result of the Harrison Act (1914), morphine came under government control and was made available only by prescription. Although morphine is addictive, *heroin*, a derivative of morphine, is much more addictive and induces a greater sense of euphoria that lasts for a longer time.

Heroin

Morphine

The structures of heroin and morphine.

Why do heroin and morphine have such powerful effects on the brain? Both drugs have been found to bind to *receptors* on the surface of the cells of the brain. The function of these receptors is to bind specific chemical signals and to direct the brain cells to respond. Yet it seemed odd that the cells of our brain should have receptors for a plant chemical. This mystery was solved in 1975, when John Hughes discovered that the brain itself synthesizes small peptide hormones with a morphinelike structure. Two of these opiate peptides are called *methionine enkephalin*, or met-enkephalin, and *leucine enkephalin*, or leu-enkephalin.

These neuropeptide hormones have a variety of effects. They inhibit intestinal motility and blood flow to the gastrointestinal tract. This explains the chronic constipation of morphine users. In addition, it is thought that these *enkephalins* play a role in pain perception, perhaps serving as a pain blockade. This is supported by the observation that they are found in higher concentrations in the bloodstream following painful stimulation. It is further suspected that they may play a role in mood and mental health. The so-called runner's high is thought to be a euphoria brought about by an excessively long or strenuous run!

Unlike morphine, the action of enkephalins is short-lived. They bind to the cellular receptor and thereby induce the cells to respond. Then they are quickly destroyed by enzymes in the brain that hydrolyze the peptide bonds of the enkephalin. Once destroyed, they are no longer able to elicit a cellular response. Morphine and heroin bind to these same receptors and induce the cells to respond. However, these drugs are not destroyed and therefore persist in the brain for long periods at concentrations high enough to continue to cause biological effects.

Many researchers are working to understand why drugs like morphine and heroin are addictive. Studies with cells in culture have suggested one mechanism for morphine tolerance and addiction. Normally, when the cell receptors bind to enkephalins, this signals the cell to decrease the production of a chemical messenger called *cyclic AMP*, or simply cAMP. (This compound is very closely related to the nucleotide adenosine-5'-monophosphate.) The decrease in cAMP level helps to block

As a result, the peptide bond is planar, and the R groups bonded to the two adjacent α-carbons lie on opposite sides of the protein chain (Figure 19.5). The hydrogen of the amide nitrogen is also *trans* to the oxygen of the carbonyl group. Almost all of the peptide bonds in proteins are planar and have a *trans* configuration. This is quite important physiologically because it makes protein structures relatively rigid. If they could not hold their shapes, they could not function.

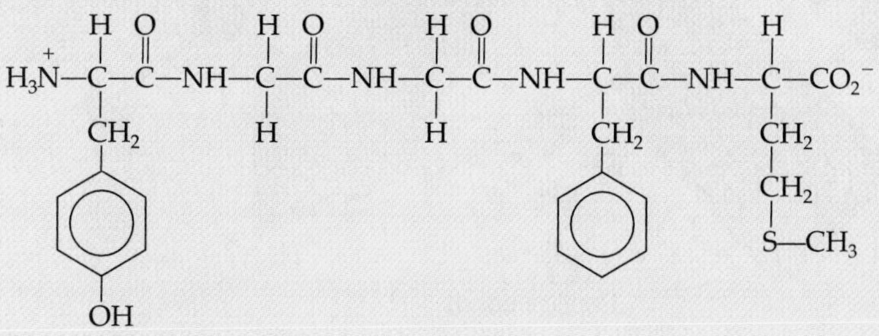

Tyr-Gly-Gly-Phe-Met
Methionine enkephalin

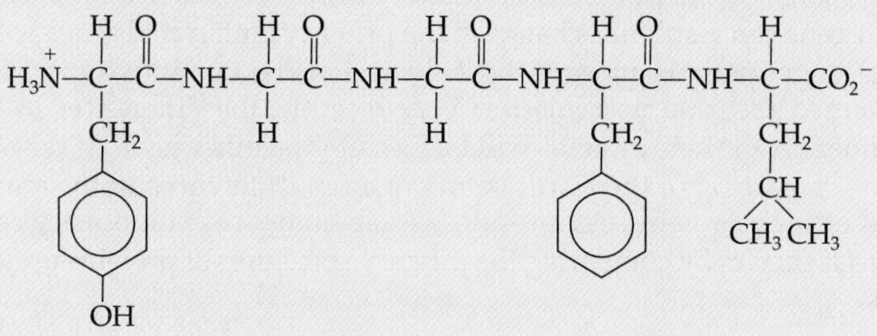

Tyr-Gly-Gly-Phe-Leu
Leucine enkephalin

Structures of the peptide opiates leucine enkephalin and methionine enkephalin. These are the body's own opiates.

pain and elevate one's mood. When morphine is applied to these cells they initially respond by decreasing cAMP levels. However, with chronic use of morphine the cells become desensitized; that is, they do not decrease cAMP production and thus behave as though no morphine were present. However, a greater amount of morphine will once again cause the decrease in cAMP levels. Thus addiction and the progressive need for more of the drug seem to result from biochemical reactions in the cells.

This logic can be extended to understand withdrawal symptoms. When an addict stops using the drug, he or she exhibits withdrawal symptoms that include excessive sweating, anxiety, and tremors. The cause may be that the high levels of morphine were keeping the cAMP levels low, thus reducing pain and causing euphoria. When morphine is removed com-

pletely, the cells overreact and produce huge quantities of cAMP. The result is all of the unpleasant symptoms known collectively as the *withdrawal syndrome.*

Clearly, morphine and heroin have demonstrated the potential for misuse and are a problem for society in several respects. Often, the money needed to support a drug habit is acquired by illegal means such as robbery, theft, and prostitution. More recently, it has become apparent that the use of shared needles for the injection of drugs is resulting in the alarming spread of the virus responsible for acquired immune deficiency syndrome (AIDS). Nonetheless, morphine remains one of the most effective painkillers known. Certainly, for people suffering from cancer, painful burns, or serious injuries the risk of addiction is far outweighed by the benefits of relief from excruciating pain.

19.4 The Primary Structure of Proteins

The **primary structure** of a protein is the amino acid sequence of the protein chain. It results from the covalent bonding between the amino acids in the chain (peptide bonds). The primary structures of proteins are translations of information contained in genes. Each protein has a different primary structure with different amino acids in different places along the chain.

Learning Goal

3

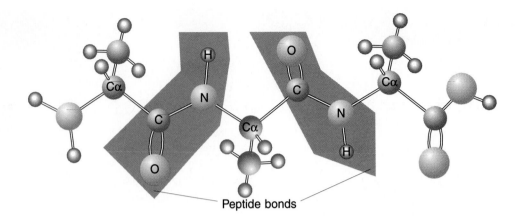

Figure 19.5

Conformation of peptide bond is planar. The C=O and N—H groups of the peptide bond are *trans* to one another. These groups are in boxes in the structure.

The genetic code and the process of protein synthesis are described in Sections 24.5 and 24.6.

Mutations and their effect on protein synthesis are discussed in Section 24.7.

Genes can change by the process of mutation during the course of evolution. A mutation in a gene can result in a change in the primary amino acid sequence of a protein. Over longer periods, more of these changes will occur. If two species of organisms diverged (became new species) very recently, the differences in the amino acid sequences of their proteins will be few. On the other hand, if they diverged millions of years ago, there will be many more differences in the amino acid sequences of their proteins. As a result, we can compare evolutionary relationships between species by comparing the primary structures of proteins present in both species.

19.5 The Secondary Structure of Proteins

The primary sequence of a protein, the chain of covalently linked amino acids, folds into regularly repeating structures that resemble designs in a tapestry. These repeating structures define the **secondary structure** of the protein. The secondary structure is the result of hydrogen bonding between the amide hydrogens and carbonyl oxygens of the peptide bonds. Many hydrogen bonds are needed to maintain the secondary structure and thereby the overall structure of the protein. Different regions of a protein chain may have different types of secondary structure. Some regions of a protein chain may have a random or non-regular structure; however, the two most common types of secondary structure are the α-helix and the β-pleated sheet because they maximize hydrogen bonding in the backbone.

α-Helix

The most common type of secondary structure is a coiled, helical conformation known as the **α-helix** (Figure 19.6). The α-helix has several important features.

- Every amide hydrogen and carbonyl oxygen associated with the peptide backbone is involved in a hydrogen bond when the chain coils into an α-helix. These hydrogen bonds lock the α-helix into place.
- Every carbonyl oxygen is hydrogen-bonded to an amide hydrogen four amino acids away in the chain.
- The hydrogen bonds of the α-helix are parallel to the long axis of the helix (see Figure 19.6).
- The polypeptide chain in an α-helix is right-handed. It is oriented like a normal screw. If you turn a screw clockwise it goes into the wall; turned counterclockwise, it comes out of the wall.
- The repeat distance of the helix, or its pitch, is 5.4 Å, and there are 3.6 amino acids per turn of the helix.

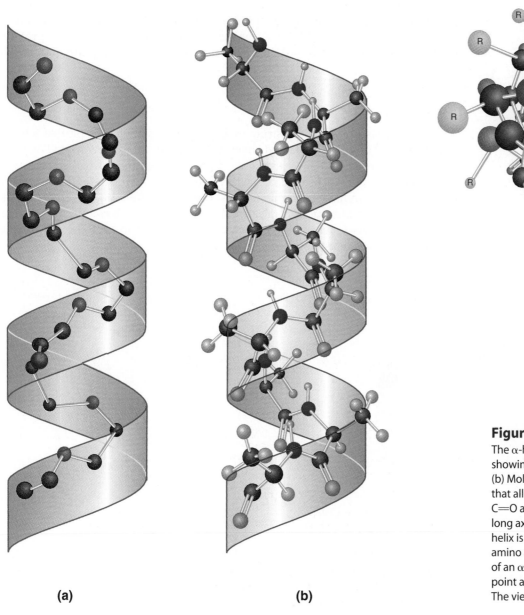

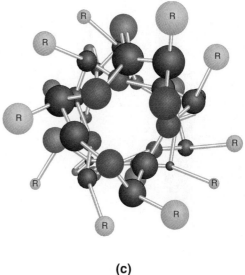

(c)

(a)   **(b)**

Figure 19.6
The α-helix. (a) Schematic diagram showing only the helical backbone. (b) Molecular model representation. Note that all of the hydrogen bonds between C=O and N—H groups are parallel to the long axis of the helix. The pitch of the helix is 5.4Å (0.54 nm), and there are 3.6 amino acid residues per turn. (c) Top view of an α-helix. The side chains of the helix point away from the long axis of the helix. The view is into the barrel of the helix.

Fibrous proteins are structural proteins arranged in fibers or sheets that have only one type of secondary structure. The **α-keratins** are fibrous proteins that form the covering (hair, wool, nails, hooves, and fur) of most land animals. Human hair provides a typical example of the structure of the α-keratins. The proteins of hair consist almost exclusively of polypeptide chains coiled up into α-helices. A single α-helix is coiled in a bundle with two other helices to give a three-stranded superstructure called a *protofibril* that is part of an array known as a *microfibril* (Figure 19.7). These structures, which resemble "molecular pigtails," possess great mechanical strength, and they are virtually insoluble in water.

The fibrous proteins of muscle are also composed of proteins that contain considerable numbers of α-helices. **Myosin,** one of the major proteins of muscle, for example, is a rodlike structure in which two α-helices form a coiled coil (Figure 19.8).

Learning Goal

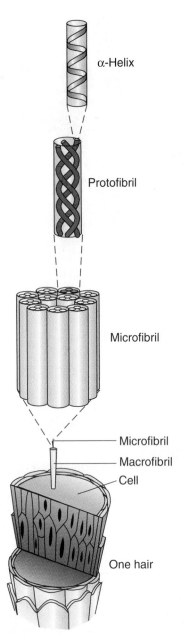

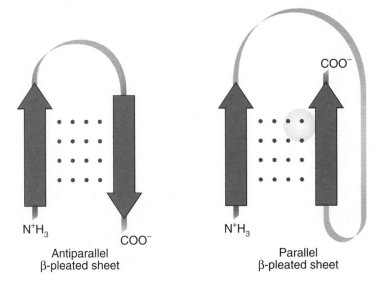

The major structural property of a coiled coil superstructure of α-helices is its great mechanical strength. This property is applied very efficiently in both the fibrous proteins of skin and those of muscle. As you can imagine, these proteins must be very strong to carry out their functions of mechanical support and muscle contraction.

β-Pleated sheet

The second common secondary structure in proteins resembles the pleated folds of drapery and is known as **β-pleated sheet** (Figure 19.9). All of the carbonyl oxygens and amide hydrogens in a β-pleated sheet are involved in hydrogen bonds, and the polypeptide chain is nearly completely extended. The polypeptide chains in a β-pleated sheet can have two orientations. If the N-termini are head to head, the structure is known as a *parallel* β-pleated sheet. And if the N-terminus of one chain is aligned with the C-terminus of a second chain (head to tail), the structure is known as an *antiparallel* β-pleated sheet.

Antiparallel
β-pleated sheet

Parallel
β-pleated sheet

Some fibrous proteins are composed of β-pleated sheets. For example, the silkworm produces *silk fibroin*, a protein whose structure is an antiparallel β-pleated sheet (Figure 19.10). The polypeptide chains of a β-pleated sheet are almost completely extended, and silk does not stretch easily. Glycine accounts for nearly half of the amino acids of silk fibroin. Alanine and serine account for most of the others. The methyl groups of alanines and the hydroxymethyl groups of serines lie on opposite sides of the sheet. Thus the stacked sheets nestle comfortably, like sheets of corrugated cardboard because the R groups are small enough to allow the stacked sheet superstructure.

Figure 19.7

Structure of the α-keratins. These proteins are assemblies of triple-helical protofibrils that are assembled in an array known as a *microfibril*. These in turn are assembled into macrofibrils. Hair is a collection of macrofibrils and hair cells.

Learning Goal

8

19.6 The Tertiary Structure of Proteins

Most fibrous proteins, such as silk, collagen, and the α-keratins, are almost completely insoluble in water. (Our skin would do us very little good if it dissolved in the rain.) The majority of cellular proteins, however, are soluble in the cell cytoplasm. Soluble proteins are usually **globular proteins.** Globular proteins have three-dimensional structures called the **tertiary structure** of the protein, which are distinct from their secondary structure. The polypeptide chain with its regions of secondary structure, α-helix and β-pleated sheet, further folds on itself to achieve the tertiary structure.

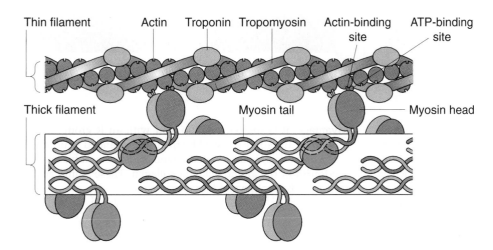

Thin filament — Actin — Troponin — Tropomyosin — Actin-binding site — ATP-binding site

Thick filament — Myosin tail — Myosin head

Figure 19.8
Schematic diagram of the structure of myosin. This muscle protein consists of a rodlike coil of α-helices with two globular heads, also composed of protein, attached to myosin at its C-terminus. In muscle, myosin molecules are assembled into thick filaments that alternate with thin filaments composed of the proteins actin, troponin, and tropomyosin. Working together, these filaments allow muscles to contract and relax.

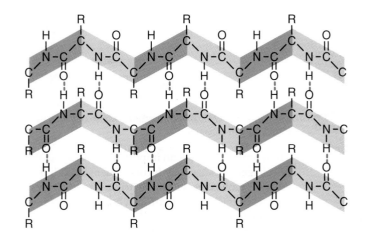

Figure 19.9
Structure of the β-pleated sheet. The polypeptide chains are nearly completely extended, and hydrogen bonds (red) between C=O and N—H groups are at right angles to the long axis of the polypeptide chains.

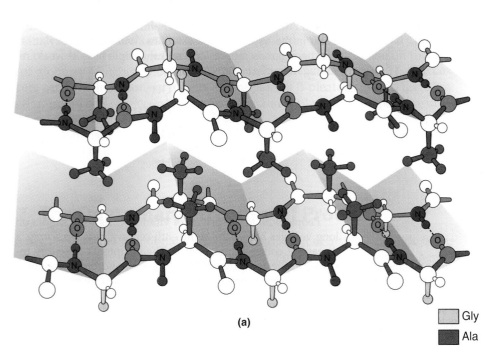

(a)

Gly
Ala

(b)

Figure 19.10
The structure of silk fibroin is almost entirely antiparallel β-pleated sheet. (a) The molecular structure of a portion of the silk fibroin protein. (b) A schematic representation of the antiparallel β-pleated sheet with the nestled R groups.

Immunoglobulins: Proteins That Defend the Body

A living organism is subjected to a constant barrage of bacterial, viral, parasitic, and fungal diseases. Without a defense against such perils we would soon perish. All vertebrates possess an *immune system*. In humans the immune system is composed of about 10^{12} cells, about as many as the brain or liver, that protect us from foreign invaders. This immune system has three important characteristics.

1. **It is highly specific.** The immune response to each infection is specific to, or directed against, only one disease organism or similar, related organisms.
2. **It has a memory.** Once the immune system has responded to an infection, the body is protected against reinfection by the same organism. This is the reason that we seldom suffer from the same disease more than once. Most of the diseases that we suffer recurrently, such as the common cold and flu, are actually caused by many different strains of the same virus. Each of these strains is "new" to the immune system.
3. **It can recognize "self" from "nonself."** When we are born, our immune system is already aware of all the antigens of our bodies. These it recognizes as "self" and will not attack. Every antigen that is not classified as "self" will be attacked by the immune system when it is encountered. Some individuals suffer from a defect of the immune response that allows it to attack the cells of one's own body. The result is an *autoimmune reaction* that can be fatal.

One facet of the immune response is the synthesis of *immunoglobulins*, or *antibodies*, that specifically bind a single macromolecule called an *antigen*. These antibodies are produced by specialized white blood cells called *B lymphocytes*. We are born with a variety of B lymphocytes that are capable of producing antibodies against perhaps a million different antigens. When a foreign antigen enters the body, it binds to the B lymphocyte that was preprogrammed to produce antibodies to destroy it. This stimulates the B cell to grow and divide. Then all of these new B cells produce antibodies that will bind to the disease agent and facilitate its destruction. Each B cell produces only one type of antibody with an absolute specificity for its target antigen. Many different B cells respond to each infection because the disease-causing agent is made up of many different antigens. Antibodies are made that bind to many of the antigens of the invader. This primary immune response is rather slow. It can take a week or two before there are enough B cells to produce a high enough level of antibodies in the blood to combat an infection.

Because the immune response has a memory, the second time we encounter a disease-causing agent the antibody response is immediate. This is why it is extremely rare to suffer from mumps, measles, or chickenpox a second time. We take advantage of this property of the immune system to protect ourselves against many diseases. In the process of *vaccination* a person can be immunized against an infectious disease by injection of a small amount of the antigens of the virus or microorganism (the vaccine). The B lymphocytes of the body then manufacture antibodies against the antigens of the infectious agent. If the individual comes into contact with the disease-causing microorganism at some later time, the sensitized B lymphocytes "remember" the antigen and very quickly produce a large amount of specific antibody to overwhelm the microorganism or virus before it can cause overt disease.

Immunoglobulin molecules contain four peptide chains that are connected by disulfide bonds and arranged in a Y-shaped quaternary structure.

Each immunoglobulin has two identical antigen-binding sites located at the tips of the Y and is therefore bivalent. Because most antigens have three or more antigen-binding sites, immunoglobulins can form large cross-linked antigen–antibody complexes that precipitate from solution.

Immunoglobulin G (IgG) is the major serum immunoglobulin. Some immunoglobulin G molecules can cross cell membranes and thus can pass between mother and fetus through the placenta, before birth. This is important because the immune system of a fetus is immature and cannot provide adequate protection from disease. Fortunately, the IgG acquired from the mother protects the fetus against most bacterial and viral infections that it might encounter before birth.

There are four additional types of antibody molecules that vary in their protein composition, but all have the same general

The oxygenation of hemoglobin in the lungs and the transfer of oxygen from hemoglobin to myoglobin in the tissues are very complex processes. We begin our investigation of these events with the inhalation of a breath of air.

The oxygenation of hemoglobin in the lungs is greatly favored by differences in the oxygen partial pressure (pO_2) in the lungs and in the blood. The pO_2 in the air in the lungs is approximately 100 mm Hg; the pO_2 in oxygen-depleted blood is only about 40 mm Hg. Oxygen diffuses from the region of high pO_2 in the lungs to

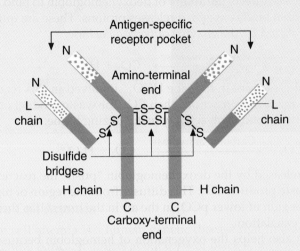

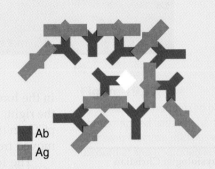

Schematic diagram of a Y-shaped immunoglobulin molecule. The binding sites for antigens are at the tips of the Y.

Schematic diagram of cross-linked immunoglobulin–antigen lattice.

Y shape. One of these is IgM, which is the first antibody produced in response to an infection. Secondarily, the B cell produces IgG molecules with the same antigen-binding region but a different protein composition in the rest of the molecule. IgA is the immunoglobulin responsible for protecting the body surfaces, such as the mucous membranes of the gut, the oral cavity, and the genitourinary tract. IgA is also found in mother's milk, protecting the newborn against diseases during the first few weeks of life. IgD is found in very small amounts and is thought to be involved in the regulation of antibody synthesis. The last type of immunoglobulin is IgE. For many years the function of IgE was unknown. It is found in large quantities in the blood of people suffering from allergies and is therefore thought to be responsible for this "overblown" immunological reaction to dust particles and pollen grains.

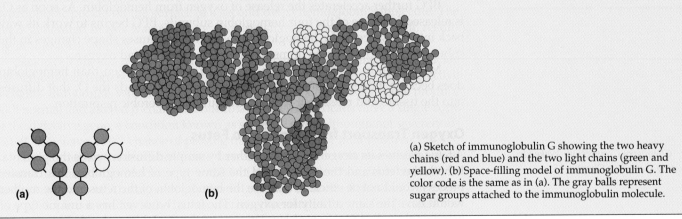

(a) Sketch of immunoglobulin G showing the two heavy chains (red and blue) and the two light chains (green and yellow). (b) Space-filling model of immunoglobulin G. The color code is the same as in (a). The gray balls represent sugar groups attached to the immunoglobulin molecule.

the region of low pO_2 in the blood. There it enters red blood cells and binds to the Fe^{2+} ions of the heme groups of deoxyhemoglobin, forming oxyhemoglobin. This binding actually helps bring more O_2 into the blood.

The events of oxygen binding are somewhat complex. Deoxyhemoglobin has a space in the center where the organic anion 2,3-bisphosphoglycerate (BPG) binds. When one of the four deoxyhemoglobin subunits binds O_2, a shape change in the protein expels the BPG. This initiates a cascade of events in which the

An Overview of Protein Structure and Function

19.47 Why is hydrogen bonding so important to protein structure?

19.48 Explain why α-keratins that have many disulfide bonds between adjacent polypeptide chains are much less elastic and much harder than those without disulfide bonds.

19.49 How does the structure of the peptide bond make the structure of proteins relatively rigid?

19.50 The primary structure of a protein known as histone H4, which tightly binds DNA, is identical in all mammals and differs by only one amino acid between the calf and pea seedlings. What does this extraordinary conservation of primary structure imply about the importance of that one amino acid?

19.51 What does it mean to say that the structure of proteins is genetically determined?

19.52 Explain why genetic mutations that result in the replacement of one amino acid with another can lead to the formation of a protein that cannot carry out its biological function.

Myoglobin and Hemoglobin

19.53 What is the function of hemoglobin?

19.54 What is the function of myoglobin?

19.55 Describe the structure of hemoglobin.

19.56 Describe the structure of myoglobin.

19.57 What is the function of heme in hemoglobin and myoglobin?

19.58 Write an equation representing the binding to and release of oxygen from hemoglobin.

19.59 Carbon monoxide binds tightly to the heme groups of hemoglobin and myoglobin. How does this affinity reflect the toxicity of carbon monoxide?

19.60 The blood of the horseshoe crab is blue because of the presence of a protein called hemocyanin. What is the function of hemocyanin?

19.61 Why does replacement of glutamic acid with valine alter hemoglobin and ultimately result in sickle cell anemia?

19.62 How do sickled red blood cells hinder circulation?

19.63 What is the difference between sickle cell disease and sickle cell trait?

19.64 How is it possible for sickle cell trait to confer a survival benefit on the person who possesses it?

Denaturation of Proteins

19.65 Define the term *denaturation*.

19.66 What is the difference between denaturation and coagulation?

19.67 Why is heat an effective means of sterilization?

19.68 As you increase the temperature of an enzyme-catalyzed reaction, the rate of the reaction initially increases. It then reaches a maximum rate and finally dramatically declines. Keeping in mind that enzymes are proteins, how do you explain these changes in reaction rate?

19.69 Why is it important that blood have several buffering mechanisms to avoid radical pH changes?

19.70 Define the term *isoelectric*.

19.71 Why do proteins become polycations at extremely low pH?

19.72 Why do proteins become polyanions at very high pH?

19.73 Yogurt is produced from milk by the action of dairy bacteria. These bacteria produce lactic acid as a by-product of their metabolism. The pH decrease causes the milk proteins to coagulate. Why are food preservatives not required to inhibit the growth of bacteria in yogurt?

19.74 Wine is made from the juice of grapes by varieties of yeast. The yeast cells produce ethanol as a by-product of their fermentation. However, when the ethanol concentration reaches 12–13%, all the yeast die. Explain this observation.

Dietary Protein and Protein Digestion

19.75 Why is it necessary to mix vegetable proteins to provide an adequate vegetarian diet?

19.76 Name some ethnic foods that apply the principle of mixing vegetable proteins to provide all of the essential amino acids.

19.77 What is the difference between essential and nonessential amino acids?

19.78 What is the difference between a complete protein and an incomplete protein?

19.79 Why must synthesis of digestive enzymes be carefully controlled?

19.80 What is the relationship between pepsin and pepsinogen?

Critical Thinking Problems

1. Calculate the length of an α-helical polypeptide that is twenty amino acids long. Calculate the length of a region of antiparallel β-pleated sheet that is forty amino acids long.

2. Proteins involved in transport of molecules or ions into or out of cells are found in the membranes of all cells. They are classified as transmembrane proteins because some regions are embedded within the lipid bilayer, whereas other regions protrude into the cytoplasm or outside the cell. Review the classification of amino acids based on the properties of their R groups. What type of amino acids would you expect to find in the regions of the proteins embedded within the membrane? What type of amino acids would you expect to find on the surface of the regions in the cytoplasm or that protrude outside the cell?

3. A biochemist is trying to purify the enzyme hexokinase from a bacterium that normally grows in the Arctic Ocean at 5°C. In the next lab, a graduate student is trying to purify the same protein from a bacterium that grows in the vent of a volcano at 98°C. To maintain the structure of the protein from the Arctic bacterium, the first biochemist must carry out all her purification procedures at refrigerator temperatures. The second biochemist must perform all his experiments in a warm room incubator. In molecular terms, explain why the same kind of enzyme from organisms with different optimal temperatures for growth can have such different thermal properties.

4. The α-keratin of hair is rich in the amino acid cysteine. The location of these cysteines in the protein chain is genetically determined; as a result of the location of the cysteines in the protein, a person may have curly, wavy, or straight hair. How can the location of cysteines in α-keratin result in these different styles of hair? Propose a hypothesis to explain how a "perm" causes straight hair to become curly.

5. Calculate the number of different pentapeptides you can make in which the amino acids phenylalanine, glycine, serine, leucine, and histidine are each found. Imagine how many proteins could be made from the twenty amino acids commonly found in proteins.

20 Enzymes

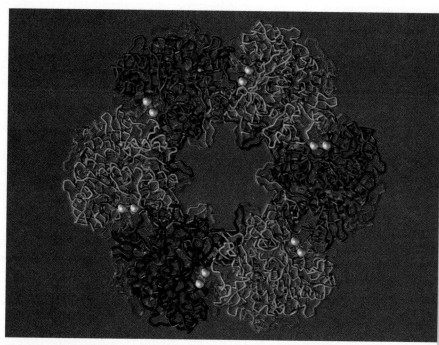

Molecular model of glutamine synthetase.

Learning Goals

1 Give examples of the correlation between an enzyme's common name and its function.

2 Classify enzymes according to the type of reaction catalyzed and the type of specificity.

3 Describe the effect that enzymes have on the activation energy of a reaction.

4 Understand the effect of substrate concentration on enzyme-catalyzed reactions.

5 Discuss the role of the active site and the importance of enzyme specificity.

6 Describe the difference between the lock-and-key model and the induced fit model of enzyme–substrate complex formation.

7 Discuss the roles of cofactors and coenzymes in enzyme activity.

8 Explain how pH and temperature affect the rate of an enzyme-catalyzed reaction.

9 Understand the mechanisms used by cells to regulate enzyme activity.

10 Discuss the mechanisms by which certain chemicals inhibit enzyme activity.

11 Discuss the role of the enzyme chymotrypsin and other serine proteases.

12 Provide examples of medical uses of enzymes.

CHEMISTRY CONNECTION

Super Hot Enzymes and the Origin of Life

Imagine the earth about four billion years ago: It was young then, not even a billion years old. Beginning as a red-hot molten sphere, slowly the earth's surface had cooled and become solid rock. But the interior, still extremely hot, erupted through the crust spewing hot gases and lava. Eventually these eruptions produced craggy land masses and an atmosphere composed of gases like hydrogen, carbon dioxide, ammonia, and water vapor. As the water vapor cooled, it condensed into liquid water, forming ponds and shallow seas.

At the dawn of biological life, the surface of the earth was still very hot and covered with rocky peaks and hot shallow oceans. The atmosphere was not very inviting either—filled with noxious gases and containing no molecular oxygen. Yet this is the environment that fostered the beginnings of life on this planet.

Some scientists think that they have found bacteria—living fossils—that may be very closely related to the first inhabitants of earth. The bacteria thrive at temperatures higher than the boiling point of water. Some need only H_2, CO_2, and H_2O for their metabolic processes and they quickly die in the presence of molecular oxygen.

But this lifestyle raises some uncomfortable questions. For instance, how do these bacteria survive at these extreme temperatures that would cook the life-forms with which we are more familiar? Researcher Mike Adams of the University of Georgia has found some of the answers. Adams and his students have studied the structure of an enzyme, a protein that acts as a biological catalyst, from one of these extraordinary bacteria. He compared the structure of the super hot enzyme with that of the same enzyme purified from an organism that grows at "normal" temperatures. The overall three-dimensional structures of the two enzymes were very similar. This makes sense because they both catalyze the same reaction.

The question, then, is why is the super hot enzyme so stable at very high temperatures, while its low temperature counterpart is not. The answer lay in the tertiary structure of the enzyme. Adams observed that the three-dimensional structure of the super hot enzyme is held together by many more R group interactions than are found in the low-temperature version. These R group interactions, along with other differences, keep the protein stable and functional even at temperatures above 100°C.

In Chapter 19 we studied the structure and properties of proteins. We are now going to apply that knowledge to the study of a group of proteins that do the majority of the work for the cell. These special proteins, the enzymes, catalyze the biochemical reactions that break down food molecules to allow the cell to harvest energy. They also catalyze the biosynthetic reactions that produce the molecules required for cellular life. In this chapter we will study the properties of this extraordinary group of proteins and learn how they dramatically speed up biochemical reactions.

Introduction

The enzymes discussed in this chapter are proteins; however, several ribonucleic acid (RNA) molecules have been demonstrated to have the ability to catalyze biological reactions. These are *ribozymes*.

An **enzyme** is a biological molecule that serves as a catalyst for a biochemical reaction. The majority of enzymes are proteins. Without enzymes to speed up biochemical reactions, life could not exist. The life of the cell depends on the simultaneous occurrence of hundreds of chemical reactions that must take place rapidly under mild conditions. It is possible, for example, to add water to an alkene. However, this reaction is usually carried out at a temperature of 100°C in aqueous sulfuric acid. Such conditions would kill a cell. The fragile cell must carry out its chemical reactions at body temperature (37°C) and in the absence of any strong acids or bases. How can this be accomplished? In Section 8.3 we saw that catalysts lower the energy of activation of a chemical reaction and thereby increase the rate of the reaction. This allows reactions to occur under milder conditions. The cell uses enzymes to solve the problem of allowing chemical reactions to occur rapidly under the mild conditions found within the cell. The enzyme *facilitates* a biochemical reaction, lowering the energy of activation and increasing the rate of the reaction. The efficient functioning of enzymes is essential for the life of the cell and of the organism.

The twin phenomena of high specificity and rapid reaction rates are the cornerstones of enzyme activity and the topic of this chapter. A typical cell contains thousands of different molecules, each of which is important to the chemistry of life processes. Each enzyme "recognizes" only one, or occasionally a few, of these molecules. One of the most remarkable features of enzymes is this *specificity*. Each can recognize and bind to a single type of *substrate* or reactant. The molecular size, shape, and charge distribution of both the enzyme and substrate must be compatible for this selective binding process to occur. The enzyme then transforms the substrate into the *product* with lightning speed. In fact, enzyme-catalyzed reactions often occur from one million to 100 million times faster than the corresponding uncatalyzed reaction.

The enzyme *catalase* provides one of the most spectacular examples of the increase in reaction rates brought about by enzymes. This enzyme is required for life in an oxygen-containing environment. In this environment the process of the aerobic (oxygen-requiring) breakdown of food molecules produces hydrogen peroxide (H_2O_2). Because H_2O_2 is toxic to the cell, it must be destroyed. One molecule of catalase converts *40 million* molecules of hydrogen peroxide to harmless water and oxygen every second:

$$2H_2O_2 \xrightarrow[\text{(an enzyme)}]{\text{Catalase}} 2H_2O + O_2$$

Reaction occurs 40 million times every second!

This is the same reaction that you witness when you pour hydrogen peroxide on a wound. The catalase released from injured cells rapidly breaks down the hydrogen peroxide. The bubbles that you see are oxygen gas released as a product of the reaction.

20.1 Nomenclature and Classification

Nomenclature of Enzymes

The common name of an enzyme is derived from the name of the **substrate** (the reactant that binds to the enzyme and is converted to product) with which the enzyme interacts and/or the type of reaction that it catalyzes. Because of this, the function of the enzyme is generally conveyed directly by its common name.

Let's look at a few examples of this simple concept. *Urea* is the substrate acted on by the enzyme *urease:*

$$\underset{\text{Substrate}}{\text{Urea}} - \text{a} + \underset{\text{Enzyme}}{\text{ase} = \text{Urease}}$$

Note that the name of this enzyme is simply the name of the substrate with the ending *-ase* added. With the exception of some historical common names, the general ending for the name of an enzyme is *-ase*. For instance, *lactose* is the substrate of *lactase:*

$$\underset{\text{Substrate}}{\text{Lactose}} - \text{ose} + \underset{\text{Enzyme}}{\text{ase} = \text{Lactase}}$$

Other enzymes may be named for the reactions they catalyze. For example,

Dehydrogenases remove hydrogen atoms, transferring them to a coenzyme.

Decarboxylases remove carboxyl groups.

The prefix *de-* indicates that a functional group is being removed. Hydrogenases and carboxylases, on the other hand, add hydrogen or carboxyl groups. Some enzyme names include *both* the names of the substrate and of the reaction type. For

Learning Goal

1

Coenzymes are molecules required by some enzymes to serve as donors or acceptors of electrons, hydrogen atoms, or other functional groups during a chemical reaction. Coenzymes are discussed in Section 20.7.

The complete name for lactate dehydrogenase is lactate: NAD oxidoreductase. This systematic name tells us the substrate, coenzyme, and type of reaction catalyzed.

example, *lactate dehydrogenase* removes hydrogen atoms from lactate ions, and *pyruvate decarboxylase* removes the carboxyl group from pyruvate.

As in other areas of chemistry, historical names, having no relationship to either substrate or reaction, continue to be used. In these cases the substrates and reactions must simply be memorized. Examples of some historical common names include catalase, pepsin, chymotrypsin, and trypsin.

Question 20.1

What is the substrate for each of the following enzymes?

a. Sucrase
b. Pyruvate decarboxylase
c. Succinate dehydrogenase

Question 20.2

What chemical reaction is mediated by each of the enzymes in Question 20.1?

Classification of Enzymes

Learning Goal

2

Enzymes may be classified according to the type of reaction in which they are involved. The six classes are as follows.

Oxidoreductases

Recall that redox reactions involve electron transfer from one substance to another (Section 9.5).

Oxidoreductases are enzymes that catalyze oxidation–reduction (redox) reactions. *Lactate dehydrogenase* is an oxidoreductase that removes hydrogen from a molecule of lactate. Other subclasses of the oxidoreductases include oxidases and reductases.

$$\underset{\text{Lactate}}{\text{HO}-\overset{\overset{\displaystyle COO^-}{|}}{\underset{\underset{\displaystyle CH_3}{|}}{C}}-H} + NAD^+ \;\rightleftharpoons\; \underset{\text{Pyruvate}}{\overset{\overset{\displaystyle COO^-}{|}}{\underset{\underset{\displaystyle CH_3}{|}}{C}}=O} + \textbf{NADH} + H^+$$

Lactate dehydrogenase

Transferases

The significance of phosphate group transfers in energy metabolism is discussed in Sections 21.1 and 21.3.

See "A Human Perspective: Amines and the Central Nervous System" in Chapter 16.

Transferases are enzymes that catalyze the transfer of functional groups from one molecule to another. For example, a *transaminase* catalyzes the transfer of an amino functional group, and a *kinase* catalyzes the transfer of a phosphate group. Kinases play a major role in energy-harvesting processes involving ATP. In the adrenal glands, norepinephrine is converted to epinephrine by the enzyme *phenylethanolamine-N-methyltransferase* (PNMT), a *transmethylase*.

Methyl group donor + HO—⬡—CHCH$_2$NH$_2$ $\overset{\text{PNMT}}{\rightleftharpoons}$ HO—⬡—CHCH$_2$NH—CH$_3$

Norepinephrine   Epinephrine

Hydrolases

Hydrolysis of esters is described in Section 15.2. The action of lipases in digestion is discussed in Section 23.1.

Hydrolases catalyze hydrolysis reactions, that is, the addition of a water molecule to a bond resulting in bond breakage. These reactions are important in the digestive process. For example, *lipases* catalyze the hydrolysis of the ester bonds in triglycerides:

$$CH_2-O-\overset{\overset{\displaystyle O}{\|}}{C}(CH_2)_nCH_3$$

$$CH-O-\overset{\overset{\displaystyle O}{\|}}{C}(CH_2)_nCH_3 \ + 3H_2O \ \xrightarrow{\text{Lipase}} \ \begin{array}{l} CH_2OH \\ CHOH \\ CH_2OH \end{array} + 3CH_3(CH_2)_nCOOH$$

$$CH_2-O-\overset{\overset{\displaystyle O}{\|}}{C}(CH_2)_nCH_3$$

Triglyceride   Glycerol   Fatty acids

Lyases

Lyases catalyze the addition of a group to a double bond or the removal of a group to form a double bond. *Citrate lyase* catalyzes the removal of an acetyl group from a molecule of citrate. The products of this reaction include oxaloacetate, acetyl CoA, ADP, and an inorganic phosphate group (P_i):

$$\begin{array}{c} COO^- \\ | \\ CH_2 \\ | \\ ^-OOC-C-OH \\ | \\ CH_2 \\ | \\ COO^- \end{array} + \ ATP \ + \ \text{Coenzyme A} \ + H_2O \ \xrightarrow{\text{Citrate lyase}}$$

Citrate

$$\begin{array}{c} COO^- \\ | \\ CH_2 \\ | \\ C=O \\ | \\ COO^- \end{array} + CH_3-\overset{\overset{\displaystyle O}{\|}}{C}\sim S-CoA + ADP + P_i$$

Oxaloacetate   Acetyl CoA

> Recall that the squiggle (~) represents a high-energy bond.

Isomerases

Isomerases rearrange the functional groups within a molecule and catalyze the conversion of one isomer into another. For example, *phosphoglyceromutase* converts one structural isomer, 3-phosphoglycerate, into another, 2-phosphoglycerate:

$$\begin{array}{c} COO^- \\ | \\ H-C-OH \\ | \\ H-C-H \\ | \\ O \\ | \\ ^-O-P=O \\ | \\ O^- \end{array} \ \underset{\text{Phosphoglyceromutase}}{\rightleftharpoons} \ \begin{array}{c} COO^- \ \ O \\ | \quad\quad \| \\ H-C-O-P-O^- \\ | \quad\quad | \\ H-C-H \ \ O^- \\ | \\ OH \end{array}$$

3-Phosphoglycerate   2-Phosphoglycerate

Ligases

Ligases are enzymes that catalyze the condensation or joining of two molecules. For example, *DNA ligase* catalyzes the joining of the hydroxyl group of a nucleotide in a DNA strand with the phosphoryl group of the adjacent nucleotide to form a phosphoester bond:

The use of DNA ligase in recombinant DNA studies is detailed in Section 24.8.

The AIDS Test

In 1981 the Centers for Disease Control in Atlanta, Georgia, recognized a new disease syndrome, acquired immune deficiency syndrome (AIDS). The syndrome is characterized by an impaired immune system, a variety of opportunistic infections and cancers, and brain damage that results in dementia. It soon became apparent that the disease was being transmitted by blood and blood products, and by sexual contact. The threat of contamination of blood supplies worldwide resulted in a multinational effort to elucidate the cause of AIDS and to develop a suitable test for the presence of the virus. As a part of this effort, Françoise Barré-Sinoussi and her colleagues at the Pasteur Institute first isolated the virus, now called *human immunodeficiency virus*, in 1983.

By April 1985 a test for virus infection was available for testing blood products. The test, called an *enzyme-linked immunosorbent assay (ELISA)*, is based on the specificity of antigen (Ag)–antibody (Ab) binding and uses a specific enzyme reaction to detect the presence of the virus in the blood. Because it is quite difficult and expensive to test for the virus itself, scientists actually test for the presence of the antibodies produced by the body in response to the virus infection.

The ELISA test is performed by coating the wells of a plastic microtiter plate with viral antigens (see the figure at right). These protein antigens are produced by growing the virus in tissue culture and purifying the viral proteins. A series of dilutions of the patient serum is prepared and placed in the wells of the microtiter plate. All samples are tested in duplicate, consistent with good analytical technique. If there are antibodies against HIV in the blood, they bind to the viral antigens on the surface of the plastic. However, this binding is invisible. How can we visualize whether the binding reaction has occurred? This involves the use of an additional antibody to which an enzyme has been covalently linked. The second antibody reacts with human IgG antibody molecules. Thus antibodies from the blood that have bound to the viral antigen on the plate will now bind to the second antibody-enzyme. Enzymes that are commonly used for this are horseradish peroxidase and alkaline phosphatase. A substrate for the enzyme is chosen that produces a colored product. For horseradish peroxidase the substrate is *ortho*-phenylenediamine, and the product is blue. For alkaline phosphatase the substrate is *para*-nitrophenylphosphate, and

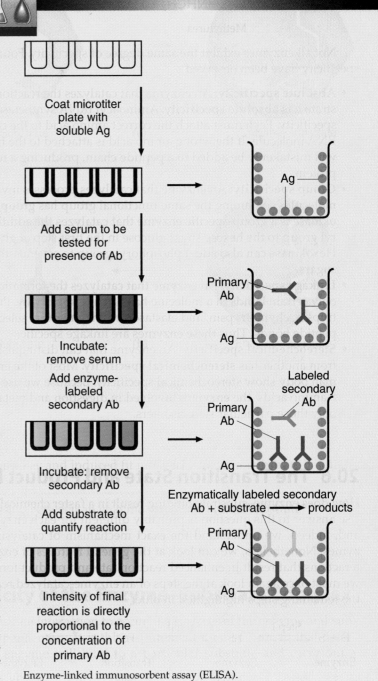

Coat microtiter
plate with
soluble Ag

Add serum to be
tested for
presence of Ab

Incubate:
remove serum

Add enzyme-
labeled
secondary Ab

Incubate: remove
secondary Ab

Add substrate to
quantify reaction

Intensity of final
reaction is directly
proportional to the
concentration of
primary Ab

Enzyme-linked immunosorbent assay (ELISA).

III). The product remains bound to the enzyme for a very brief time, then in step IV the product and enzyme dissociate from one another, leaving the enzyme completely unchanged.

What kinds of transition state changes might occur in the substrate that would make a reaction proceed more rapidly?

1. The enzyme might put "stress" on a bond and thereby facilitate bond breakage. Consider the hydrolysis of the sugar sucrose by the enzyme sucrase. The enzyme

the product is yellow. If the second antibody-enzyme has been bound by the sample, a colored product appears when the substrate is added, and it can be concluded that the test is positive for the presence of HIV infection. If no color change is observed, the patient has not been exposed to HIV, and the test is negative. The more intense the color observed, the greater the amount of HIV antibodies in the test serum.

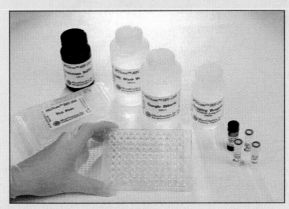

ELISA test for the presence of HIV antibodies in the blood serum. The more intense the color reaction, the greater the concentration of antibodies in the serum.

It would seem that the advent of this test would remove all threat of a contaminated blood supply, but this is unfortunately not the case. Sometimes, individuals who are not infected with the virus show positive results with the test: false positives. This is caused by other antibodies in the blood of the individual that react with other antigens that contaminate the HIV protein preparation. Other individuals who are infected demonstrate a negative result: false negatives. One reason for a false negative is the fact that it can take up to six months before the body produces antibodies against the virus. (The longest lag reported was forty-two months.) During this period the individual tests negative but is infectious. Such an individual could donate contaminated blood that would then be available for transfusion.

Because of the incidence of false positive results, any blood that tests positive is tested, in duplicate, a second time. If the result is positive in the second test, then a more accurate test is

done to determine with certainty whether the subject has been infected with HIV. This test is called a *Western blot* and relies on the use of gel electrophoresis to separate HIV proteins according to their size. The proteins are then detected by using antigen–antibody binding and enzyme assays, as with the ELISA test. Proteins of the HIV virus are electrophoresed and transferred to a membrane. The membrane is then covered with the serum of the test subject. If antibodies are present that can bind to the individual HIV proteins on the membrane, antigen–antibody complexes form. The membrane is then treated with an antibody–enzyme complex similar to that used in the ELISA test. Finally, the enzyme substrate is added. At any position on the membrane where the necessary antigen-antibody–antibody-enzyme complexes have formed, a colored band appears, as seen in the following figure.

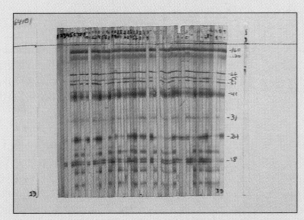

Western blot analysis of serum for the presence of HIV antibodies.

A number of new tests are currently available that can be done on blood, saliva, and urine samples. There are even blood tests that can be done at home. All of these new tests are based on the principles of the ELISA and Western blot tests. The urine and saliva tests have two advantages: No needles are involved, and both urine and saliva have very low concentrations of HIV and are considered low-risk body fluids. These factors minimize the risk to the health care worker collecting the samples.

catalyzes the hydrolysis of the disaccharide sucrose into the monosaccharides glucose and fructose. The formation of the enzyme–substrate complex (Figures 20.4a and 20.4b) results in a change in the shape of the enzyme. This, in turn, may stretch or put pressure on one of the bonds of the substrate. Such a stress weakens the bond, allowing it to be broken much more easily than in the absence of the enzyme. This is represented in Figure 20.4c as the bending of the O-glycosidic bond between the fructose and the glucose. In the transition state the substrate

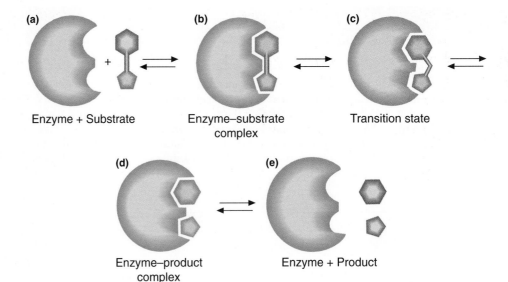

Figure 20.4

Bond breakage is facilitated by the enzyme as a result of stress on a bond. (a, b) The enzyme–substrate complex is formed. (c) In the transition state, the enzyme changes shape and thereby puts stress on the O-glycosidic linkage holding the two monosaccharides together. This lowers the energy of activation of this reaction. (d, e) The bond is broken, and the products are released.

has a molecular form resembling both the disaccharide, the original substrate, and the two monosaccharides, the eventual products. Clearly, the stress placed on the bond weakens it, and much less energy is required to break the bond to form products (Figures 20.4d and 20.4e). This also has the effect of speeding up the reaction.

2. An enzyme may facilitate a reaction by bringing two reactants into close proximity and in the proper orientation for reaction to occur. Consider now the dehydration reaction between glucose and fructose to produce sucrose (Figure 20.5a). Each of the sugars has five hydroxyl groups that could undergo condensation to produce a disaccharide. But the purpose is to produce sucrose, not some other disaccharide. By random molecular collision there is a one in twenty-five chance that the two molecules will collide in the proper orientation to produce sucrose. The probability that the two will react is actually much less than that because of a variety of conditions in addition to orientation that must be satisfied for the reaction to occur. For example, at body temperature, most molecular collisions will not have a sufficient amount of energy to overcome the energy of activation, even if the molecules are in the proper orientation. The enzyme can facilitate the reaction by bringing the two molecules close together in the correct alignment (Figure 20.5b), thereby forcing the desired reactive groups of the two molecules together in the transition state and greatly speeding up the reaction.

3. The active site of an enzyme may modify the pH of the microenvironment surrounding the substrate. To accomplish this, the enzyme may, for example, serve as a donor or an acceptor of H^+. As a result, there would be a change in the pH in the vicinity of the substrate without disturbing the normal pH elsewhere in the cell.

Question 20.7

Summarize three ways in which an enzyme might lower the energy of activation of a reaction.

Question 20.8

What is the transition state in an enzyme-catalyzed reaction?

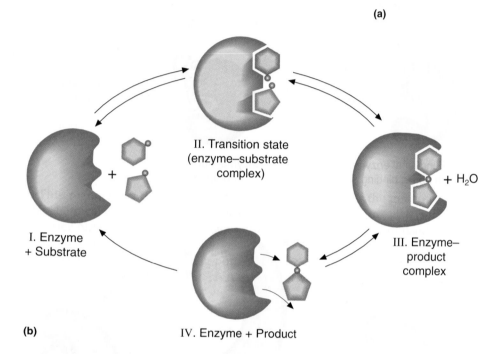

(a)

(b)

I. Enzyme + Substrate

II. Transition state (enzyme–substrate complex)

III. Enzyme–product complex

IV. Enzyme + Product

Figure 20.5

An enzyme may lower the energy of activation required for a reaction by holding the substrates in close proximity and in the correct orientation. (a) A dehydration reaction in which glucose and fructose are joined in O-glycosidic linkage to produce sucrose. (b) The enzyme–substrate complex forms, bringing the two monosaccharides together with the hydroxyl groups involved in the linkage extended toward one another.

20.7 Cofactors and Coenzymes

In Section 19.7 we saw that some proteins, the conjugated proteins, require an additional nonprotein prosthetic group to function. The same is true of some enzymes. The polypeptide portion of such an enzyme is called the **apoenzyme,** and the nonprotein prosthetic group is called the **cofactor.** Together they form the active enzyme called the **holoenzyme.** Cofactors are generally metal ions that must be bound to the enzyme to maintain the correct configuration of the enzyme active site (Figure 20.6). When the cofactor is bound and the active site is in the proper conformation, the enzyme can bind the substrate and catalyze the reaction.

Other enzymes require the temporary binding of a **coenzyme.** Such binding is generally mediated by weak interactions like hydrogen bonds. The coenzymes are organic molecules that generally serve as carriers of electrons or chemical groups. In chemical reactions they may either donate groups to the substrate or serve as recipients of groups that are removed from the substrate (Figure 20.7).

Often coenzymes contain modified vitamins as part of their structure. A **vitamin** is an organic substance that is required in the diet in only small amounts. Of the water soluble vitamins, only vitamin C has not been associated with a coenzyme. Table 20.1 is a summary of some coenzymes and the water soluble vitamins from which they are made.

Learning Goal

7

Water soluble vitamins are discussed in greater detail in Appendix F.

Regulation of Enzyme Activity

20.55 a. Why is it important for cells to regulate the level of enzyme activity?
   b. Why must synthesis of digestive enzymes be carefully controlled?
20.56 What is an allosteric enzyme?
20.57 What is the difference between positive and negative allosterism?
20.58 a. Define feedback inhibition.
   b. Describe the role of allosteric enzymes in feedback inhibition.
   c. Is this positive or negative allosterism?
20.59 What is a zymogen?
20.60 Three zymogens that are involved in digestion of proteins in the stomach and intestines are pepsinogen, chymotrypsinogen, and trypsinogen. What is the advantage of producing these enzymes as inactive peptides?

Inhibition of Enzyme Activity

20.61 Define *competitive enzyme inhibition.*
20.62 How do the sulfa drugs selectively kill bacteria while causing no harm to humans?
20.63 Describe the structure of a structural analog.
20.64 How can structural analogs serve as enzyme inhibitors?
20.65 Define *irreversible enzyme inhibition.*
20.66 Why are irreversible enzyme inhibitors often called *poisons?*
20.67 Suppose that a certain drug company manufactured a compound that had nearly the same structure as a substrate for a certain enzyme but that could not be acted upon chemically by the enzyme. What type of interaction would the compound have with the enzyme?
20.68 The addition of phenylthiourea to a preparation of the enzyme polyphenoloxidase completely inhibits the activity of the enzyme.
   a. Knowing that phenylthiourea binds all copper ions, what conclusion can you draw about whether polyphenoloxidase requires a cofactor?
   b. What kind of inhibitor is phenylthiourea?

Proteolytic Enzymes

20.69 What do the similar structures of chymotrypsin, trypsin, and elastase suggest about their evolutionary relationship?
20.70 What properties are shared by chymotrypsin, trypsin, and elastase?
20.71 Draw the complete structural formula for the peptide tyr-lys-ala-phe. Show which bond would be broken when this peptide is reacted with chymotrypsin.
20.72 Repeat Question 20.71 for the peptide trp-pro-gly-tyr.
20.73 The sequence of a peptide that contains ten amino acid residues is as follows:

   ala-gly-val-leu-trp-lys-ser-phe-arg-pro

   Indicate with arrows and label the peptide bond(s) that are cleaved by elastase, trypsin, and chymotrypsin.
20.74 What structural features of trypsin, chymotrypsin, and elastase account for their different specificities?

Uses of Enzymes in Medicine

20.75 List the enzymes whose levels are elevated in blood serum following a myocardial infarction.
20.76 List the enzymes whose levels are elevated as a result of hepatitis or cirrhosis of the liver.

Critical Thinking Problems

1. Ethylene glycol is a poison that causes about fifty deaths a year in the United States. Treating people who have drunk ethylene glycol with massive doses of ethanol can save their lives. Suggest a reason for the effect of ethanol.

2. Generally speaking, feedback inhibition involves regulation of the first step in a pathway. Consider the following hypothetical pathway:

$$
\begin{array}{c}
C \\
\downarrow E_2 \\
A \xrightarrow{E_1} B \xrightarrow{E_4} E \xrightarrow{E_5} F \xrightarrow{E_6} G \\
\uparrow E_3 \\
D
\end{array}
$$

   Which step in this pathway do you think should be regulated? Explain your reasoning.

3. In an amplification cascade, each step greatly increases the amount of substrate available for the next step, so that a very large amount of the final product is made. Consider the following hypothetical amplification cascade:

$$
\begin{array}{c}
A_{active} \\
\downarrow \\
B_{inactive} \longrightarrow B_{active} \\
\qquad\qquad \downarrow \\
C_{inactive} \longrightarrow C_{active} \\
\qquad\qquad\qquad \downarrow \\
D_{inactive} \longrightarrow D_{active}
\end{array}
$$

   If each active enzyme in the pathway converts 100 molecules of its substrate to active form, how many molecules of D will be produced if the pathway begins with one molecule of A?

4. L-1-(*p*-toluenesulfonyl)-amido-2-phenylethylchloromethyl ketone (TPCK, shown below) inhibits chymotrypsin, but not trypsin. Propose a hypothesis to explain this observation.

5. A graduate student is trying to make a "map" of a short peptide so that she can eventually determine the amino acid sequence. She digested the peptide with several proteases and determined the sizes of the resultant digestion products.

Enzyme	M.W. of Digestion Products
Trypsin	2000, 3000
Chymotrypsin	500, 1000, 3500
Elastase	500, 1000, 1500, 2000

   Suggest experiments that would allow the student to map the order of the enzyme digestion sites along the peptide.

21

Carbohydrate Metabolism

Some familiar fermentation products.

Learning Goals

1 Discuss the importance of ATP in cellular energy transfer processes.

2 Describe the three stages of catabolism of dietary proteins, carbohydrates, and lipids.

3 Discuss glycolysis in terms of its two major segments.

4 Looking at an equation representing any of the chemical reactions that occur in glycolysis, describe the kind of reaction that is occurring and the significance of that reaction to the pathway.

5 Describe the mechanism of regulation of the rate of glycolysis. Discuss particular examples of that regulation.

6 Discuss the practical and metabolic roles of fermentation reactions.

7 List several products of the pentose phosphate pathway that are required for biosynthesis.

8 Compare glycolysis and gluconeogenesis.

9 Summarize the regulation of blood glucose levels by glycogenesis and glycogenolysis.

22 Aerobic Respiration and Energy Production

Long distance runners have a great capacity for aerobic respiration.

Learning Goals

1 Name the regions of the mitochondria and the function of each region.

2 Describe the reaction that results in the conversion of pyruvate to acetyl CoA, describing the location of the reaction and the components of the pyruvate dehydrogenase complex.

3 Summarize the reactions of aerobic respiration.

4 Looking at an equation representing any of the chemical reactions that occur in the citric acid cycle, describe the kind of reaction that is occurring and the significance of that reaction to the pathway.

5 Explain the mechanisms for the control of the citric acid cycle.

6 Describe the process of oxidative phosphorylation.

7 Describe the conversion of amino acids to molecules that can enter the citric acid cycle.

8 Explain the importance of the urea cycle and describe its essential steps.

9 Discuss the cause and effect of hyperammonemia.

10 Summarize the role of the citric acid cycle in catabolism and anabolism.

CHEMISTRY CONNECTION

Mitochondria from Mom

In this chapter we will be studying the amazing, intricate set of reactions that allow us to completely degrade fuel molecules such as sugars and amino acids. These oxygen-requiring reactions occur in cellular organelles called *mitochondria*.

We are used to thinking of the organelles as a collection of membrane-bound structures that are synthesized under the direction of the genetic information in the nucleus of the cell. Not so with the mitochondria. These organelles have their own genetic information and are able to make some of their own proteins. They grow and multiply in a way very similar to the simple bacteria. This, along with other information on the structure and activities of mitochondria, has led researchers to conclude that the mitochondria are actually the descendants of bacteria captured by eukaryotic cells millions of years ago.

Recent studies of the mitochondrial genetic information (DNA) have revealed fascinating new information. For instance, although each of us inherited half our genetic information from our mothers and half from our fathers, each of us inherited all of our mitochondria from our mothers. The reason for this is that when the sperm fertilizes the egg, only the sperm nucleus enters the cell.

The observation that all of our mitochondria are inherited from our mothers led Dr. A. Wilson to study the mitochondrial DNA of thousands of women around the world. He thought that by looking for similarities and differences in the mitochondrial DNA he would be able to identify a "Mitochondrial Eve"—the mother of all humanity. He didn't really think that he could identify a single woman who would have lived tens of thousands of years ago. But he hoped to determine the location

of the first population of human women to help answer questions about the origin of humankind. Although the idea was a good one, the study had a number of experimental flaws. Currently, a hot debate is going on among hundreds of scientists about the Mitochondrial Eve. This controversy should encourage better experiments and analysis to help us identify our origins and to better understand the workings of the mitochondria.

Like the mitochondria themselves, some genetic diseases of energy metabolism are maternally inherited. One such disease, Leber's hereditary optic neuropathy (LHON), causes blindness and heart problems. People with LHON have a reduced ability to make ATP. As a result, sensitive tissues that demand a great deal of energy eventually die. LHON sufferers eventually lose their sight because the optic nerve dies from lack of energy.

Researchers have identified and cloned a mutant mitochondrial gene that is responsible for LHON. The defect is a mutant form of *NADH dehydrogenase*. NADH dehydrogenase is a huge, complex enzyme that accepts electrons from NADH and sends them on through an electron transport system. Passage of electrons through the electron transport system allows the synthesis of ATP. If NADH dehydrogenase is defective, passage of electrons through the electron transport system is less efficient, and less ATP is made. In LHON sufferers the result is eventual blindness.

In this chapter and the next, we will study some of the important biochemical reactions that occur in the mitochondria. A better understanding of the function of healthy mitochondria will eventually allow us to help those suffering from LHON and other mitochondrial genetic diseases.

Introduction

As we have seen, the anaerobic glycolysis pathway begins the breakdown of glucose and produces a small amount of ATP and NADH. But it is aerobic catabolic pathways that complete the oxidation of glucose to CO_2 and H_2O and provide most of the ATP needed by the body. In fact, this process, called *aerobic respiration*, produces thirty-six ATP molecules using the energy harvested from each glucose molecule that enters glycolysis. These reactions occur in metabolic pathways located in *mitochondria*, the cellular "power plants." Mitochondria are a type of membrane-enclosed cell *organelle*.

Here, in the mitochondria, the final oxidations of carbohydrates, lipids, and proteins occur. Here, also, the electrons that are harvested in these oxidation reactions are used to make ATP. In these remarkably efficient reactions, nearly 40% of the potential energy of glucose is stored as ATP.

An organelle is a compartment within the cytoplasm that has a specialized function.

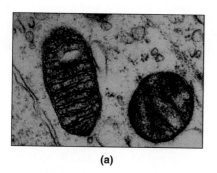 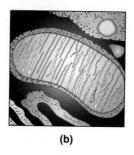

(a)   (b)

Figure 22.1
Structure of the mitochondrion.
(a) Electron micrograph of mitochondria.
(b) Schematic drawing of the mitochondrion.

22.1 The Mitochondria

Mitochondria are football-shaped organelles that are roughly the size of a bacterial cell. They are bounded by an **outer mitochondrial membrane** and an **inner mitochondrial membrane** (Figure 22.1). The region between the two membranes is known as the **intermembrane space,** and the region enclosed by the inner membrane is known as the **matrix space.** The enzymes of the citric acid cycle, of the β-oxidation pathway for the breakdown of fatty acids, and for the degradation of amino acids are all found in the mitochondrial matrix space.

Learning Goal

1

Structure and Function

The outer mitochondrial membrane is freely permeable to substances of molar mass less than 10,000 g/mol. Small molecules to be oxidized for the production of ATP, such as the pyruvate produced by glycolysis, can freely enter the intermembrane space through pores in the outer membrane.

The inner membrane is highly folded, which greatly increases the surface area. The folded membranes are known as **cristae.** The inner mitochondrial membrane is almost completely impermeable to most substances. For this reason it has many transport proteins to bring particular fuel molecules into the matrix space. Also embedded within the inner mitochondrial membrane are the protein electron carriers of the *electron transport system* and *ATP synthase.* ATP synthase is a large complex of many proteins that catalyzes the synthesis of ATP.

Origin of the Mitochondria

Not only are mitochondria roughly the size of bacteria, they have several other features that have led researchers to suspect that they may once have been free-living bacteria that were "captured" by eukaryotic cells. They have their own genetic information (DNA). They also make their own ribosomes that are very similar to those of bacteria. These ribosomes allow the mitochondria to synthesize some of their own proteins. Finally, mitochondria are actually self-replicating; they grow in size and divide to produce new mitochondria. All of these characteristics suggest that the mitochondria that produce the majority of the ATP for our cells evolved from bacteria "captured" perhaps as long as 1.5×10^9 years ago.

As we will see in Chapter 24, ribosomes are complexes of protein and RNA that serve as small platforms for protein synthesis.

What is the function of the mitochondria?

Q u e s t i o n 22.1

How do the mitochondria differ from the other components of eukaryotic cells?

Q u e s t i o n 22.2

Draw a schematic diagram of a mitochondrion, and label the parts of this organelle.

Q u e s t i o n 22.3

A HUMAN PERSPECTIVE

Exercise and Energy Metabolism

The Olympic sprinters get set in the blocks. The gun goes off, and roughly ten seconds later the 100-m dash is over. Elsewhere, the marathoners line up. They will run 26 miles and 385 yards in a little over two hours. Both sports involve running, but they utilize very different sources of energy.

Let's look at the sprinter first. The immediate source of energy for the sprinter is stored ATP. But the quantity of stored ATP is very small, only about three ounces. This allows the sprinter to run as fast as he or she can for about three seconds. Obviously, another source of stored energy must be tapped, and that energy store is *creatine phosphate:*

$$\begin{array}{c} O \quad H \quad NH \\ \parallel \quad | \quad \parallel \\ {}^-O-P-N-C-N-CH_2-C \\ | \qquad\qquad\qquad\quad \diagdown O^- \\ O^- \qquad\quad CH_3 \qquad\quad O \end{array}$$

The structure of creatine phosphate.

Creatine phosphate, stored in the muscle, donates its high-energy phosphate to ADP to produce new supplies of ATP.

This will keep our runner in motion for another five or six seconds before the store of creatine phosphate is also depleted. This is almost enough energy to finish the 100-m dash, but in reality, all the runners are slowing down, owing to energy depletion, and the winner is the sprinter who is slowing down the least!

Consider a longer race, the 400-m or the 800-m. These runners run at maximum capacity for much longer. When they have depleted their ATP and creatine phosphate stores, they must synthesize more ATP. Of course, the cells have been making ATP all the time, but now the demand for energy is much greater. To supply this increased demand, the anaerobic energy-generating reactions (glycolysis and lactate fermentation, Chapter 21) and aerobic processes (citric acid cycle and oxidative phosphorylation) begin to function much more rapidly. Often, however, these athletes are running so strenuously that they cannot provide enough oxygen to the exercising muscle to allow oxidative phosphorylation to function efficiently. When this happens, the muscles must rely on glycolysis and lactate fermentation to provide *most* of the energy requirement. The

$$\begin{array}{c} O \quad H \quad NH \\ \parallel \quad | \quad \parallel \\ {}^-O-P-N-C-N-CH_2-C \\ | \qquad\qquad\qquad\quad \diagdown O^- \\ O^- \qquad\quad CH_3 \qquad\quad O \end{array} + ADP \xrightarrow{\text{Creatine kinase}} \begin{array}{c} NH \\ \parallel \\ H_2N-C-N-CH_2-C \\ \qquad\quad | \qquad\qquad \diagdown O^- \\ \qquad\quad CH_3 \qquad\quad O \end{array} + ATP$$

Phosphoryl group transfer from creatine phosphate to ADP is catalyzed by the enzyme creatine kinase.

Question 22.4

Describe the evidence that suggests that mitochondria evolved from free-living bacteria.

22.2 Conversion of Pyruvate to Acetyl CoA

Learning Goal

As we saw in Chapter 21, under anaerobic conditions, glucose is metabolized to two pyruvate molecules that are then converted to a stable fermentation product. This limited degradation of glucose releases very little of the potential energy of glucose. Under aerobic conditions the cells can use oxygen and completely oxidize glucose to CO_2 in a metabolic pathway called the *citric acid cycle.*

This pathway is often referred to as the *Krebs cycle* in honor of Sir Hans Krebs who worked out the steps of this cyclic pathway from his own experimental data and that of other researchers. It is also called the *tricarboxylic acid (TCA) cycle* because several of the early intermediates in the pathway have three carboxyl groups.

chemical by-product of these anaerobic processes, lactate, builds up in the muscle and diffuses into the bloodstream. However, the concentration of lactate inevitably builds up in the working muscle and causes muscle fatigue and, eventually, muscle failure. Thus exercise that depends primarily on anaerobic ATP production cannot continue for very long.

The marathoner presents us with a different scenario. This runner will deplete his or her stores of ATP and creatine phosphate as quickly as a short-distance runner. The anaerobic glycolytic pathway will begin to degrade glucose provided by the blood at a more rapid rate, as will the citric acid cycle and oxidative phosphorylation. The major difference in ATP production between the long-distance runner and the short- or middle-distance runner is that the muscles of the long-distance runner derive almost all the energy through aerobic pathways. These individuals continue to run long distances at a pace that allows them to supply virtually all the oxygen needed by the exercising muscle. In fact, only aerobic pathways can provide a constant supply of ATP for exercise that goes on for hours. Theoretically, under such conditions our runner could run indefinitely, utilizing first his or her stored glycogen and eventually stored lipids. Of course, in reality, other factors such as dehydration and fatigue place limits on the athlete's ability to continue.

We have seen, then, that long-distance runners must have a great capacity to produce ATP aerobically, in the mitochondria, whereas short- and middle-distance runners need a great capacity to produce energy anaerobically, in the cytoplasm of the muscle cells. It is interesting to note that the muscles of these runners reflect these diverse needs.

When one examines muscle tissue that has been surgically removed, one finds two predominant types of muscle fibers.

Fast twitch muscle fibers are large, relatively plump, pale cells. They have only a few mitochondria but contain a large reserve of glycogen and high concentrations of the enzymes that are needed for glycolysis and lactate fermentation. These muscle fibers fatigue rather quickly because fermentation is inefficient, quickly depleting the cell's glycogen store and causing the accumulation of lactate.

Slow twitch muscle fiber cells are about half the diameter of fast twitch muscle cells and are red. The red color is a result of the high concentrations of myoglobin in these cells. Recall that myoglobin stores oxygen for the cell (Section 19.9) and facilitates rapid diffusion of oxygen throughout the cell. In addition, slow twitch muscle fiber cells are packed with mitochondria. With this abundance of oxygen and mitochondria these cells have the capacity for extended ATP production via aerobic pathways—ideal for endurance sports like marathon racing.

It is not surprising, then, that researchers have found that the muscles of sprinters have many more fast twitch muscle fibers and those of endurance athletes have many more slow twitch muscle fibers. One question that many researchers are trying to answer is whether the type of muscle fibers an individual has is a function of genetic makeup or training. Is a marathon runner born to be a long-distance runner, or are his or her abilities due to the type of training the runner undergoes? There is no doubt that the training regimen for an endurance runner does indeed increase the number of slow twitch muscle fibers and that of a sprinter increases the number of fast twitch muscle fibers. But there is intriguing new evidence to suggest that the muscles of endurance athletes have a greater proportion of slow twitch muscle fibers before they ever begin training. It appears that some of us truly were born to run.

We now turn our attention to the production of the molecule that carries two carbon fragments, acetyl groups, from pyruvate into the aerobic energy-harvesting pathways. This molecule is **acetyl CoA.**

The structure of acetyl CoA is seen in Figure 22.2. The **coenzyme A** portion of this molecule is derived from ATP and the vitamin pantothenic acid. It serves as an acceptor of acetyl groups, shown in red in Figure 22.2, that are linked to the thiol group of the molecule by a thioester bond. We can consider acetyl CoA to be an "activated" form of the acetyl group.

The reaction that converts pyruvate to acetyl CoA is shown in Figure 22.3. The pyruvate is decarboxylated, which liberates a molecule of CO_2. It is also oxidized, and the hydride anion that is removed is accepted by NAD^+, which is thus reduced. Finally, the remaining acetyl group, $CH_3CO—$, is linked to coenzyme A by a thioester bond. This very complex reaction is carried out by three enzymes and five coenzymes. To economize the process, these enzymes and coenzymes are all localized in a single bundle called the **pyruvate dehydrogenase complex** (see Figure 22.3). In this way the substrate can be passed from one enzyme to the next as each modification occurs. A schematic representation of this "disassembly line" is shown in Figure 22.3b.

Coenzyme A is described in Sections 13.9 and 15.4.

Thioester bonds are discussed in Section 15.4.

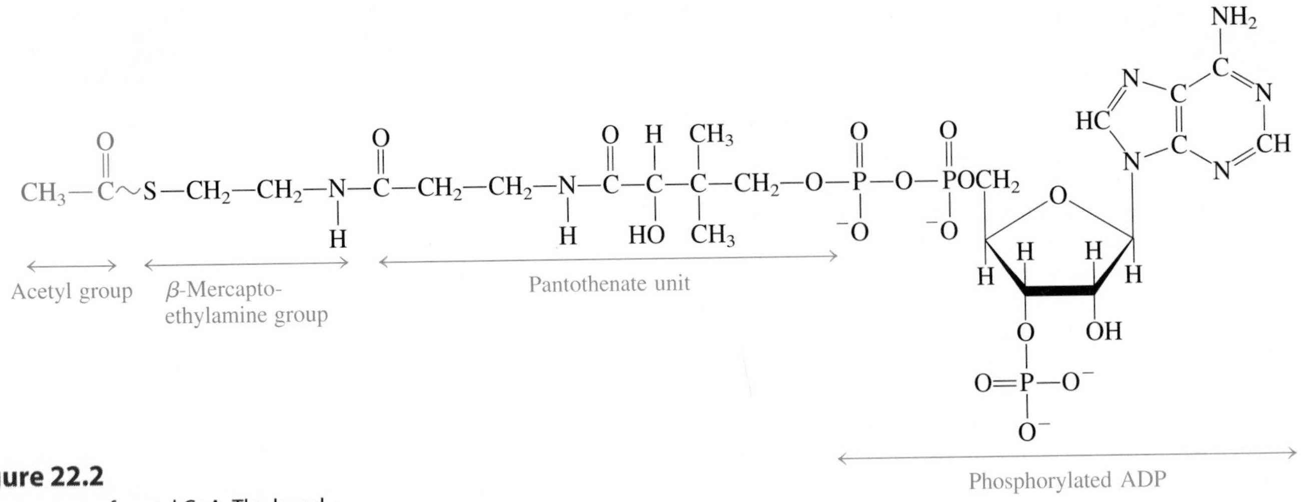

Figure 22.2
The structure of acetyl CoA. The bond between the acetyl group and coenzyme A is a high-energy thioester bond.

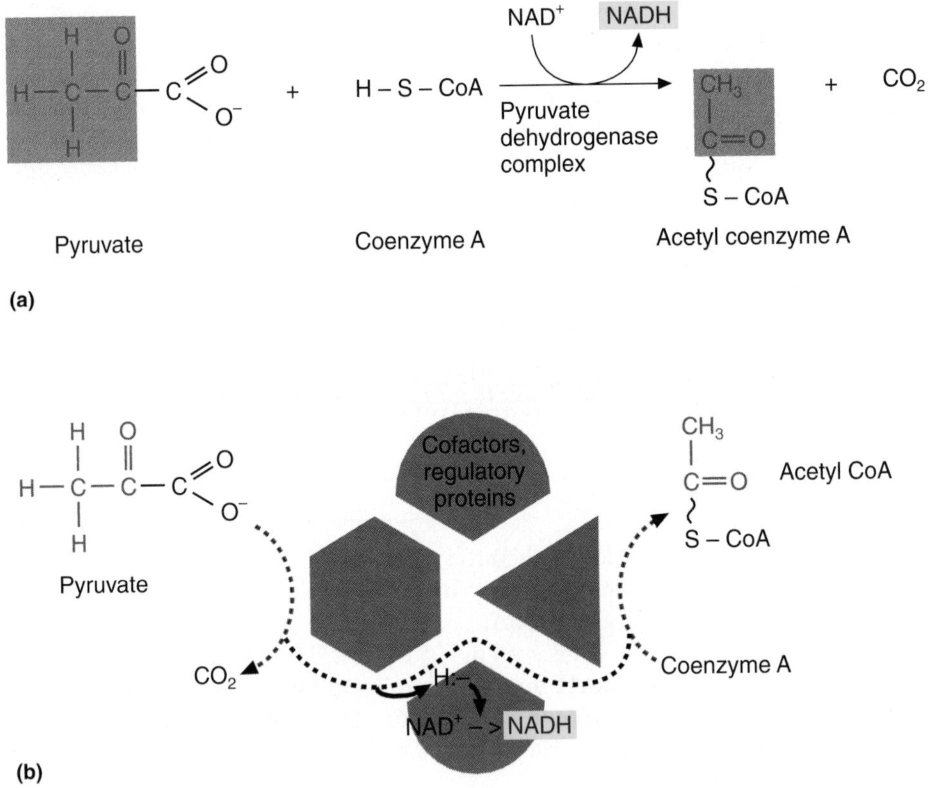

Acetyl coenzyme A
(acetyl CoA)

(a)

Figure 22.3
The decarboxylation and oxidation of pyruvate to produce acetyl CoA. (a) The overall reaction in which CO_2 and an H:$^-$ are removed from pyruvate and the remaining acetyl group is attached to coenzyme A. This requires the concerted action of three enzymes and five coenzymes. (b) The pyruvate dehydrogenase complex that carries out this reaction is actually a cluster of enzymes and coenzymes. The substrate is passed from one enzyme to the next as the reaction occurs.

(b)

This single reaction requires four coenzymes made from four different vitamins, in addition to the coenzyme lipoamide. These are thiamine pyrophosphate, derived from thiamine (Vitamin B_1); FAD, derived from riboflavin (Vitamin B_2); NAD$^+$, derived from niacin; and coenzyme A, derived from pantothenic acid. Obviously, a deficiency in any of these vitamins would seriously reduce the amount of acetyl CoA that our cells could produce. This, in turn, would limit the amount of ATP that the body could make and would contribute to vitamin-deficiency

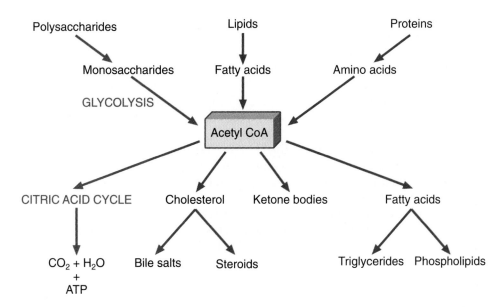

Figure 22.4
The central role of acetyl CoA in cellular metabolism.

diseases. Fortunately, a well-balanced diet provides an adequate supply of these and other vitamins.

In Figure 22.4 we see that acetyl CoA is a central character in cellular metabolism. It is produced by the degradation of glucose, fatty acids, and some amino acids. The major function of acetyl CoA in energy-harvesting pathways is to carry the acetyl group to the citric acid cycle, in which it will be used to produce large amounts of ATP. In addition to these catabolic duties, the acetyl group of acetyl CoA can also be used for *anabolic* or biosynthetic reactions to produce cholesterol and fatty acids. It is through this intermediate, acetyl CoA, that all the energy sources (fats, proteins, and carbohydrates) are interconvertible.

What vitamins are required for acetyl CoA production from pyruvate?

Q u e s t i o n **22.5**

What is the major role of coenzyme A in catabolic reactions?

Q u e s t i o n **22.6**

22.3 An Overview of Aerobic Respiration

The structure of the mitochondrion is very complex, but each region of the mitochondrion has an important role to play in the process of **aerobic respiration.** Aerobic respiration is the oxygen-requiring breakdown of food molecules and production of ATP.

The enzymes for the citric acid cycle are found in the mitochondrial matrix space. The first of these enzymes catalyzes a reaction that joins the acetyl group of acetyl CoA (two carbons) to a four-carbon molecule (oxaloacetate) to produce citrate (six carbons). The remaining enzymes catalyze a series of rearrangements, decarboxylations (removal of CO_2), and oxidation–reduction reactions. The eventual products of this cyclic pathway are two CO_2 molecules and oxaloacetate—the molecule we began with.

At several steps in the citric acid cycle a substrate is oxidized. In three of these steps a pair of electrons is transferred from the substrate to NAD^+, producing NADH (three NADH molecules per turn of the cycle). At another step a pair of

Learning Goal

Remember (Section 20.7) that it is really the hydride anion with its pair of electrons ($H{:}^-$) that is transferred to NAD^+ to produce NADH. Similarly, a pair of hydrogen atoms (and thus two electrons) are transferred to FAD to produce $FADH_2$.

electrons is transferred from a substrate to FAD, producing FADH$_2$ (one FADH$_2$ molecule per turn of the cycle).

The electrons are passed from NADH or FADH$_2$, through an electron transport system located in the inner mitochondrial membrane, and finally to the terminal electron acceptor, molecular oxygen (O$_2$). The transfer of electrons through the electron transport system causes protons (H$^+$) to be pumped from the mitochondrial matrix into the intermembrane compartment. The result is a high-energy H$^+$ reservoir.

In the final step the energy of the H$^+$ reservoir is used to make ATP. This last step is carried out by the enzyme complex ATP synthase. As protons flow back into the mitochondrial matrix through a pore in the ATP synthase complex, the enzyme catalyzes the synthesis of ATP.

This long, involved process is called *oxidative phosphorylation*, because the energy of electrons from the *oxidation* of substrates in the citric acid cycle is used to *phosphorylate* ADP and produce ATP. The details of each of these steps will be examined in upcoming sections.

Question 22.7	What is meant by the term *oxidative phosphorylation*?
Question 22.8	What does the term *aerobic respiration* mean?

22.4 The Citric Acid Cycle (The Krebs Cycle)

Reactions of the Citric Acid Cycle

Learning Goal

The **citric acid cycle** is sometimes called the *Krebs cycle*, in honor of its discoverer, Sir Hans Krebs. It is the final stage of the breakdown of carbohydrates, fats, and amino acids released from dietary proteins (Figure 22.5).

To understand this important cycle, let's follow the fate of the acetyl group of an acetyl CoA as it passes through the citric acid cycle. The numbered steps listed below correspond to the steps in the citric acid cycle that are summarized in Figure 22.5.

The formation of acetyl CoA was described in Section 22.2.

Reaction 1. The acetyl group of acetyl CoA is transferred to oxaloacetate in a reaction catalyzed by the enzyme *citrate synthase*. The product that is formed is citrate:

$$\text{Oxaloacetate} + \text{Acetyl CoA} + H_2O \xrightarrow{\text{Citrate synthase}} \text{Citrate} + \text{HS—CoA} + H^+$$

Oxaloacetate   Acetyl CoA   Citrate   Coenzyme A

Reaction 2. The enzyme *aconitase* catalyzes the dehydration of citrate, producing *cis*-aconitate. The same enzyme, aconitase, then catalyzes addition of a water molecule to the *cis*-aconitate, converting it to isocitrate. The net effect of these two steps is the isomerization of citrate to isocitrate:

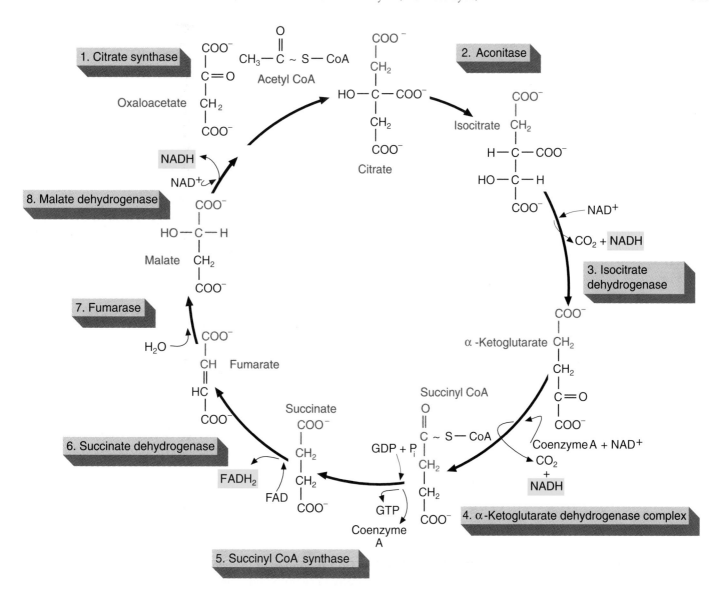

Figure 22.5
The reactions of the citric acid cycle.

Reaction 3. The first oxidative step of the citric acid cycle is catalyzed by *isocitrate dehydrogenase.* It is a complex reaction in which three things happen:
   a. the hydroxyl group of isocitrate is oxidized to a ketone,
   b. carbon dioxide is released, and
   c. NAD^+ is reduced to NADH.

The structure of NAD^+ and its reduction to NADH are shown in Figure 20.8.

A HUMAN PERSPECTIVE

Brown Fat: The Fat That Makes You Thin?

Humans have two types of fat, or adipose, tissue. *White fat* is distributed throughout the body and is composed of aggregations of cells having membranous vacuoles containing stored triglycerides. The size and number of these storage vacuoles determine whether a person is overweight or not. The other type of fat is *brown fat*. Brown fat is a specialized tissue for heat production, called *nonshivering thermogenesis*. As the name suggests, this is a means of generating heat in the absence of the shivering response. The cells of brown fat look nothing like those of white fat. They do contain small fat vacuoles; however, the distinguishing feature of brown fat is the huge number of mitochondria within the cytoplasm. In addition, brown fat tissue contains a great many blood vessels. These provide oxygen for the thermogenic metabolic reactions.

Brown fat is most pronounced in newborns, cold-adapted mammals, and hibernators. One major difficulty faced by a newborn is temperature regulation. The baby leaves an environment in which he or she was bathed in fluid of a constant 37°C, body temperature. Suddenly, the child is thrust into a world that is much colder and in which he or she must generate his or her own warmth internally. By having a good reserve of active brown fat to generate that heat, the newborn is protected against cold shock at the time of birth. However, this thermogenesis literally burns up most of the brown fat tissue, and adults typically have so little brown fat that it can be found only by using a special technique called *thermography*, which detects temperature differences throughout a body. However, in some individuals, brown fat is very highly developed. For instance, the Korean diving women who spend 6–7 hours every day diving for pearls in cold water have a massive amount of brown fat to warm them by nonshivering thermogenesis. Thus development of brown fat is a mechanism of cold adaptation.

When it was noticed that such cold-adapted individuals were seldom overweight, a correlation was made between the amount of brown fat in the body and the tendency to become overweight. Studies done with rats suggest that, to some degree, fatness is genetically determined. In other words, you are as lean as your genes allow you to be. In these studies, cold-adapted and non-

cold-adapted rats were fed cafeteria food—as much as they wanted—and their weight gain was monitored. In every case the cold-adapted rats, with their greater quantity of brown fat, gained significantly less weight than their non–cold-adapted counterparts, despite the fact that they ate as much as the non–cold-adapted rats. This and other studies led researchers to conclude that brown fat burns excess fat in a highly caloric diet.

How does brown fat generate heat and burn excess calories? For the answer we must turn to the mitochondrion. In addition to the ATP synthase and the electron transport system proteins that are found in all mitochondria, there is a protein in the inner mitochondrial membrane of brown fat tissue called *thermogenin*. This protein has a channel in the center through which the protons (H^+) of the intermembrane space could pass back into the mitochondrial matrix. Under normal conditions this channel is plugged by a GDP molecule so that it remains closed and the proton gradient can continue to drive ATP synthesis by oxidative phosphorylation.

When brown fat is "turned on," by cold exposure or in response to certain hormones, there is an immediate increase in the rate of glycolysis and β-oxidation of the stored fat (Chapter 23). These reactions produce acetyl CoA, which then fuels the citric acid cycle. The citric acid cycle, of course, produces NADH and $FADH_2$, which carry electrons to the electron transport system. Finally, the electron transport system pumps protons into the intermembrane space. Under usual conditions the energy of the proton gradient would be used to synthesize ATP. However, when brown fat is stimulated, the GDP that had plugged the pore in thermogenin is lost. Now protons pass freely back into the matrix space, and the proton gradient is dissipated. The energy of the gradient, no longer useful for generating ATP, is released as *heat*, the heat that warms and protects newborns and cold-adapted individuals.

Brown fat is just one of the body's many systems for maintaining a constant internal environment regardless of the conditions in the external environment. Such mechanisms, called *homeostatic mechanisms*, are absolutely essential to allow the body to adapt to and survive in an ever-changing environment.

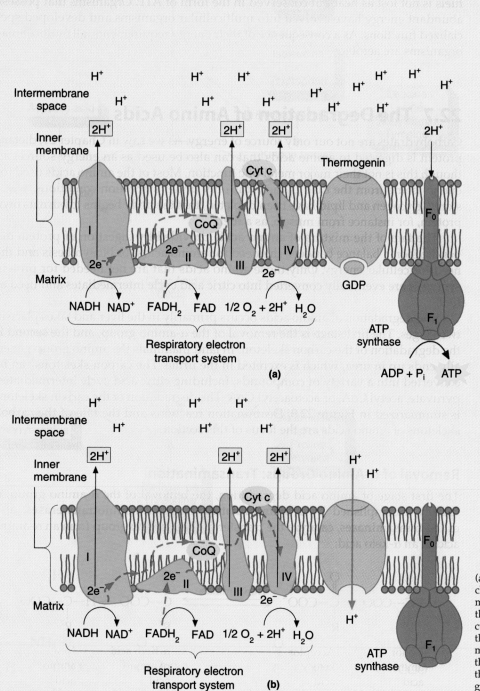

(a) The inner membrane of brown fat mitochondria contains thermogenin. In the normal state the pore in the center of thermogenin is plugged by a GDP molecule. (b) When brown fat is activated for thermogenesis, the GDP molecule is removed from the pore, and the protons from the H^+ reservoir are free to flow back into the matrix of the mitochondrion. As the gradient dissipates, heat energy is released.

23 Fatty Acid Metabolism

A tasty source of energy.

Learning Goals

1 Summarize the digestion and storage of lipids.

2 Describe the degradation of fatty acids by β-oxidation.

3 Explain the role of acetyl CoA in fatty acid metabolism.

4 Understand the role of ketone body production in β-oxidation.

5 Compare β-oxidation of fatty acids and fatty acid biosynthesis.

6 Describe the regulation of lipid and carbohydrate metabolism in relation to the liver, adipose tissue, muscle tissue, and the brain.

7 Summarize the antagonistic effects of glucagon and insulin.

CHEMISTRY CONNECTION

Obesity: A Genetic Disorder?

Approximately a third of all Americans are obese; that is, they are more than 20% overweight. One million are morbidly obese; they carry so much extra weight that it threatens their health. Many obese people simply eat too much and exercise too little, but others actually gain weight even though they eat fewer calories than people of normal weight. This observation led many researchers to the hypothesis that obesity in some people is a genetic disorder.

This hypothesis was supported by the 1950 discovery of an obesity mutation in mice. Selective breeding produced a strain of genetically obese mice from the original mutant mouse. The hypothesis was further strengthened by the results of experiments performed in the 1970s by Douglas Coleman. Coleman connected the circulatory systems of a genetically obese mouse and a normal mouse. The obese mouse started eating less and lost weight. Coleman concluded that there was a substance in the blood of normal mice that signals the brain to decrease the appetite. Obese mice, he hypothesized, can't produce this "satiety factor," and thus they continue to eat and gain weight.

In 1987, Jeffrey Friedman assembled a team of researchers to map and then clone the obesity gene that was responsible for appetite control. In 1994, after seven years of intense effort, the scientists achieved their goal, but they still had to demonstrate that the protein encoded by the cloned obesity gene did, indeed, have a metabolic effect. The gene was modified to be compatible with the genetic system of bacteria so that they could be used to manufacture the protein. When the engineered gene was then introduced into bacteria, they produced an abundance of the protein product. The protein was then purified in preparation for animal testing.

The researchers calculated that a normal mouse has about 12.5 mg of the protein in its blood. They injected that amount into each of ten mice that were so fat they couldn't squeeze into the feeding tunnels used for normal mice. The day after the first injection, graduate student Jeff Halaas observed that the mice had eaten less food. Injections were given daily, and each day the obese mice ate less. After two weeks of treatment, each of the ten mice had lost about 30% of its weight. In addition, the mice had become more active and their metabolisms had speeded up.

When normal mice underwent similar treatment, their body fat fell from 12.2% to 0.67%, which meant that these mice had no extra fat tissue. The 0.67% of their body weight represented by fat was accounted for by the membranes that surround each of the cells of their bodies! Because of the dramatic results, Friedman and his colleagues called the protein leptin, from the Greek word *leptos*, meaning slender.

The leptin protein is a hormone, and ongoing research is aimed at understanding how leptin works to control metabolism and food intake. Friedman has hypothesized that it is a signal in a metabolic thermostat. Fat cells produce leptin and secrete it into the bloodstream. As a result, the leptin concentration in a normal person is proportional to the amount of body fat. The blood concentration of the hormone is monitored by a center in the brain, probably the hypothalamus, a region known to control appetite and set metabolic rates. When the concentration reaches a certain level, it triggers the hypothalamus to suppress the appetite. If a genetically obese person, or mouse, produces no leptin or only small amounts of it, the hypothalamus "thinks" that the individual has too little body fat or is starving. Under these circumstances it does not send a signal to suppress hunger and the individual continues to eat.

The human leptin gene also has been cloned and shown to correct genetic obesity in mice. But what about obesity in humans? Leptin has been tested in a small number of obese individuals. Unfortunately, the dramatic results achieved with mice were *not* observed with humans. Why? It seems that the obese volunteers already produced an abundance of leptin. But remember that the leptin must bind to its receptor to suppress the appetite. Apparently the majority of the genetically obese humans have a defect in the leptin receptor gene, not the gene for the hormone itself. The leptin receptor gene has been cloned and a great deal of research is focusing on the interaction between the hormone and its receptor.

Lipid metabolism in animals is a complex process and is not yet fully understood. The discovery of the leptin gene, and the hormone it produces, is just one part of the story. In this chapter we will study other aspects of lipid metabolism: the pathways for fatty acid degradation and biosynthesis and the processes by which dietary lipids are digested and excess lipids are stored.

Introduction

The metabolism of fatty acids and lipids revolves around the fate of acetyl CoA. We saw in Chapter 22 that, under aerobic conditions, pyruvate is converted to acetyl CoA, which feeds into the citric acid cycle. Fatty acids are also degraded to acetyl CoA and oxidized by the citric acid cycle, as are certain amino acids. Moreover, acetyl CoA is itself the starting material for the biosynthesis of fatty acids, fully half of the amino acids, cholesterol, and steroid hormones. Acetyl CoA is thus one of the major metabolites of intermediary metabolism.

23.1 Lipid Metabolism in Animals

Digestion and Absorption of Dietary Triglycerides

Triglycerides are highly hydrophobic ("water fearing"). Because of this they must be processed before they can be digested, absorbed, and metabolized. Because processing of dietary lipids occurs in the small intestine, the water soluble **lipases,** enzymes that hydrolyze triglycerides, that are found in the stomach and in the saliva are not very effective. In fact, most dietary fat arrives in the duodenum, the first part of the small intestine, in the form of fat globules. These fat globules stimulate the secretion of bile from the gallbladder. **Bile** is composed of micelles of lecithin, cholesterol, protein, bile salts, inorganic ions, and bile pigments. **Micelles** (Figure 23.1) are aggregations of molecules having a polar region and a nonpolar region. The nonpolar ends of bile salts tend to bunch together when placed in water. The hydrophilic ("water loving") regions of these molecules interact with water. Bile salts are made in the liver and stored in the gallbladder, awaiting the stimulus to be secreted into the duodenum. The major bile salts in humans are cholate and chenodeoxycholate (Figure 23.2).

Cholesterol is almost completely insoluble in water, but the conversion of cholesterol to bile salts creates *detergents* whose polar heads make them soluble in the aqueous phase of the cytoplasm and whose hydrophobic tails bind triglycerides. After a meal is eaten, bile flows through the common bile duct into the duodenum, where bile salts emulsify the fat globules into tiny droplets. This increases the surface area of the lipid molecules, allowing them to be more easily hydrolyzed by lipases (Figure 23.3).

Much of the lipid in these droplets is in the form of **triglycerides,** or triacylglycerols, which are fatty acid esters of glycerol. A protein called **colipase** binds to the surface of the lipid droplets and helps pancreatic lipases to stick to the surface and hydrolyze the ester bonds between the glycerol and fatty acids of the triglycerides (Figure 23.4). In this process, two of the three fatty acids are liberated, and the monoglycerides and free fatty acids produced mix freely with the micelles of bile. These micelles are readily absorbed through the membranes of the intestinal epithelial cells (Figure 23.5).

Learning Goal

See Sections 15.1 and 18.2 for a discussion of micelles.

Triglycerides are described in Section 18.3.

To tho
duct a
blood

for nea
cle, cor
by mit

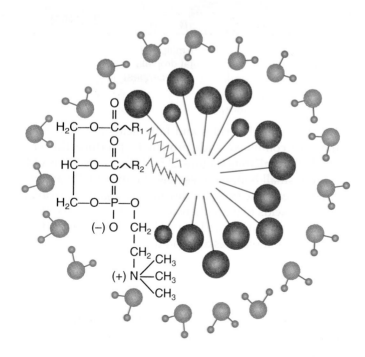

Figure 23.1

The structure of a micelle formed from the phospholipid lecithin. The straight lines represent the long hydrophobic fatty acid tails, and the spheres represent the hydrophilic heads of the phospholipid.

Biochemical analysis showed that he had only one-fifth of the normal amount of carnitine in his muscle cells.

What effect will carnitine deficiency have on β-oxidation? What effect will carnitine deficiency have on glucose metabolism?

6. Acetyl CoA carboxylase catalyzes the formation of malonyl CoA from acetyl CoA and the bicarbonate anion, a reaction that requires the hydrolysis of ATP. Write a balanced equation showing this reaction.

The reaction catalyzed by acetyl CoA carboxylase is the rate-limiting step in fatty acid biosynthesis. The malonyl group is transferred from coenzyme A to acyl carrier protein; similarly, the acetyl group is transferred from coenzyme A to acyl carrier protein. This provides the two beginning substrates of fatty acid biosynthesis shown in Figure 23.11.

Consider the following case study. A baby boy was brought to the emergency room with severe respiratory distress. Examination revealed muscle pathology, poor growth, and severe brain damage. A liver biopsy revealed that the child didn't make acetyl CoA carboxylase. What metabolic pathway is defective in this child? How is this defect related to the respiratory distress suffered by the baby?

24

Introduction to Molecular Genetics

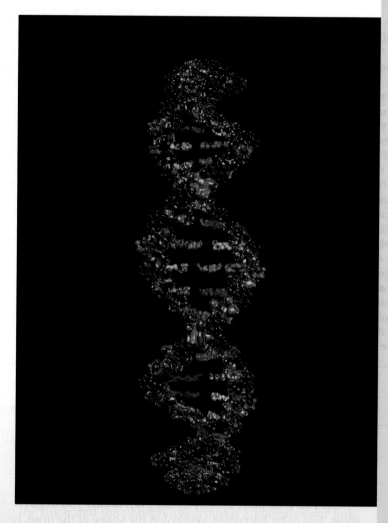

Computer-generated model of DNA.

Learning Goals

1 Draw the general structure of DNA and RNA nucleotides.

2 Describe the structure of DNA and compare it with RNA.

3 Explain DNA replication.

4 List three classes of RNA molecules and describe their functions.

5 Explain the process of transcription.

6 List and explain the three types of post-transcriptional modifications of eukaryotic mRNA.

7 Describe the essential elements of the genetic code, and develop a "feel" for its elegance.

8 Describe the process of translation.

9 Define mutation and understand how mutations cause cancer and cell death.

10 Describe the tools used in the study of DNA and in genetic engineering.

11 Describe the process of polymerase chain reaction and discuss potential uses of the process.

A CLINICAL PERSPECTIVE

Fooling the AIDS Virus with "Look-Alike" Nucleotides

The virus that is responsible for the acquired immune deficiency syndrome (AIDS) is called the *human immunodeficiency virus,* or *HIV.* HIV is a member of a family of viruses called *retroviruses,* all of which have single-stranded RNA as their genetic material. The RNA is copied by a viral enzyme called *reverse transcriptase* into a double-stranded DNA molecule. This process is the opposite of the central dogma, which states that the flow of genetic information is from DNA to RNA. But these viruses reverse that flow, RNA to DNA. For this reason these viruses are called retroviruses, which literally means "backward viruses." The process of producing a DNA copy of the RNA is called *reverse transcription.*

Because our genetic information is DNA and it is expressed by the classical DNA → RNA → protein pathway, our cells have no need for a reverse transcriptase enzyme. Thus the HIV reverse transcriptase is a good target for antiviral chemotherapy because inhibition of reverse transcription should kill the virus but have no effect on the human host. Many drugs have been tested for the ability to selectively inhibit HIV reverse transcription. Among these is the DNA chain terminator 3'-azido-2', 3'-dideoxythymidine, commonly called *AZT* or *zidovudine.*

How does AZT work? It is one of many drugs that looks like one of the normal nucleosides. These are called *nucleoside analogs.* A nucleoside is just a nucleotide without any phosphate groups attached. The analog is phosphorylated by the cell and then tricks a polymerase, in this case viral reverse transcriptase, into incorporating it into the growing DNA chain in place of the normal phosphorylated nucleoside. AZT is a nucleoside analog that looks like the nucleoside thymidine except that in the 3'position of the deoxyribose sugar there is an azido group (—N₃)

Comparison of the structures of the normal nucleoside, 2'-deoxythymidine, and the nucleoside analog, 3'-azido-2', 3'-dideoxythymidine.

rather than the 3'-OH group. Compare the structures of thymidine and AZT shown in the accompanying figure. The 3'-OH group is necessary for further DNA polymerization because it is there that the phosphoester linkage must be made between the growing DNA strand and the next nucleotide. If an azido group or some other group is present at the 3' position, the nucleotide analog can be incorporated into the growing DNA strand, but further chain elongation is blocked, as shown in the following figure. If the viral RNA cannot be reverse transcribed into the DNA form, the virus will not be able to replicate and can be considered to be dead.

AZT is particularly effective because the HIV reverse transcriptase actually prefers it over the normal nucleotide, thymidine. Nonetheless, AZT is not a cure. At best it prolongs the life of a person with AIDS for a year or two. Eventually, however, AZT has a negative effect on the body. The cells of our bone marrow are constantly dividing to produce new blood cells: red

The two strands of DNA are held together by hydrogen bonds between the nitrogenous bases in the center of the helix. Adenine forms two hydrogen bonds with thymine, and cytosine forms three hydrogen bonds with guanine (Figures 24.4 and 24.5). These are called **base pairs.** The two strands of DNA are **complementary strands** because the sequence of bases on one automatically determines the sequence of bases on the other. When there is an adenine on one strand, there will always be a thymine in the same location on the opposite strand.

The diameter of the double helix is 2.0 nm. This is dictated by the dimensions of the purine–pyrimidine base pairs. The helix completes one turn every ten base pairs. One complete turn is 3.4 nm. Thus each base pair advances the helix by 0.34 nm.

One last important feature of the DNA double helix is that the two strands are **antiparallel strands,** as this example shows:

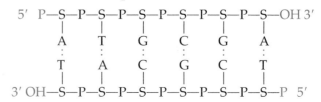

blood cells to carry oxygen to the tissues, white blood cells of the immune system, and platelets for blood clotting. For cells to divide, they must replicate their DNA. The DNA polymerases of these dividing cells also accidentally incorporate AZT into the growing DNA chains with the result that cells of the bone marrow begin to die. This can result in anemia and even further depression of the immune response.

Another problem that has arisen with prolonged use of AZT is that AZT-resistant mutants of the virus appear. It is well known that HIV is a virus that mutates rapidly. Some of these mutant forms of the virus have an altered reverse transcriptase that will no longer use AZT. When these mutants appear, AZT is no longer useful in treating the infection.

It is hoped that research with other nucleoside analogs, alternative types of antiviral treatments, and combinations of drugs will provide a means of effectively treating HIV infection. Such a therapy must have fewer toxic side effects while stopping the replication of the virus and the progress of the disease.

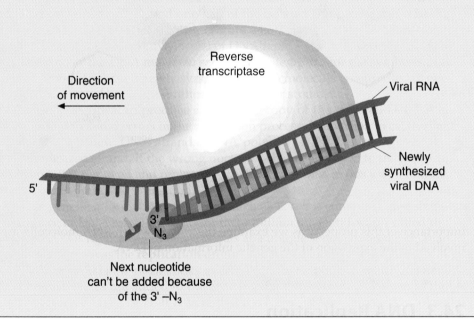

The mechanism by which AZT inhibits HIV reverse transcriptase. Incorporation of AZT into the growing HIV DNA strand in place of deoxythymidine results in DNA chain termination; the azido group on the 3' carbon of the sugar cannot react to produce the phosphoester linkage required to add the next nucleotide.

In other words, the two strands of the helix run in opposite directions (see Figure 24.4). Only when the two strands are antiparallel can the base pairs form the hydrogen bonds that hold the two strands together.

RNA Structure

The sugar–phosphate backbone of RNA consists of ribonucleotides, also linked by 3'–5' phosphodiester bonds. These phosphodiester bonds are identical to those found in DNA. However, RNA molecules differ from DNA molecules in three basic properties.

- RNA molecules are usually single-stranded.
- The sugar–phosphate backbone of RNA consists of *ribonucleotides* linked by 3'–5' phosphodiester bonds. Thus the sugar ribose is found in place of 2'-deoxyribose.
- The nitrogenous base uracil (U) replaces thymine (T).

Although RNA molecules are single-stranded, base pairing between uracil and adenine and between guanine and cytosine can still occur. We will show the

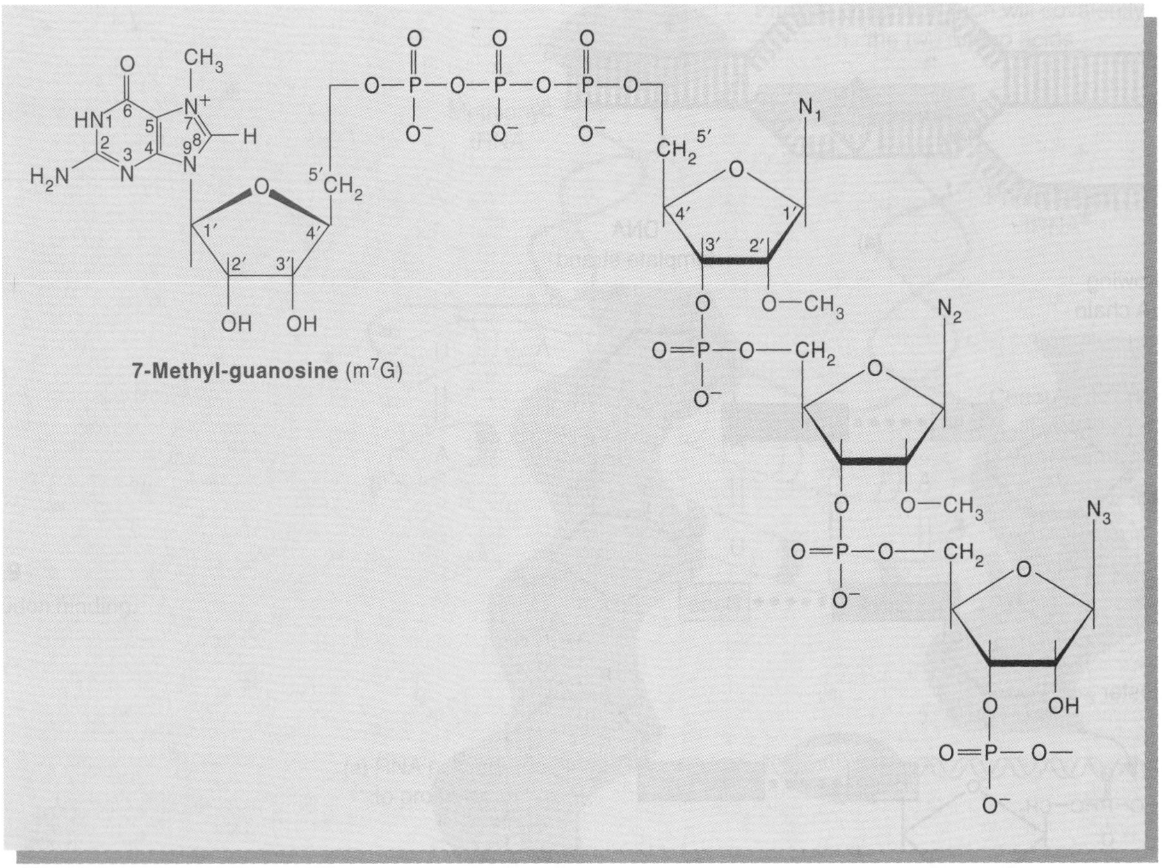

Figure 24.12
The 5'-methylated cap structure of
eukaryotic mRNA.

The second modification is the enzymatic addition of a **poly(A) tail** to the 3' end of the transcript. *Poly(A) polymerase* uses ATP and catalyzes the stepwise polymerization of 100–200 adenosine nucleotides on the 3' end of the RNA. The poly(A) tail protects the 3' end of the mRNA from enzymatic degradation and thus prolongs the lifetime of the mRNA.

The third modification, **RNA splicing,** involves the removal of portions of the primary transcript that are not protein coding. Bacterial genes are continuous; all the nucleotide sequences of the gene are found in the mRNA. However, study of the gene structure of eukaryotes revealed a fascinating difference. Eukaryotic genes are discontinuous; there are *extra* DNA sequences within these genes that do not encode any amino acid sequences for the protein. These sequences are called *intervening sequences* of **introns.** The primary transcript contains both the introns and the protein coding sequences, called **exons.** The presence of introns in the mRNA would make it impossible for the process of translation to synthesize the correct protein. Therefore they must be removed, which is done by the process of RNA splicing.

As you can imagine, RNA splicing must be very precise. If too much, or too little, RNA is removed, the mRNA will not carry the correct code for the protein. Thus there are "signals" in the DNA to mark the boundaries of the introns. The sequence GpU is always found at the intron's 5' boundary and the sequence ApG is found at the 3' boundary.

Recognition of the splice boundaries and stabilization of the splicing complex requires the assistance of particles called *spliceosomes.* Spliceosomes are composed of a variety of *small nuclear ribonucleoproteins* (snRNPs, read "snurps"). Each snRNP consists of a small RNA and associated proteins. The RNA components of different snRNPs are complementary to different sequences involved in splicing.

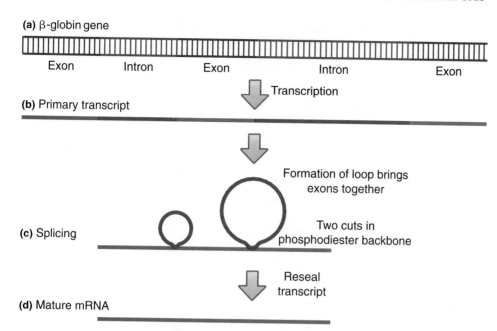

(a) β-globin gene

Exon   Intron   Exon   Intron   Exon

Transcription

(b) Primary transcript

Formation of loop brings exons together

(c) Splicing

Two cuts in phosphodiester backbone

 Reseal transcript

(d) Mature mRNA

Figure 24.13

Schematic diagram of mRNA splicing. (a) The β-globin gene contains protein coding exons, as well as noncoding sequences called *introns*. (b) The primary transcript of the DNA carries both the introns and the exons. (c) The introns are looped out, the phosphodiester backbone of the mRNA is cut twice, and the pieces are tied together. (d) The final mature mRNA now carries only the coding sequences (exons) of the gene.

By hydrogen bonding to a splice boundary or intron sequences the snRNPs recognize and bring together the sequences involved in the splicing reactions.

One of the first eukaryotic genes shown to contain introns was the gene for the β subunit of adult hemoglobin (Figure 24.13). On the DNA the gene for β-hemoglobin is 1200 nucleotides long, but only 438 nucleotides carry the genetic information for protein. The remaining sequences are found in two introns of 116 and 646 nucleotides that are removed by splicing before translation. It is interesting that the larger intron is longer than the final β-hemoglobin mRNA! In the genes that have been studied, introns have been found to range in size from 50 to 20,000 nucleotides in length, and there may be many throughout a gene. Thus a typical human gene might be 10–30 times longer than the final mRNA.

24.5 The Genetic Code

The mRNA carries the genetic code for a protein. But what is the nature of this code? In 1954, George Gamow proposed that because there are only four "letters" in the DNA alphabet (A, T, G, and C) and because there are twenty amino acids, the genetic code must contain words made of at least three letters taken from the four letters in the DNA alphabet. How did he come to this conclusion? He reasoned that a code of two-letter words constructed from any combination of the four letters has a "vocabulary" of only sixteen words (4^2). In other words, there are only sixteen different ways to put A, T, C, and G together two bases at a time (AA, AT, AC, AG, TT, TA, etc.). That is not enough to encode all twenty amino acids. A code of four-letter words gives 256 words (4^4), far more than are needed. A code of three-letter words, however, has a possible vocabulary of sixty-four words (4^3), sufficient to encode the twenty amino acids but not too excessive.

A series of elegant experiments proved that Gamow was correct by demonstrating that the genetic code is, indeed, a triplet code. Mutations were introduced into the DNA of a bacterial virus. These mutations inserted (or deleted) one, two, or three nucleotides into a gene. The researchers then looked for the protein encoded by that gene. When one or two nucleotides were inserted, no protein was produced. However, when a third base was inserted, the sense of the mRNA was

Learning Goal

7

restored, and the protein was made. You can imagine this experiment by using a sentence composed of only three-letter words. For instance,

THE CAT RAN OUT

What happens to the "sense" of the sentence if we insert one letter?

THE FCA TRA NOU T

The reading frame of the sentence has been altered, and the sentence is now nonsense. Can we now restore the sense of the sentence by inserting a second letter?

THE FAC ATR ANO UT

No, we have not restored the sense of the sentence. Once again, we have altered the reading frame, but because our code has only three-letter words, the sentence is still nonsense. If we now insert a third letter, it should restore the correct reading frame:

THE FAT CAT RAN OUT

Indeed, by inserting three new letters we have restored the sense of the message by restoring the reading frame. This is exactly the way in which the message of the mRNA is interpreted. Each group of three nucleotides in the sequence of the mRNA is called a *codon,* and each codes for a single amino acid. If the sequence is interrupted or changed, it can change the amino acid composition of the protein that is produced or even result in the production of no protein at all.

As we noted, a three-letter genetic code contains sixty-four words, called *codons,* but there are only twenty amino acids. Thus there are forty-four more codons than are required to specify all of the amino acids found in proteins. Three of the codons—UAA, UAG, and UGA—specify termination signals for the process of translation. But this still leaves us with forty-one additional codons. What is the function of the "extra" code words? Crick (recall Watson and Crick and the double helix) proposed that the genetic code is a **degenerate code.** The term *degenerate* is used to indicate that different triplet codons may serve as code words for the same amino acid.

The complete genetic code is shown in Figure 24.14. We can make several observations about the genetic code. First, methionine and tryptophan are the only amino acids that have a single codon. All others have at least two codons, and serine and leucine have six codons each. The genetic code is also somewhat mutation-resistant. For those amino acids that have multiple codons the first two bases are often identical and thus identify the amino acid, and only the third position is variable. Mutations—changes in the nucleotide sequence—in the third position therefore often have no effect on the amino acid that is incorporated into a protein.

Question 24.7   Why is the genetic code said to be degenerate?

Question 24.8   Why is the genetic code said to be mutation-resistant?

24.6 Protein Synthesis

Learning Goal

The process of protein synthesis is called *translation.* It involves translating the genetic information from the sequence of nucleotides into the sequence of amino acids in the primary structure of a protein. Translation is carried out on **ribosomes,** which are complexes of ribosomal RNA (rRNA) and proteins. Each ribosome is made up of two subunits: a small and a large ribosomal subunit (Figure 24.15a). In eukaryotic cells the small ribosomal subunit contains one rRNA molecule and

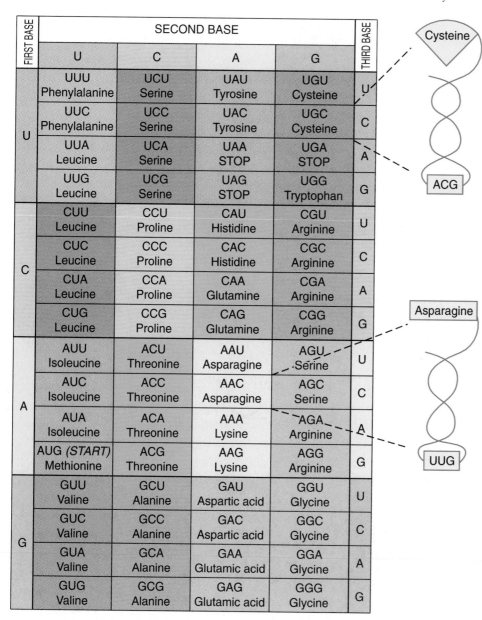

FIRST BASE	SECOND BASE				THIRD BASE
	U	C	A	G	
U	UUU Phenylalanine	UCU Serine	UAU Tyrosine	UGU Cysteine	U
	UUC Phenylalanine	UCC Serine	UAC Tyrosine	UGC Cysteine	C
	UUA Leucine	UCA Serine	UAA STOP	UGA STOP	A
	UUG Leucine	UCG Serine	UAG STOP	UGG Tryptophan	G
C	CUU Leucine	CCU Proline	CAU Histidine	CGU Arginine	U
	CUC Leucine	CCC Proline	CAC Histidine	CGC Arginine	C
	CUA Leucine	CCA Proline	CAA Glutamine	CGA Arginine	A
	CUG Leucine	CCG Proline	CAG Glutamine	CGG Arginine	G
A	AUU Isoleucine	ACU Threonine	AAU Asparagine	AGU Serine	U
	AUC Isoleucine	ACC Threonine	AAC Asparagine	AGC Serine	C
	AUA Isoleucine	ACA Threonine	AAA Lysine	AGA Arginine	A
	AUG (START) Methionine	ACG Threonine	AAG Lysine	AGG Arginine	G
G	GUU Valine	GCU Alanine	GAU Aspartic acid	GGU Glycine	U
	GUC Valine	GCC Alanine	GAC Aspartic acid	GGC Glycine	C
	GUA Valine	GCA Alanine	GAA Glutamic acid	GGA Glycine	A
	GUG Valine	GCG Alanine	GAG Glutamic acid	GGG Glycine	G

Figure 24.14

The genetic code. The table shows the possible codons found in mRNA. To read the universal biological language from this chart, find the first base in the column on the left, the second base from the row across the top, and the third base from the column to the right. This will direct you to one of the sixty-four squares in the matrix. Within that square you will find the codon and the amino acid that it specifies. In the cell this message is decoded by tRNA molecules like those shown to the right of the table.

thirty-three different ribosomal proteins, and the large subunit contains three rRNA molecules and about forty-nine different proteins.

Protein synthesis involves the simultaneous action of many ribosomes on a single mRNA molecule. These complexes of many ribosomes along a single mRNA are known as *polyribosomes* or **polysomes** (Figure 24.15b). Each ribosome is synthesizing one copy of the protein molecule encoded by the mRNA. Thus many copies of a protein are simultaneously produced.

The Role of Transfer RNA

The codons of mRNA must be read if the genetic message is to be translated into protein. The molecule that decodes the information in the mRNA molecule into the primary structure of a protein is transfer RNA (tRNA). To decode the genetic message into the primary sequence of a protein, the tRNA must faithfully perform two functions.

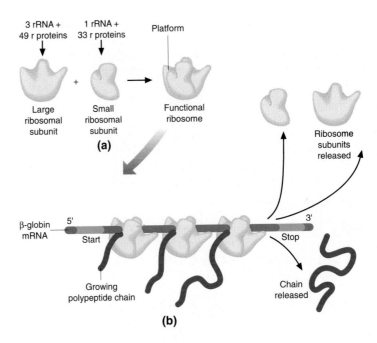

Figure 24.15

Structure of the ribosome. (a) The large and small subunits form the functional complex in association with an mRNA molecule. (b) A polyribosome translating the mRNA for a β-globin chain of hemoglobin.

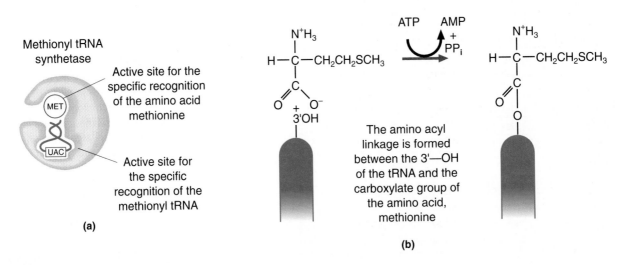

Figure 24.16

Methionyl tRNA synthetase. (a) The enzyme specifically recognizes the amino acid methionine in one region of the active site and the methionyl tRNA in another. (b) The acylation reaction that results in a covalent linkage of the amino acid to the tRNA.

First, the tRNA must covalently bind one, and only one, specific amino acid. There is at least one transfer RNA for each amino acid. As shown in Figure 24.8, all tRNA molecules have the sequence CCA at their 3′ ends. This is the site where the amino acid will be covalently attached to the tRNA molecule. Each tRNA is specifically recognized by the active site of an enzyme called an **aminoacyl tRNA synthetase.** This enzyme also recognizes the correct amino acid and covalently links the amino acid to the 3′ end of the tRNA molecule. Figure 24.16a shows the recognition of the amino acid methionine and its tRNA by the methionyl tRNA synthetase. The resulting structure is called an **aminoacyl tRNA,** in this case methionyl tRNA. In Figure 24.16b the reaction that results in the attachment of the aminoacyl group to the tRNA is shown. The covalently bound amino acid will be transferred from the tRNA to a growing polypeptide chain during protein synthesis.

Second, the tRNA must be able to recognize the appropriate codon on the mRNA that calls for that amino acid. This is mediated through a sequence of three bases called the *anticodon,* which is located at the bottom of the tRNA cloverleaf (refer to Figure

24.8). The anticodon sequence for each tRNA is complementary to the codon on the mRNA that specifies a particular amino acid. As you can see in Figure 24.9, the anticodon–codon complementary hydrogen bonding will bring the correct amino acid to the site of protein synthesis.

How are codons related to anticodons?

If the sequence of a codon on the mRNA is 5′-AUG-3′, what will the sequence of the anticodon be? Remember that the hydrogen bonding rules require antiparallel strands. It is easiest to write the anticodon first 3′ → 5′ and then reverse it to the 5′ → 3′ order.

The Process of Translation

Initiation

The first stage of protein synthesis is *initiation.* Proteins called **initiation factors** are required to mediate the formation of a translation complex composed of an mRNA molecule, the small and large ribosomal subunits, and the initiator tRNA. This initiator tRNA recognizes the codon AUG and carries the amino acid methionine.

The ribosome has two sites for binding tRNA molecules. The first site, called the **peptidyl tRNA binding site (P-site),** holds the peptidyl tRNA, the growing peptide bound to a tRNA molecule. The second site, called the **aminoacyl tRNA binding site (A-site),** holds the aminoacyl tRNA carrying the next amino acid to be added to the peptide chain. Each of the tRNA molecules is hydrogen bonded to the mRNA molecule by codon–anticodon complementarity. The entire complex is further stabilized by the fact that the mRNA is also bound to the ribosome. Figure 24.17a shows the series of events that result in the formation of the initiation complex. The initiator methionyl tRNA occupies the P-site in this complex.

Learning Goal

8

Chain Elongation

The second stage of translation is *chain elongation.* This occurs in three steps that are repeated until protein synthesis is complete. We enter the action after a tetrapeptide has already been assembled, and a peptidyl tRNA occupies the P-site (Figure 24.17b).

The first event is binding of an aminoacyl-tRNA molecule to the empty A-site. Next, peptide bond formation occurs. This is catalyzed by an enzyme on the ribosome called *peptidyl transferase.* Now the peptide chain is shifted to the tRNA that occupies the A-site. Finally, the tRNA in the P-site falls away, and the ribosome changes positions so that the next codon on the mRNA occupies the A-site. This movement of the ribosome is called **translocation.** The process shifts the new peptidyl tRNA from the A-site to the P-site. The chain elongation stage of translation requires the hydrolysis of GTP to GDP and P_i. Several **elongation factors** are also involved in this process.

Recent evidence indicates that the peptidyl transferase is a catalytic region of the 28S ribosomal RNA.

Termination

The last stage of translation is *termination.* There are three **termination codons**—UAA, UAG, and UGA—for which there are no corresponding tRNA molecules. When one of these "stop" codons is encountered, translation is terminated. A **release factor** binds the empty A-site. The peptidyl transferase that had previously catalyzed peptide bond formation hydrolyzes the ester bond between the peptidyl tRNA and the last amino acid of the newly synthesized protein (Figure 24.17c). At this point the tRNA, the newly synthesized peptide, and the two ribosomal subunits are released.

hemoglobin the sixth amino acid is valine. How did this amino acid substitution arise? The answer lies in examination of the codons for glutamic acid and valine:

Glutamic acid: GAA or GAG

Valine: GUG, GUC, GUA, or GUU

A point mutation of A → U in the second nucleotide changes some codons for glutamic acid into codons for valine:

GAA   ⟶   GUA

GAG   ⟶   GUG

Glutamic acid codon   Valine codon

This mutation in a single codon leads to the change in amino acid sequence at position 6 in the β-chain of human hemoglobin from glutamic acid to valine. The result of this seemingly minor change is sickle cell anemia in individuals who inherit two copies of the mutant gene.

Question 24.13

The sequence of a gene on the mRNA is normally AUGCCCGACUUU. A point mutation in the gene results in the mRNA sequence AUGCGCGACUUU. What are the amino acid sequences of the normal and mutant proteins? Would you expect this to be a silent mutation?

Question 24.14

The sequence of a gene on the mRNA is normally AUGCCCGACUUU. A point mutation in the gene results in the mRNA sequence AUGCCGGACUUU. What are the amino acid sequences of the normal and mutant proteins? Would you expect this to be a silent mutation?

Mutagens and Carcinogens

Any chemical that causes a change in the DNA sequence is called a *mutagen*. Often, mutagens are also **carcinogens,** cancer-causing chemicals. Most cancers result from mutations in a single normal cell. These mutations result in the loss of normal growth control, causing the abnormal cell to proliferate. If that growth is not controlled or destroyed, it will result in the death of the individual. We are exposed to many carcinogens in the course of our lives. Sometimes we are exposed to a carcinogen by accident, but in some cases it is by choice. There are about 3000 chemical components in cigarette smoke, and several are potent mutagens. As a result, people who smoke have a much greater chance of lung cancer than those who don't.

Ultraviolet Light Damage and DNA Repair

Ultraviolet (UV) light is another agent that causes damage to DNA. Absorption of UV light by DNA causes adjacent pyrimidine bases to become covalently linked. The product (Figure 24.18) is called a **pyrimidine dimer.** As a result of pyrimidine dimer formation, there is no hydrogen bonding between these pyrimidine molecules and the complementary bases on the other DNA strand. This stretch of DNA cannot be replicated or transcribed!

Bacteria such as *Escherichia coli* have four different mechanisms to repair ultraviolet light damage. However, even a repair process can make a mistake. Mutations occur when the UV damage repair system makes an error and causes a change in the nucleotide sequence of the DNA.

In medicine the pyrimidine dimerization reaction is used to advantage in hospitals where germicidal (UV) light is used to kill bacteria in the air and on

A CLINICAL PERSPECTIVE

The Ames Test for Carcinogens

Each day we come into contact with a variety of chemicals, including insecticides, food additives, hair dyes, automobile emissions, and cigarette smoke. Some of these chemicals have the potential to cause cancer. How do we determine whether these agents are harmful? More particularly, how do we determine whether they cause cancer?

If we consider the example of cigarette smoke, we see that it can be years, even centuries, before a relationship is seen between a chemical and cancer. Europeans and Americans have been smoking since Sir Walter Raleigh introduced tobacco into England in the seventeenth century. However, it was not until three centuries later that physicians and scientists demonstrated the link between smoking and lung cancer. Obviously, this epidemiological approach takes too long, and too many people die. Alternatively, we can test chemicals by treating laboratory animals, such as mice, and observing them for various kinds of cancer. However, this, too, can take years, is expensive, and requires the sacrifice of many laboratory animals. How, then, can chemicals be tested for carcinogenicity (the ability to cause cancer) quickly and inexpensively? In the 1970s it was recognized that most carcinogens are also mutagens. That is, they cause cancer by causing mutations in the DNA, and the mutations cause the cells of the body to lose growth control. Bruce Ames, a biochemist and bacterial geneticist, developed a

test using mutants of the bacterium *Salmonella typhimurium* that can demonstrate in 48–72 hours whether a chemical is a mutagen and thus a suspected carcinogen.

Ames chose several mutants of *S. typhimurium* that cannot grow unless the amino acid histidine is added to the growth medium. The Ames test involves subjecting these bacteria to a chemical and determining whether the chemical causes reversion of the mutation. In other words, the researcher is looking for a mutation that reverses the original mutation. When a reversion occurs, the bacteria will be able to grow in the absence of histidine.

The details of the Ames test are shown in the accompanying figure. Both an experimental and a control test are done. The control test contains no carcinogen and will show the number of spontaneous revertants that occur in the culture. If there are many colonies on the surface of the experimental plate and only a few colonies on the negative control plate, it can be concluded that the chemical tested is a mutagen. It is therefore possible that the chemical is also a carcinogen.

The Ames test has greatly accelerated our ability to test new compounds for mutagenic and possibly carcinogenic effects. However, once the Ames test identifies a mutagenic compound, testing in animals must be done to show conclusively that the compound also causes cancer.

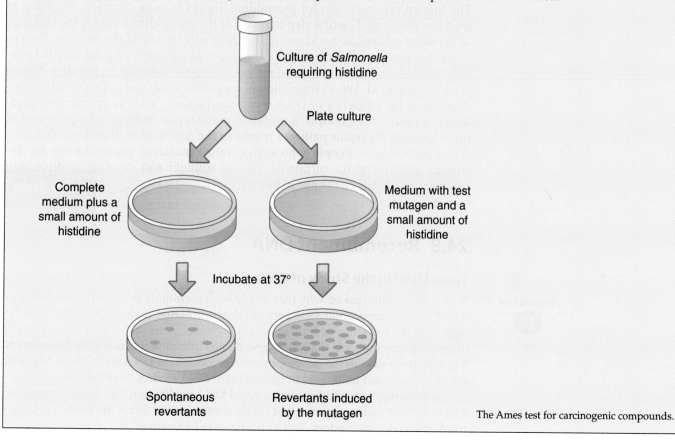

The Ames test for carcinogenic compounds.

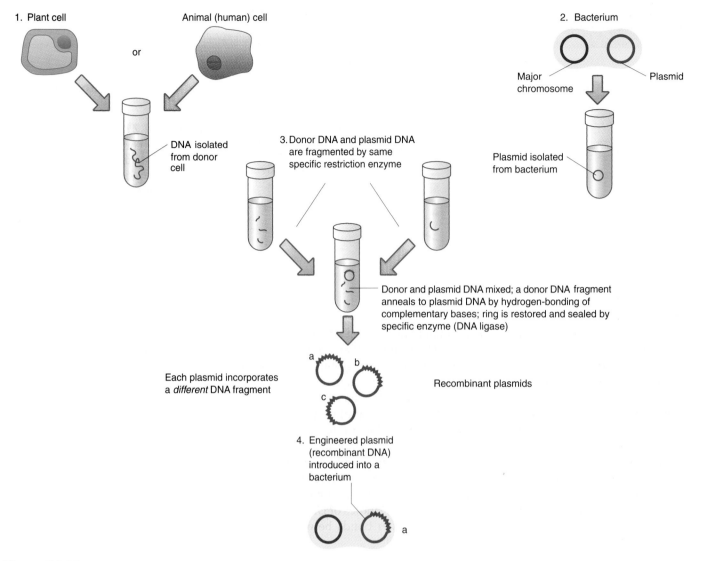

1. Plant cell   Animal (human) cell

or

DNA isolated from donor cell

2. Bacterium

Major chromosome   Plasmid

Plasmid isolated from bacterium

3. Donor DNA and plasmid DNA are fragmented by same specific restriction enzyme

Donor and plasmid DNA mixed; a donor DNA fragment anneals to plasmid DNA by hydrogen-bonding of complementary bases; ring is restored and sealed by specific enzyme (DNA ligase)

Each plasmid incorporates a *different* DNA fragment

a   b

c

Recombinant plasmids

4. Engineered plasmid (recombinant DNA) introduced into a bacterium

a

Figure 24.22
Cloning of eukaryotic DNA into a plasmid cloning vector.

radioactive probe will hybridize only to the complementary sequences of the β-globin gene. When the membrane filter is exposed to X-ray film, a "spot" will appear on the developed film only at the site of a colony carrying the desired clone. By going back to the original plate, we can select cells from that colony and grow the cells for further study (Figure 24.23).

The same procedure can be used to clone the β-chain gene of sickle cell hemoglobin. Then the two can be studied and compared to determine the nature of the genetic defect.

This simple example makes it appear that all gene cloning is very easy and straightforward. This has proved to be far from the truth. Genetic engineers have had to overcome many obstacles to clone eukaryotic genes of particular medical interest. One of the first obstacles encountered was the presence of introns within eukaryotic genes. Bacteria that are used for cloning lack the enzymatic machinery to splice out introns. Molecular biologists found that a DNA copy of a eukaryotic mRNA could be made by using the enzyme reverse transcriptase from a family of viruses called retroviruses (see "A Clinical Perspective: Fooling the AIDS Virus with 'Look-Alike' Nucleotides"). Such a DNA copy of the mRNA carries all the protein-coding sequences of a gene but none of the intron sequences. Thus bacteria are able to transcribe and translate the cloned DNA and produce valuable products for use in medicine and other applications.

This is only one of the many technical problems that have been overcome by the amazing developments in recombinant DNA technology. A brief but impressive list of medically important products of genetic engineering is presented in Table 24.4.

Although great progress has been made in applying genetic engineering to important medical problems, most of these products represent only a treatment, not a cure. The ultimate dream for the future is to intervene more absolutely, to remove "bad" genes from the human population entirely. Replacing a mutant gene in a fertilized egg with a normal gene is one approach to this goal. However, realization of that dream in humans is years, perhaps decades, away. Meanwhile, molecular biologists continue to work toward that goal by studying the normal structure and expression of DNA and by working to develop safe delivery systems for the introduction of new DNA into cells such as the fragile human fertilized egg.

24.9 Polymerase Chain Reaction

A bacterium originally isolated from a hot spring in Yellowstone National Park provides the key to a powerful molecular tool for the study of DNA. Polymerase chain reaction (PCR) allows scientists to produce unlimited amounts of any gene of interest and the bacterium *Thermus aquaticus* produces a heat-stable DNA polymerase (Taq polymerase) that allows the process to work.

The human genome consists of approximately three billion base pairs of DNA. But suppose you are interested in studying only one gene, perhaps the gene responsible for muscular dystrophy or cystic fibrosis. It's like looking for a needle in a haystack. Using PCR, a scientist can make millions of copies of the gene you want to study, while ignoring the 100,000–200,000 other genes on human chromosomes.

The secret to this specificity is the synthesis of a DNA primer, a short piece of single-stranded DNA that will specifically hybridize to the beginning of a particular gene. DNA polymerases require a primer for initiation of DNA synthesis

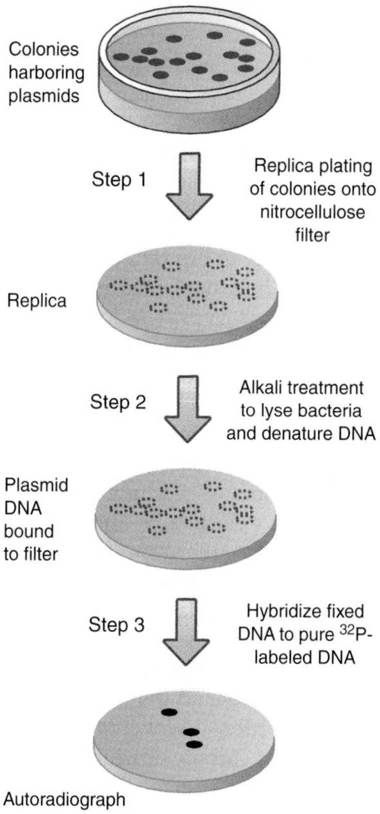

Figure 24.23
Colony blot hybridization for detection of cells carrying a plasmid clone of the β-chain gene of hemoglobin.

Table 24.4	A Brief List of Medically Important Proteins Produced by Genetic Engineering
Protein	**Medical Condition Treated**
Insulin	Insulin-dependent diabetes
Human growth hormone	Pituitary dwarfism
Factor VIII	Type A hemophilia
Factor IX	Type B hemophilia
Tissue plasminogen factor	Stroke, myocardial infarction
Streptokinase	Myocardial infarction
Interferon	Cancer, some virus infection
Interleukin-2	Cancer
Tumor necrosis factor	Cancer
Atrial natriuretic factor	Hypertension
Erythropoietin	Anemia
Thymosin α-1	Stimulate immune system
Hepatitis B virus (HBV) vaccine	Prevent HBV viral hepatitis
Influenza vaccine	Prevent influenza infection

base pairs with thymine, and cytosine base pairs with guanine. The two strands of DNA in the helix are antiparallel to one another. RNA is single-stranded.

24.3   DNA Replication

DNA replication involves synthesis of a faithful copy of the DNA molecule. It is *semiconservative;* each daughter molecule consists of one parental strand and one newly synthesized strand. *DNA polymerase* "reads" each parental strand and synthesizes the complementary daughter strand according to the rules of base pairing.

24.4   Information Flow in Biological Systems

The *central dogma* states that the flow of biological information in cells is DNA → RNA → protein. There are three classes of RNA: *messenger RNA, transfer RNA,* and *ribosomal RNA. Transcription* is the process by which RNA molecules are synthesized. *RNA polymerase* catalyzes the synthesis of RNA. Transcription occurs in three stages: initiation, elongation, and termination. Eukaryotic genes contain *introns,* sequences that do not encode protein. These are removed from the primary transcript by the process of *RNA splicing.* The final mRNA contains only the protein coding sequences or *exons.* This final mRNA also has an added 5′ cap structure and 3′ poly(A) tail.

24.5   The Genetic Code

The genetic code is a triplet code. Each code word is called a *codon* and consists of three nucleotides. There are sixty-four codons in the genetic code. Of these, three are *termination codons* (UAA, UAG, and UGA), and the remaining sixty-one specify an amino acid. Most amino acids have several codons. As a result, the genetic code is said to be *degenerate.*

24.6   Protein Synthesis

The process of protein synthesis is called *translation.* The genetic code words on the mRNA are decoded by tRNA. Each tRNA has an *anticodon* that is complementary to a codon on the mRNA. In addition the tRNA is covalently linked to its correct amino acid. Thus hydrogen bonding between codon and anticodon brings the correct amino acid to the site of protein synthesis. Translation also occurs in three stages called initiation, chain elongation, and termination.

24.7   Mutation, Ultraviolet Light, and DNA Repair

Any change in a DNA sequence is a *mutation.* Mutations are classified according to the type of DNA alteration, including *point mutations, deletion mutations,* and *insertion mutations.* Ultraviolet light (UV) causes formation of

pyrimidine dimers. Mistakes can be made during pyrimidine dimer repair, causing UV-induced mutations. Germicidal (UV) lamps are used to kill bacteria on environmental surfaces. UV damage to skin can result in skin cancer.

24.8   Recombinant DNA

Several tools are required for genetic engineering, including *restriction enzymes, agarose gel electrophoresis, hybridization,* and *cloning vectors.* Cloning a DNA fragment involves digestion of the target and vector DNA with a restriction enzyme. DNA ligase joins the target and vector DNA covalently, and the recombinant DNA molecules are introduced into bacterial cells by transformation. The desired clone is located by using antibiotic selection and hybridization. Many eukaryotic genes have been cloned for the purpose of producing medically important proteins.

24.9   Polymerase Chain Reaction

Using a heat-stable DNA polymerase produced by the bacterium *Thermus aquaticus* and specific DNA primers, polymerase chain reaction allows the amplification of DNA sequences that are present in small quantities. This technique is useful in genetic screening, diagnosis of viral or bacterial disease, and forensic science.

Key Terms

aminoacyl tRNA (24.6)
aminoacyl tRNA binding site of ribosome (A-site) (24.6)
aminoacyl tRNA synthetase (24.6)
anticodon (24.4)
antiparallel strands (24.2)
base pairs (24.2)
cap structure (24.4)
carcinogen (24.7)
central dogma (24.4)
cloning vector (24.8)
codon (24.4)
complementary strands (24.2)
degenerate code (24.5)
deletion mutation (24.7)
deoxyribonucleic acid (DNA) (24.1)
deoxyribonucleotide (24.1)
DNA polymerase (24.3)
double helix (24.2)

elongation factor (24.6)
exon (24.4)
hybridization (24.8)
initiation factor (24.6)
insertion mutation (24.7)
intron (24.4)
messenger RNA (mRNA) (24.4)
mutagen (24.7)
mutation (24.7)
nucleotide (24.1)
peptidyl tRNA binding site of ribosome (P-site) (24.6)
point mutation (24.7)
poly(A) tail (24.4)
polysome (24.6)
post-transcriptional modification (24.4)
primary transcript (24.4)
promoter (24.4)
purine (24.1)
pyrimidine (24.1)
pyrimidine dimer (24.7)

release factor (24.6)
replication fork (24.3)
replication origin (24.3)
restriction enzyme (24.8)
ribonucleic acid (RNA)
  (24.1)
ribonucleotide (24.1)
ribosomal RNA (rRNA)
  (24.4)
ribosome (24.6)
RNA polymerase (24.4)

RNA splicing (24.4)
semiconservative
  replication (24.3)
silent mutation (24.7)
termination codon (24.6)
transcription (24.4)
transfer RNA (tRNA)
  (24.4)
translation (24.4)
translocation (24.6)

Questions and Problems

The Structure of the Nucleotide

24.15 Draw the structure of the purine ring, and indicate the nitrogen that is bonded to sugars in nucleotides.

24.16 **a.** Draw the ring structure of the pyrimidines.
  b. In a nucleotide, which nitrogen atom of pyrimidine rings is bonded to the sugar?

24.17 ATP is the universal energy currency of the cell. What components make up the ATP nucleotide?

24.18 One of the energy-harvesting steps of the citric acid cycle results in the production of GTP. What is the structure of the GTP nucleotide?

The Structure of DNA and RNA

24.19 The two strands of a DNA molecule are antiparallel. What is meant by this description?

24.20 List three differences between DNA and RNA.

24.21 How many hydrogen bonds link the adenine–thymine base pair?

24.22 How many hydrogen bonds link the guanine–cytosine base pair?

24.23 Write the structure that results when deoxycytosine-5'-monophosphate is linked by a 3' → 5' phosphodiester bond to thymidine-5'-monophosphate.

24.24 Write the structure that results when adenosine-5'-monophosphate is linked by a 3' → 5' phosphodiester bond to uridine-5'-monophosphate.

DNA Replication

24.25 What is meant by semiconservative DNA replication?

24.26 Draw a diagram illustrating semiconservative DNA replication.

24.27 What are the two primary functions of DNA polymerase?

24.28 **a.** Why is DNA polymerase said to be template-directed?
  b. Why is DNA replication a self-correcting process?

24.29 If a DNA strand had the nucleotide sequence

5'-ATGCGGCTAGAATATTCCA-3'

what would the sequence of the complementary daughter strand be?

24.30 If the sequence of a double-stranded DNA is

5'-GAATTCCTTAAGGATCGATC-3'
   |||||||||||||||||||||
3'-CTTAAGGAATTCCTAGCTAG-5'

what would the sequence of the two daughter DNA molecules be after DNA replication? Indicate which strands are newly synthesized and which are parental.

24.31 What is the replication origin of a DNA molecule?

24.32 What is occurring at the replication fork?

Information Flow in Biological Systems

24.33 What is the central dogma of molecular biology?

24.34 What are the roles of DNA, RNA, and protein in information flow in biological systems?

24.35 On what molecule is the anticodon found?

24.36 On what molecule is the codon found?

24.37 If a gene had the nucleotide sequence

5'-TACCTAGCTCTGGTCATTAAGGCAGTA-3'

what would the sequence of the mRNA be?

24.38 If a mRNA had the nucleotide sequence

5'-AUGCCCUUUCAUUACCCGGUA-3'

what was the sequence of the DNA strand that was transcribed?

24.39 What is meant by the term *RNA splicing*?

24.40 The following is the unspliced transcript of a eukaryotic gene:

exon 1   intron A   exon 2   intron B   exon 3   intron C   exon 4

What would the structure of the final mature mRNA look like, and which of the above sequences would be found in the mature mRNA?

24.41 List the three classes of RNA molecules.

24.42 What is the function of each of the classes of RNA molecules?

24.43 What is the function of the spliceosome?

24.44 What are snRNPs? How do they facilitate RNA splicing?

24.45 What is a poly(A) tail?

24.46 What is the purpose of the poly(A) tail on eukaryotic mRNA?

24.47 What is the cap structure?

24.48 What is the function of the cap structure on eukaryotic mRNA?

The Genetic Code

24.49 How many codons constitute the genetic code?

24.50 What is meant by a triplet code?

24.51 What is meant by the reading frame of a gene?

24.52 What happens to the reading frame of a gene if a nucleotide is deleted?

24.53 Which two amino acids are encoded by only one codon?

24.54 Which amino acids are encoded by six codons?

24.55 An essential gene has the codon 5'-UUU-3' in a critical position. If this codon is mutated to the sequence 5'-UUA-3', what is the expected consequence for the cell?

24.56 An essential gene has the codon 5'-UUA-3' in a critical position. If this codon is mutated to the sequence 5'-UUG-3', what is the expected consequence for the cell?

Protein Synthesis

24.57 What is the function of ribosomes?

24.58 What are the two tRNA binding sites on the ribosome?

in which the symbol $\propto$ is shorthand for the words *proportional to;* it reads: "Volume is proportional to temperature." Use of a *proportionality constant, k,* and an *equals sign* to replace $\propto$ results in a valid mathematical equation:

$$V = kT$$

In this example, k is the Charles's law constant. Graphical representation of a direct proportion results in a straight-line (linear) relationship between variables:

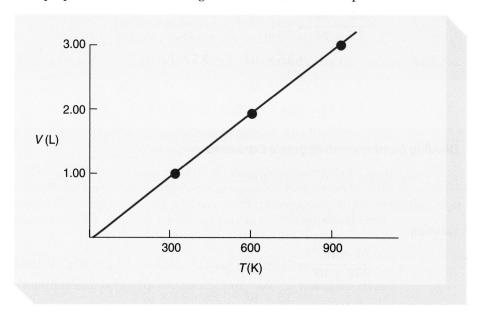

Graphs of this type allow prediction of the volume at *any* temperature within the range in which Charles's law applies.

Inverse Proportionality

Boyle's law resulted from the observation that gas volumes decrease as the pressure of the gas increases when the temperature and number of moles of the gas remain constant. The following data were obtained to illustrate Boyle's law.

Experiment	P (atm)	V of Helium (L)
1	2.00	6.00
2	4.00	3.00
3	6.00	2.00

Doubling the pressure (from 2 atm to 4 atm) causes the volume of helium to decrease by a factor of one-half. Tripling the pressure (from 2 atm to 6 atm) decreases the volume to one-third of the original value (from 6 L to 2 L).

Pressure and volume are *inversely proportional.* This relationship is expressed as

$$V \propto \frac{1}{P}$$

in which again the symbol $\propto$ is short for the words *proportional to;* it reads: "Volume is inversely proportional to pressure." Use of a *proportionality constant, k,* and an *equals sign* to replace $\propto$ results in a valid mathematical equation:

$$V = \frac{k}{P}$$

The proportionality constant, *k*, is the Boyle's law constant. Graphically we may represent the data as

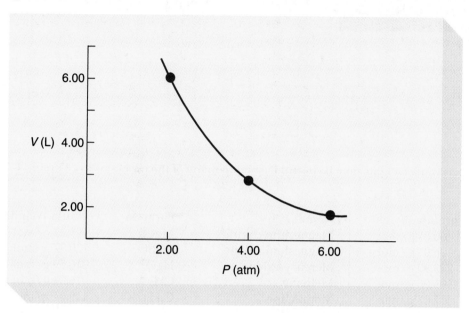

A curved relationship, such as that shown, is not ideal for predicting other pairs of values from the graph. However, regraphing the data as

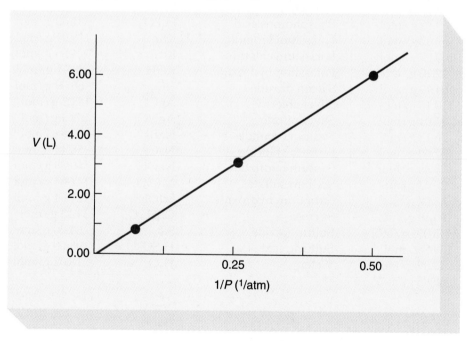

results in a linear relationship that is useful for predicting the volume at *any* pressure within the limits of Boyle's law.

Appendix B

Table of Formula Weights

The following list of names, formulas, and formula weights may be useful in solving many of the problems in Chapters 4–7.

Name	Formula	Formula Weight	Name	Formula	Formula Weight
Acetic acid	CH_3COOH	60.05 g/mol	Magnesium chloride	$MgCl_2$	95.21 g/mol
Acetylene	C_2H_2	26.04 g/mol	Magnesium sulfate	$MgSO_4$	120.37 g/mol
Aluminum carbonate	$Al_2(CO_3)_3$	233.99 g/mol	Mercury oxide	HgO	216.59 g/mol
Aluminum oxide	Al_2O_3	101.96 g/mol	Methane	CH_4	16.04 g/mol
Ammonia	NH_3	17.03 g/mol	Methionine	$C_5H_{11}NO_2S$	149.21 g/mol
Ammonium chloride	NH_4Cl	53.49 g/mol	Nitrogen	N_2	28.02 g/mol
Ammonium nitrate	NH_4NO_3	80.05 g/mol	Nitrous oxide	N_2O	44.02 g/mol
Aspirin	$C_9H_8O_4$	180.15 g/mol	Octane	C_8H_{18}	114.2 g/mol
Barium carbonate	$BaCO_3$	197.35 g/mol	Oxygen	O_2	32.00 g/mol
Boron oxide	B_2O_3	69.62 g/mol	Phosphorous acid	H_3PO_3	82.00 g/mol
Bromine	Br_2	159.82 g/mol	Potassium bromide	KBr	119.01 g/mol
Calcium carbonate	$CaCO_3$	100.09 g/mol	Potassium chloride	KCl	74.55 g/mol
Calcium hydride	CaH_2	42.10 g/mol	Potassium hydroxide	KOH	56.11 g/mol
Calcium nitrate	$Ca(NO_3)_2$	164.10 g/mol	Silicon dioxide	SiO_2	60.09 g/mol
Calcium phosphate	$Ca_3(PO_4)_2$	310.18 g/mol	Silver chloride	$AgCl$	143.3 g/mol
Carbon dioxide	CO_2	44.01 g/mol	Silver nitrate	$AgNO_3$	169.9 g/mol
Carbon disulfide	CS_2	76.13 g/mol	Sodium bromide	$NaBr$	102.9 g/mol
Chromium(III) oxide	Cr_2O_3	152.00 g/mol	Sodium chloride	$NaCl$	58.44 g/mol
Chromium(III) chloride	$CrCl_3$	158.35 g/mol	Sodium hydroxide	$NaOH$	40.00 g/mol
Diborane	B_2H_6	27.67 g/mol	Sodium sulfate	Na_2SO_4	142.04 g/mol
Ethyl alcohol (Ethanol)	C_2H_5OH	46.07 g/mol	Strontium hydroxide	$Sr(OH)_2$	121.64 g/mol
Glucose	$C_6H_{12}O_6$	180.16 g/mol	Sucrose	$C_{12}H_{22}O_{11}$	342.3 g/mol
Hydrogen	H_2	2.016 g/mol	Sulfur dioxide	SO_2	64.06 g/mol
Iron(III) oxide	Fe_2O_3	159.7 g/mol	Sulfuric acid	H_2SO_4	98.08 g/mol
Lithium chloride	$LiCl$	42.39 g/mol	Water	H_2O	18.02 g/mol
Lithium nitrate	$LiNO_3$	68.95 g/mol			

Appendix C

Determination of Composition and Formulas of Compounds

C.1 Percentage Composition of Compounds

The chemical formula provides information about the composition of a compound in terms of moles. For example, 1 mol of glucose, $C_6H_{12}O_6$, contains 6 mol each of both carbon and oxygen atoms and 12 mol of hydrogen atoms.

Percentage composition of a compound, on the other hand, provides us with the composition of the compound in terms of mass. If we calculate the percentage composition of glucose, it will tell us the relative masses of C, H, and O that are present in any amount of the compound. The relative masses are expressed as percentages of the whole.

EXAMPLE C.1

Calculating Percentage Composition

Calculate the percentage composition of glucose ($C_6H_{12}O_6$).

Solution

First of all, calculate the molar mass of glucose:

$$6 \text{ mol C} \times \frac{12.01 \text{ g C}}{1 \text{ mol C}} = 72.06 \text{ g C}$$

$$12 \text{ mol H} \times \frac{1.008 \text{ g H}}{1 \text{ mol H}} = 12.10 \text{ g H}$$

$$6 \text{ mol O} \times \frac{16.00 \text{ g O}}{1 \text{ mol O}} = 96.00 \text{ g O}$$

and

$$72.06 \text{ g} + 12.10 \text{ g} + 96.00 \text{ g} = 180.16 \text{ g}$$

The percentage of carbon in glucose is the mass of carbon in glucose divided by the mass of the compound (in terms of grams in one mole of glucose). This quantity is multiplied by 100% to express the answer as a percentage:

$$\frac{72.06 \text{ g C}}{180.16 \text{ g glucose}} \times 100\% = 40.00\% \text{ C}$$

Similarly,

$$\frac{12.10 \text{ g H}}{180.16 \text{ g glucose}} \times 100\% = 6.72\% \text{ H}$$

and

alkane (11.2) a hydrocarbon that contains only carbon and hydrogen and is bonded together through carbon-hydrogen and carbon-carbon single bonds; a saturated hydrocarbon with the general molecular formula C_nH_{2n+2}

alkene (11.1, 12.1) a hydrocarbon that contains one or more carbon-carbon double bonds; an unsaturated hydrocarbon with the general formula C_nH_{2n}

alkyl group (11.2) a hydrocarbon group that results from the removal of one hydrogen from the original hydrocarbon (e.g., methyl, CH_3—; ethyl, CH_3CH_2—)

alkyl halide (11.6) a substituted hydrocarbon with the general structure R—X, in which R— represents any alkyl group and X = a halogen (F—, Cl—, Br—, or I—)

alkylammonium ion (16.1) the ion formed when the lone pair of electrons of the nitrogen atom of an amine is shared with a proton (H^+) from a water molecule

alkyne (11.1, 12.1) a hydrocarbon that contains one or more carbon-carbon triple bonds; an unsaturated hydrocarbon with the general formula C_nH_{2n-2}

allosteric enzyme (20.9) an enzyme that has an effector binding site and an active site; effector binding changes the shape of the active site, rendering it either active or inactive

alpha particle (10.1) a particle consisting of two protons and two neutrons; the alpha particle is identical to a helium nucleus

amide bond (16.3) the bond between the carbonyl carbon of a carboxylic acid and the amino nitrogen of an amine

amides (16.3) the family of organic compounds formed by the reaction between a carboxylic acid derivative and an amine and characterized by the amide group

amines (16.1) the family of organic molecules with the general formula RNH_2, R_2NH, or R_3N (R— can represent either an alkyl or aryl group); they may be viewed as substituted ammonia molecules in which one or more of the ammonia hydrogens has been substituted by a more complex organic group

α-amino acid (19.2) the subunits of proteins composed of an α-carbon bonded to a carboxylate group, a protonated amino group, a hydrogen atom, and a variable R group

aminoacyl group (16.4) the functional group that is characteristic of an amino acid; the aminoacyl group has the following general structure:

$$\overset{+}{H_3}N-\overset{\overset{\displaystyle H}{|}}{\underset{\underset{\displaystyle R}{|}}{C}}-\overset{\overset{\displaystyle O}{\|}}{C}-$$

aminoacyl tRNA (24.6) the transfer RNA covalently linked to the correct amino acid

aminoacyl tRNA binding site of ribosome (A-site) (24.6) a pocket on the surface of a ribosome that holds the amino acyl tRNA during translation

aminoacyl tRNA synthetase (24.6) an enzyme that recognizes one tRNA and covalently links the appropriate amino acid to it

aminotransferase (22.7) an enzyme that catalyzes the transfer of an amino group from one molecule to another

amorphous solid (6.3) a solid with no organized, regular structure

amphibolic pathway (22.9) a metabolic pathway that functions in both anabolism and catabolism

amylopectin (17.4) a highly branched form of amylose; the branches are attached to the C-6 hydroxyl by $\alpha(1 \rightarrow 6)$ glycosidic linkage; a component of starch

amylose (17.4) a linear polymer of α-D-glucose molecules bonded in $\alpha(1 \rightarrow 4)$ glycosidic linkage that is a major component of starch; a polysaccharide storage form

anabolism (21.1, 22.9) all of the cellular energy-requiring biosynthetic pathways

anaerobic threshold (21.4) the point at which the level of lactate in the exercising muscle inhibits glycolysis and the muscle, deprived of energy, ceases to function

analgesic (16.2) any drug that acts as a painkiller, e.g., aspirin, acetaminophen

anaplerotic reaction (22.9) a reaction that replenishes a substrate needed for a biochemical pathway

anesthetic (16.2) a drug that causes a lack of sensation in part of the body (local anesthetic) or causes unconsciousness (general anesthetic)

angular structure (4.4) a planar molecule with bond angles other than 180°

anion (2.2, 3.3) a negatively charged atom or group of atoms

anode (2.3, 9.5) the positively charged electrode in an electrical cell

antibodies (19.1) immunoglobulins; specific glycoproteins produced by cells of the immune system in response to invasion by infectious agents

anticodon (24.4) a sequence of three ribonucleotides on a tRNA that are complementary to a codon on the mRNA; codon-anticodon binding results in delivery of the correct amino acid to the site of protein synthesis

antigen (19.1) any substance that is able to stimulate the immune system; generally a protein or large carbohydrate

antiparallel (24.2) a term describing the polarities of the two strands of the DNA double helix; on one strand the sugar–phosphate backbone advances in the $5' \rightarrow 3'$ direction; on the opposite, complementary strand the sugar–phosphate backbone advances in the $3' \rightarrow 5'$ direction

apoenzyme (20.7) the protein portion of an enzyme that requires a cofactor to function in catalysis

aqueous solution (7.3) any solution in which the solvent is water

arachidonic acid (18.2) a fatty acid derived from linoleic acid; the precursor of the prostaglandins

aromatic hydrocarbon (11.1, 12.5) an organic compound that contains the benzene ring or a derivative of the benzene ring

Arrhenius theory (9.1) a theory that describes an acid as a substance that dissociates to produce H^+ and a base as a substance that dissociates to produce OH^-

artificial radioactivity (10.6) radiation that results from the conversion of a stable nucleus to another, unstable nucleus

asymmetric carbon (17.2) a chiral carbon; a carbon bonded to four different groups

atherosclerosis (18.4) deposition of excess plasma cholesterol and other lipids and proteins on the walls of arteries, resulting in decreased artery diameter and increased blood pressure

atom (2.2) the smallest unit of an element that retains the properties of that element

atomic mass (2.2) the mass of an atom expressed in atomic mass units

atomic mass unit (5.1) 1/12 of the mass of a ^{12}C atom, equivalent to 1.661×10^{-24}

atomic number (2.2) the number of protons in the nucleus of an atom; it is a characteristic identifier of an element

atomic orbital (2.4) a specific region of space where an electron may be found

ATP synthase (22.6) a multiprotein complex within the inner mitochondrial membrane that uses the energy of the proton (H^+) gradient to produce ATP

autoionization (9.2) also known as *self-ionization*, the reaction of a substance, such as water, with itself to produce a positive and a negative ion

Avogadro's law (6.1) a law that states that the volume is directly proportional to the number of moles of gas particles, assuming that the pressure and temperature are constant

Avogadro's number (5.1) 6.022×10^{23} particles of matter contained in 1 mol of a substance

axial atom (11.4) an atom that lies above or below a cycloalkane ring

B

background radiation (10.7) the radiation that emanates from natural sources

barometer (6.1) a device for measuring pressure

base (9.1) a substance that behaves as a proton acceptor

base pair (24.2) a hydrogen-bonded pair of bases within the DNA double helix; the standard base pairs always involve a purine and a pyrimidine; in particular, adenine always base pairs with thymine and cytosine with guanine

Benedict's reagent (17.2) a buffered solution of Cu^{2+} ions that can be used to test for reducing sugars or to

distinguish between aldehydes and ketones

Benedict's test (14.4) a test used to determine the presence of reducing sugars or to distinguish between aldehydes and ketones; it requires a buffered solution of Cu^{2+} ions that are reduced to Cu^+, which precipitates as brick-red Cu_2O

beta particle (10.1) an electron formed in the nucleus by the conversion of a neutron into a proton

bile (23.1) micelles of lecithin, cholesterol, bile salts, protein, inorganic ions, and bile pigments that aid in lipid digestion by emulsifying fat droplets

binding energy (10.3) the energy required to break down the nucleus into its component parts

boiling point (4.3) the temperature at which the vapor pressure of a liquid is equal to the atmospheric pressure

bond energy (4.4) the amount of energy necessary to break a chemical bond

Boyle's law (6.1) a law stating that the volume of a gas varies inversely with the pressure exerted if the temperature and number of moles of gas are constant

breeder reactor (10.4) a nuclear reactor that produces its own fuel in the process of providing electrical energy

Brønsted–Lowry theory (9.1) a theory that describes an acid as a proton donor and a base as a proton acceptor

buffer capacity (9.4) a measure of the ability of a solution to resist large changes in pH when a strong acid or strong base is added

buffer solution (9.4) a solution containing a weak acid or base and its salt that is resistant to large changes in pH upon addition of strong acids or bases

buret (9.3) a device calibrated to deliver accurately known volumes of liquid, as in a titration

C

C-terminal amino acid (19.3) the amino acid in a peptide that has a free α-CO_2^- group; the last amino acid in a peptide

calorimetry (5.2) the measurement of heat energy changes during a chemical reaction

cap structure (24.4) a 7-methylguanosine unit covalently bonded to the 5′ end of a mRNA by a 5′–5′ triphosphate bridge

carbinol carbon (13.4) that carbon in an alcohol to which the hydroxyl group is attached

carbohydrate (Chapter 17, Introduction) generally sugars and polymers of sugars; the primary source of energy for the cell

carbonyl group (Chapter 14, Introduction) the functional group that contains a carbon-oxygen double bond: —C=O; the functional group found in aldehydes and ketones

carboxyl group (15.1) the —COOH functional group; the functional group found in carboxylic acids

carboxylic acid (15.1) a member of the family of organic compounds that contain the —COOH functional group

carboxylic acid derivative (15.2) any of several families of organic compounds, including the esters and amides, that are derived from carboxylic acids and have the general formula

Z = —OR or OAr for the esters, and Z = —NH$_2$ for the amides

carcinogen (24.7) any chemical or physical agent that causes mutations in the DNA that lead to uncontrolled cell growth or cancer

catabolism (21.1, 22.9) the degradation of fuel molecules and production of ATP for cellular functions

catalyst (8.3) any substance that increases the rate of a chemical reaction (by lowering the activation energy of the reaction) and that is not destroyed in the course of the reaction

cathode (2.3, 9.5) the negatively charged electrode in an electrical cell

cathode rays (2.3) a stream of electrons that are given off by the cathode (negative electrode) in a cathode ray tube

cation (2.2, 3.3) a positively charged atom or group of atoms

cellulose (17.4) a polymer of β-D-glucose linked by β(1 → 4) glycosidic bonds

central dogma (24.4) a statement of the directional transfer of the genetic information in cells: DNA → RNA → Protein

chain reaction (10.4) the reaction in a fission reactor that involves neutron production and causes subsequent reactions accompanied by the production of more neutrons in a continuing process

chair conformation (11.4) the most energetically favorable conformation for a six-member cycloalkane; so-called for its resemblance to a lawn chair

Charles's law (6.1) a law stating that the volume of a gas is directly proportional to the temperature of the gas, assuming that the pressure and number of moles of the gas are constant

chemical bond (4.1) the attractive force holding two atomic nuclei together in a chemical compound

chemical change (2.1) the conversion of one type of substance into another through a reorganization of the atoms; this process is represented by a chemical equation

chemical equation (5.4) a record of chemical change, showing the conversion of reactants to products

chemical formula (5.2) the representation of a compound or ion in which elemental symbols represent types of atoms and subscripts show the relative numbers of atoms

chemical property (2.1) properties of a substance that relate to the substance's participation in a chemical reaction

chemical reaction (2.1) a process in which atoms are rearranged to produce new combinations

chemistry (1.1) the study of matter and the changes that matter undergoes

chiral molecule (17.2) molecule capable of existing in mirror-image forms

cholesterol (18.4) a twenty-seven-carbon steroid ring structure that serves as the precursor of the steroid hormones

chylomicron (18.5, 23.1) a plasma lipoprotein (aggregate of protein and triglycerides) that carries triglycerides from the intestine to all body tissues via the bloodstream

***cis-trans* isomers** (11.3) isomers that differ from one another in the placement of substituents on a double bond or ring

citric acid cycle (22.4) a cyclic biochemical pathway that is the final stage of degradation of carbohydrates, fats, and amino acids. It results in the complete oxidation of acetyl groups derived from these dietary fuels

cloning vector (24.8) a DNA molecule that can carry a cloned DNA fragment into a cell and that has a replication origin that allows the DNA to be replicated abundantly within the host cell

coagulation (19.10) the process by which proteins in solution are denatured and aggregate with one another to produce a solid

codon (24.4) a group of three ribonucleotides on the mRNA that specifies the addition of a specific amino acid onto the growing peptide chain

coenzyme (20.7) an organic group required by some enzymes; it generally serves as a donor or acceptor of electrons or a functional group in a reaction

coenzyme A (22.2) a molecule derived from ATP, the vitamin pantothenic acid, and β-mercaptoethylamine; coenzyme A functions in the transfer of acetyl groups in lipid and carbohydrate metabolism

cofactor (20.7) an inorganic group, usually a metal ion, that must be bound to an apoenzyme to maintain the correct configuration of the active site

colipase (23.1) a protein that aids in lipid digestion by binding to the surface of lipid droplets and facilitating binding of pancreatic lipase

colligative property (7.6) property of a solution that is dependent only on the concentration of solute particles

colloidal suspension (7.3) a heterogeneous mixture of solute particles in a solvent; distribution of solute particles is not uniform because of the size of the particles

combination reaction (7.1) a reaction in which two substances join to form another substance

combined gas law (6.1) an equation that describes the behavior of a gas when volume, pressure, and temperature may change simultaneously

combustion (11.5) the oxidation of hydrocarbons by burning in the presence of air to produce carbon dioxide and water

competitive inhibitor (20.10) a structural analog; a molecule that has a structure very similar to the natural substrate of an enzyme, competes with the natural substrate for binding to the enzyme active site, and inhibits the reaction

complementary strands (24.2) the opposite strands of the double helix are hydrogen-bonded to one another such that adenine and thymine or guanine and cytosine are always paired

complete protein (19.11) a protein source that contains all the essential and nonessential amino acids

complex lipid (18.5) a lipid bonded to other types of molecules

compound (2.1) a substance that is characterized by constant composition and that can be chemically broken down into elements

concentration (1.6, 7.4) a measure of the quantity of a substance contained in a specified volume of solution

condensation (6.2) the conversion of a gas to a liquid

condensed formula (11.2) a structural formula showing all of the atoms in a molecule and placing them in a sequential arrangement that details which atoms are bonded to each other; the bonds themselves are not shown

conformations, conformers (11.4) discrete, distinct isomeric structures that may be converted, one to the other, by rotation about the bonds in the molecule

conjugate acid (9.1) substance that has one more proton than the base from which it is derived

conjugate acid-base pair (9.1) two species related to each other through the gain or loss of a proton

conjugate base (9.1) substance that has one fewer proton than the acid from which it is derived

conjugated protein (19.7) a protein that is functional only when it carries other chemical groups attached by covalent linkages or by weak interactions

constitutional isomers (11.2) two molecules having the same molecular formulas, but different chemical structures

conversion factor (1.3) an equivalence statement or multiplier consisting of a ratio of two equivalent quantities in different units, used to convert a quantity from one unit to another

Cori Cycle (21.6)   a metabolic pathway in which the lactate produced by working muscle is taken up by cells in the liver and converted back to glucose by gluconeogenesis

corrosion (9.5)   the unwanted oxidation of a metal

covalent bond (4.1)   a pair of electrons shared between two atoms

covalent solid (6.3)   a collection of atoms held together by covalent bonds

crenation (7.6)   the shrinkage of red blood cells caused by water loss to the surrounding medium

cristae (22.1)   the folds of the inner membrane of the mitochondria

crystal lattice (4.2)   a unit of a solid characterized by a regular arrangement of components

crystalline solid (6.3)   a solid having a regular repeating atomic structure

curie (10.9)   the quantity of radioactive material that produces 3.7×10^{10} nuclear disintegrations per second

cycloalkane (11.3)   a cyclic alkane; a saturated hydrocarbon that has the general formula C_nH_{2n}

D

Dalton's law (6.1)   also called the law of partial pressures; states that the total pressure exerted by a gas mixture is the sum of the partial pressures of the component gases

data (1.2)   a group of facts resulting from an experiment

decomposition reaction (7.1)   the breakdown of a substance into two or more substances

degenerate code (24.5)   a term used to describe the fact that several triplet codons may be used to specify a single amino acid in the genetic code

dehydration (of alcohols) (13.5)   a reaction that involves the loss of a water molecule, in this case the loss of water from an alcohol and the simultaneous formation of an alkene

deletion mutation (24.7)   a mutation that results in the loss of one or more nucleotides from a DNA sequence

denaturation (19.10)   the process by which the organized structure of a

protein is disrupted, resulting in a completely disorganized, nonfunctional form of the protein

density (1.6)   mass per unit volume of a substance

deoxyribonucleic acid (DNA) (24.1)   the nucleic acid molecule that carries all of the genetic information of an organism; the DNA molecule is a double helix composed of two strands, each of which is composed of phosphate groups, deoxyribose, and the nitrogenous bases thymine, cytosine, adenine, and guanine

deoxyribonucleotide (24.1)   a nucleoside phosphate or nucleotide composed of a nitrogenous base in β-*N*-glycosidic linkage to the 1' carbon of the sugar 2'-deoxyribose and with one, two, or three phosphoryl groups esterified at the hydroxyl of the 5' carbon

diabetes mellitus (23.3)   a disease caused by the production of insufficient levels of insulin and characterized by the appearance of very high levels of glucose in the blood and urine

dialysis (7.8)   the removal of waste material via transport across a membrane

diglyceride (18.3)   the product of esterification of glycerol at two positions

dipole-dipole interactions (6.2)   attractive forces between polar molecules

disaccharide (17.1, 17.3)   a sugar composed of two monosaccharides joined through an oxygen atom bridge

dissociation (4.3)   production of positive and negative ions when an ionic compound dissolves in water

disulfide (13.9)   an organic compound that contains a disulfide group (—S—S—)

DNA polymerase (24.3)   the enzyme that catalyzes the polymerization of daughter DNA strands using the parental strand as a template

double bond (4.4)   a bond in which two pairs of electrons are shared by two atoms

double helix (24.2)   the spiral staircase-like structure of the DNA molecule characterized by two sugar–phosphate backbones wound around the outside and nitrogenous bases extending into the center

double-replacement reaction (7.1)   a chemical change in which cations and anions "exchange partners"

dynamic equilibrium (5.6, 8.4)   the state that exists when the rate of change in the concentration of products and reactants is equal, resulting in no net concentration change

E

eicosanoid (18.2)   any of the derivatives of twenty-carbon fatty acids, including the prostaglandins, leukotrienes, and thromboxanes

electrolysis (9.5)   an electrochemical process that uses electrical energy to cause nonspontaneous oxidation–reduction reactions to occur

electrolyte (4.3, 7.3)   a material that dissolves in water to produce a solution that conducts an electrical current

electrolytic solution (4.3)   a solution composed of an electrolytic solute dissolved in water

electromagnetic radiation (2.3)   energy that is propagated as waves at the speed of light

electromagnetic spectrum (2.3)   the complete range of electromagnetic waves

electron (2.2)   a negatively charged particle outside of the nucleus of an atom

electron affinity (3.4)   the energy released when an electron is added to an isolated atom

electron configuration (3.2)   the arrangement of electrons around a nucleus of an atom, ion, or a collection of nuclei of a molecule

electron density (2.4)   the probability of finding the electron in a particular location

electron transport system (22.6)   the series of electron transport proteins embedded in the inner mitochondrial membrane that accept high-energy electrons from NADH and $FADH_2$ and transfer them in stepwise fashion to molecular oxygen (O_2)

electronegativity (4.1)   a measure of the tendency of an atom in a molecule to attract shared electrons

electronic transitions (2.3)   involves the movement of an electron from one energy level to another within an atom

Answers to Odd-Numbered Problems

Chapter 1

1.1 **a.** 1.0×10^3 mL
b. 1.0×10^6 μL
c. 1.0×10^{-3} kL
d. 1.0×10^2 cL
e. 1.0×10^{-1} daL

1.3 **a.** 1.3×10^{-2} m
b. 0.71 L
c. 2.00 oz
d. 1.5×10^{-4} m^2

1.5 **a.** Three
b. Three
c. Four
d. Two
e. Three

1.7 **a.** 2.4×10^{-3}
b. 1.80×10^{-2}
c. 2.24×10^2

1.9 **a.** 8.09
b. 5.9
c. 20.19

1.11 **a.** 51
b. 8.0×10^1
c. 1.6×10^2

1.13 **a.** 61.4
b. 6.17
c. 6.65×10^{-2}

1.15 **a.** 0°C
b. 273 K

1.17 23.7 g alcohol

1.19 **a.** Chemistry is the study of matter and the changes that matter undergoes.
b. Matter is the material component of the universe.
c. Energy is the ability to do work.

1.21 **a.** Precision is the degree of agreement among replicate measurements of the same quantity.
b. Accuracy is the nearness of an experimental value to the true value.
c. Data are a group of observations resulting from an experiment.

1.23 **a.** gram (or kilogram)
b. liter
c. meter

1.25 Weight is the force exerted on a body by gravity; mass is a quantity of matter. Mass is an independent quantity whereas weight is dependent on gravity, which may differ from location to location.

1.27 Density is mass per volume. Specific gravity is the ratio of the density of a substance to the density of water at 4°C or any specified temperature.

1.29 The scientific method is an organized way of doing science. It uses carefully planned experimentation to study our surroundings.

1.31 **a.** 32 oz
b. 1.0×10^{-3} t
c. 9.1×10^2 g
d. 9.1×10^5 mg
e. 9.1×10^1 da

1.33 **a.** 6.6×10^{-3} lb
b. 1.1×10^{-1} oz
c. 3.0×10^{-3} kg
d. 3.0×10^2 cg
e. 3.0×10^3 mg

1.35 **a.** 10°C
b. 283 K

1.37 **a.** 293 K
b. 68°F

1.39 4 L

1.41 101°F

1.43 **a.** Three
b. Three
c. Three
d. Four
e. Four
f. Three

1.45 **a.** 3.87×10^{-3}
b. 5.20×10^{-2}
c. 2.62×10^{-3}
d. 2.43×10^1
e. 2.40×10^2
f. 2.41×10^0

1.47 **a.** 1.5×10^4
b. 2.41×10^{-1}
c. 5.99
d. 1139.42
e. 7.21×10^3

1.49 **a.** 1.23×10^1
b. 5.69×10^{-2}
c. -1.527×10^3
d. 7.89×10^{-7}
e. 9.2×10^7
f. 5.280×10^{-3}
g. 1.279×10^0
h. -5.3177×10^2

1.51 **a.** 3240
b. 0.000150
c. 0.4579
d. −683,000
e. −0.0821
f. 299,790,000

g. 1.50
h. 602,200,000,000,000,000,000,000

1.53 6.00 g/mL

1.55 1.08×10^3 g

1.57 9.8×10^{-1} g/cm³, teak

1.59 0.789

1.61 $d_{lead} = 7.9$ g/cm³
$d_{uranium} = 19$ g/cm³
$d_{platinum} = 21.4$ g/cm³
Lead has the lowest density and platinum has the greatest density.

Chapter 2

2.1 **a.** physical property
b. chemical property
c. physical property
d. physical property
e. physical property

2.3 **a.** pure substance
b. heterogeneous mixture
c. homogeneous mixture
d. pure substance

2.5 **a.** sixteen protons, sixteen electrons, sixteen neutrons
b. eleven protons, eleven electrons, twelve neutrons

2.7 20.18 amu

2.9 DeBroglie considered electrons to have both wave and particle properties.

2.11 A physical property is a characteristic of a substance that can be observed without the substance undergoing a change in chemical composition.

2.13 **a.** chemical reaction
b. physical change
c. physical change

2.15 **a.** physical
b. chemical

2.17 flammability and toxicity

2.19 A pure substance has constant composition with only a single substance whereas a mixture is composed of two or more substances.

2.21 A homogeneous mixture has uniform composition whereas a heterogeneous mixture has nonuniform composition.

2.23 A gas is made up of particles that are widely separated. A gas will expand to fill any container and has no definite shape or volume.

2.25 An intensive property is a characteristic of a substance that is independent of the quantity of the substance. An extensive property depends on the quantity of the substance.

2.27 An element is a pure substance that cannot be changed into a simpler form of matter by any chemical reaction. An atom is the smallest unit of an element that retains the properties of that element.

2.29 **a.** eight protons, eight electrons, eight neutrons
b. fifteen protons, fifteen electrons, sixteen neutrons

2.31 Isotopes are atoms of the same element that differ in mass because they contain different numbers of neutrons.

2.33

Particle	Mass	Charge
a. electron	5.4×10^{-4} amu	−1
b. proton	1.00 amu	+1
c. neutron	1.00 amu	0

2.35 **a.** An ion is a charged atom or group of atoms formed by the loss or gain of electrons.
b. A loss of electrons by a neutral species results in a cation.
c. A gain of electrons by a neutral species results in an anion.

2.37

Atomic Symbol	# Protons	# Neutrons	# Electrons	Charge
a. $^{23}_{11}Na$	11	12	11	0
b. $^{32}_{16}S^{2-}$	16	16	18	2−
c. $^{16}_{8}O$	8	8	8	0
d. $^{24}_{12}Mg^{2+}$	12	12	10	2+
e. $^{39}_{19}K^{+}$	19	20	18	1+

2.39 All matter consists of tiny particles called atoms. Atoms cannot be created, divided, destroyed, or converted to any other type of atom. All atoms of a particular element have identical properties. Atoms of different elements have different properties. Atoms combine in simple whole-number ratios. Chemical change involves joining, separating, or rearranging atoms.

2.41 **a.** Chadwick demonstrated the existence of the neutron.
b. DeBroglie theorized that electrons had wavelike and particlelike properties.
c. Geiger provided the basic experimental evidence for the existence of a nucleus.
d. The Bohr theory describes electron arrangement in atoms.

2.43 Geiger bombarded a piece of gold foil with alpha particles, and observed that some alpha particles passed straight through the foil, others were deflected, and some simply bounced back. This led Rutherford to propose that the atom consisted of a small, dense nucleus (alpha particles bounced back), surrounded by a cloud of electrons (some alpha particles were deflected). The size of the nucleus is small when compared to the volume of the atom (alpha particles were able to pass through the foil).

2.45 Electrons are found in orbits at discrete distances from the nucleus. The orbits are quantitized—they are of discrete energies. Electrons can only be found in these orbits, never in between (they are able to jump instantaneously from orbit to orbit). Electrons can undergo transitions—if an electron absorbs energy, it will jump to a higher orbit; when the electron falls back down to a lower orbit, it will release energy.

2.47 Crookes used the cathode ray tube. He observed particles emitted by the cathode and traveling toward the anode. This ray was deflected by an electric field. Thomson measured the curvature of the ray influenced by the electric field. This measurement provided the mass-to-charge ratio of the negative particle. Thomson also gave the particle the name, electron.

2.49 A cathode ray is the negatively charged particle formed in a cathode ray tube. It was characterized as an electron, with a very small mass and a charge of −1.

2.51 **a.** neutrons
b. protons
c. protons, neutrons
d. ion
e. nucleus, negative

2.53 radiowave
microwave
infrared
visible   Increasing
ultraviolet   Wavelength
X-ray
gamma ray

2.55 The deBroglie hypothesis stated that the electron has both particlelike and wavelike properties.

2.57 Bohr's atomic model was the first to successfully account for electronic properties of atoms, specifically, the interaction of atoms and light (spectroscopy).

Chapter 3

3.1 **a.** Zr (zirconium)
 b. 22.99
 c. Cr (chromium)
 d. Bi (bismuth)
3.3 **a.** helium, atomic number = 2, mass = 4.00 amu
 b. fluorine, atomic number = 9, mass = 19.00 amu
 c. manganese, atomic number = 25, mass = 54.94 amu
3.5 **a.** Total electrons = 11, valence electrons = 1
 b. Total electrons = 12, valence electrons = 2
 c. Total electrons = 16, valence electrons = 6
 d. Total electrons = 17, valence electrons = 7
 e. Total electrons = 18, valence electrons = 8
3.7 **a.** Sulfur: $1s^2, 2s^2, 2p^6, 3s^2, 3p^4$
 b. Calcium: $1s^2, 2s^2, 2p^6, 3s^2, 3p^6, 4s^2$
3.9 **a.** [Ne] $3s^2, 3p^4$
 b. [Ar] $4s^2$
3.11 **a.** Ca^{2+} and Ar are isoelectronic
 b. Sr^{2+} and Kr are isoelectronic
 c. S^{2-} and Ar are isoelectronic
 d. Mg^{2+} and Ne are isoelectronic
 e. P^{3-} and Ar are isoelectronic
3.13 **a.** (smallest) F, N, Be (largest)
 b. (lowest) Be, N, F (highest)
 c. (lowest) Be, N, F, (highest)
3.15 **a.** elemental properties are periodic as a function of their atomic number
 b. a horizontal row across the periodic table
 c. a vertical column on the periodic table
 d. a charged unit resulting from the gain or loss of electrons from a neutral atom or group of atoms
3.17 **a.** true
 b. true
3.19 **a.** Na, Ni, Al
 b. Na, Al
 c. Na, Ni, Al
 d. Ar
3.21 **a.** sodium
 b. potassium
 c. magnesium
3.23 Group 1A (or 1): lithium, sodium, potassium, rubidium, cesium, and francium
3.25 Group VIIA (or 17): fluorine, chlorine, bromine, iodine, and astatine
3.27 The early periodic table contained many fewer elements and was arranged by atomic weight.
3.29 An element along the "staircase" boundary between metals and nonmetals; metalloids exhibit both metallic and nonmetallic properties.
3.31 **a.** one
 b. one
 c. three
 d. seven
 e. zero (or eight)
 f. zero (or two)
3.33 One valence electron located in an s orbital and an outermost electron configuration of ns^1.
3.35 $2n^2$

3.37 A principal energy level is designated $n = 1, 2, 3$, and so forth. It is similar to Bohr's orbits in concept. A sublevel is a part of a principal energy level and is designated $s, p, d,$ and f.
3.39

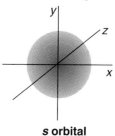

s orbital

The s orbital represents the probability of finding an electron in a region of space surrounding the nucleus.
3.41 Three
3.43 A $3p$ orbital is a higher energy orbital than a $2p$ orbital because it is a part of a higher energy principal energy level.
3.45 $2e^-$ for $n = 1$
 $8e^-$ for $n = 2$
 $18e^-$ for $n = 3$
3.47 **a.** $3p$
 b. $3s$
 c. $3d$
 d. $4s$
 e. $3d$
 f. $3p$
3.49 **a.** Not possible; $n = 1$ level can have only s-level orbitals
 b. Possible; it is the electron configuration of C.
 c. Not possible; $n = 2$ level can contain only s and p orbitals
 d. Not possible; an s orbital cannot contain $3e^-$
3.51 **a.** Li^+
 b. O^{2-}
 c. Ca^{2+}
 d. Br^-
 e. S^{2-}
 f. Al^{3+}
3.53 **a.** O^{2-}, $10e^-$; Ne, $10e^-$ Isoelectronic
 b. S^{2-}, $18e-$; Cl^-, $18e^-$ Isoelectronic
3.55 Group IA metals form only a 1+ ion because the loss of one electron produces an electron configuration similar to their nearest noble gas. Group IIA metals form only a 2+ ion because the loss of two electrons produces an electron configuration similar to their nearest noble gas.
3.57 **a.** Na^+
 b. S^{2-}
 c. Cl^-
3.59 **a.** $1s^2, 2s^2, 2p^6, 3s^2, 3p^6$
 b. $1s^2, 2s^2, 2p^6$
3.61 **a.** (Smallest) F, O, N (Largest)
 b. (Smallest) Li, K, Cs (Largest)
 c. (Smallest) Cl, Br, I (Largest)
3.63 Cl
3.65 **a.** (Smallest) O, N, F (Largest)
 b. (Smallest) Cs, K, Li (Largest)
 c. (Smallest) I, Br, Cl (Largest)
3.67 A positive ion is always smaller than its parent atom because the positive charge of the nucleus is shared among fewer electrons in the ion. As a result, each electron is pulled closer to the nucleus and the volume of the ion decreases.
3.69 The fluoride ion has a completed octet of electrons and an electron configuration resembling its nearest noble gas.

b. 1.20×10^{24} O atoms

c. 32.00 g O

d. 10.7 g O

5.59 7.39 g O_2

5.61 6.14×10^4 g O_2

5.63 70.6 g $C_{10}H_{22}$

5.65 9.13×10^2 g N_2

5.67 92.6%

5.69 6.85×10^2 g N_2

Chapter 6

6.1 a. 0.954 atm

b. 0.382 atm

c. 0.730 atm

6.3 a. 38 atm

b. 25 atm

6.5 a. 3.76 L

b. 3.41 L

c. 2.75 L

6.7 0.200 atm

6.9 4.46 mol H_2

6.11 9.00 L

6.13 0.223 mol N_2

6.15 1 atm

6.17 5 L-atm

6.19 5.23 atm

6.21 Charles's law states that the volume of a gas varies directly with the absolute temperature if pressure and number of moles of gas are constant.

6.23 The volume increases from 2.00 L to 2.96 L.

6.25 1.51 L

6.27 *increase*

6.29 $V_f = \dfrac{P_i V_i T_f}{P_f T_i}$

6.31 1.82×10^{-2} L

6.33 6.00 L

6.35 Avogadro's law states that equal volumes of a gas contain the same number of moles if measured under the same conditions of temperature and pressure.

6.37 5.94×10^{-2} L

6.39 22.4 L

6.41 9.08×10^3 L

6.43 172°C

6.45 Gases exhibit more ideal behavior at low pressures. At low pressures, gas particles are more widely separated and therefore the attractive forces between particles are less. The ideal gas model assumes negligible attractive forces between gas particles.

6.47 The kinetic molecular theory states that the average kinetic energy of the gas particles increases as the temperature increases. Kinetic energy is proportional to (velocity)². Therefore, as the temperature increases the gas particle velocity increases and the rate of mixing increases as well.

6.49 Dalton's law states that the total pressure of a mixture of gases is the sum of the partial pressures of the component gases.

6.51 0.74 atm

6.53 Intermolecular forces in liquids are considerably stronger than intermolecular forces in gases. Particles are, on average, much closer together in liquids and the strength of attraction is inversely proportional to the distance of separation.

6.55 The vapor pressure of a liquid increases as the temperature of the liquid increases.

6.57 Evaporation is the conversion of a liquid to a gas at a temperature lower than the boiling point of the liquid. Condensation is the conversion of a gas to a liquid at a temperature lower than the boiling point of the liquid.

6.59 Viscosity is the resistance to flow caused by intermolecular attractive forces. Complex molecules may become entangled and not slide smoothly across one another.

6.61 Solids are essentially incompressible because the average distance of separation among particles in the solid state is small. There is literally no space for the particles to crowd closer together.

6.63 a. high melting temperature, brittle

b. high melting temperature, hard

6.65 Beryllium; metabolic solids are good electrical conductors. Carbon forms covalent solids, which are poor electrical conductors.

Chapter 7

7.1 a. DR

b. SR

c. DR

d. D

7.3 a. $KCl(aq) + AgNO_3(aq) \rightarrow KNO_3(aq) + AgCl(s)$
A precipitation reaction occurs.

b. $CH_3COOK(aq) + AgNO_3(aq) \rightarrow$ no reaction
No precipitation reaction occurs.

7.5 16.7% NaCl

7.7 7.50% KCl

7.9 2.56×10^{-2}%

7.11 20.0%

7.13 0.125 mol HCl

7.15 Dilute 1.7×10^{-2} L of 12 M HCl with sufficient water to produce 1.0×10^2 mL of total solution.

7.17 1.0×10^{-2} osmol

7.19 0.24 atm

7.21 CO is more soluble than CO_2. CO is a polar molecule as is water and "like dissolves like."

7.23 0.0154 mol/L

7.25 a. Heating an alkaline earth metal carbonate,
$MgCO_3(s) \xrightarrow{\Delta} MgO(s) + CO_2(g)$

b. The replacement of copper by zinc in copper sulfate,
$Zn(s) + CuSO_4(aq) \rightarrow ZnSO_4(aq) + Cu(s)$

7.27 Reaction of two soluble substances to form an insoluble product
$2NaOH(aq) + FeCl_2(aq) \rightarrow Fe(OH)_2(s) + 2NaCl(aq)$

7.29 a. $2C_2H_6(g) + 7O_2(g) \rightarrow 4CO_2(g) + 6H_2O(g)$

b. $6K_2O(s) + P_4O_{10}(s) \rightarrow 4K_3PO_4(s)$

c. $MgBr_2(aq) + H_2SO_4(aq) \rightarrow 2HBr(g) + MgSO_4(aq)$

7.31 a. $Ca(s) + F_2(g) \rightarrow CaF_2(s)$

b. $2Mg(s) + O_2(g) \rightarrow 2MgO(s)$

c. $3H_2(g) + N_2(g) \rightarrow 2NH_3(g)$

7.33 a. 2.00% NaCl

b. 6.60% $C_6H_{12}O_6$

7.35 a. 5.00% ethanol

b. 10.0% ethanol

7.37 a. 21.0% NaCl

b. 3.75% NaCl

7.39 a. 2.25 g NaCl

b. 3.13 g CH_3COONa

7.41 a. 0.342 M NaCl
   b. 0.367 M $C_6H_{12}O_6$
7.43 a. 1.46 g NaCl
   b. 9.00 g $C_6H_{12}O_6$
7.45 0.146 M $C_{12}H_{22}O_{11}$
7.47 5.00×10^{-2} L
7.49 20.0 M
7.51 A colligative property is a solution property that depends on the concentration of solute particles rather than the identity of the particles.
7.53 Salt is an ionic substance that dissociates in water to produce positive and negative ions. These ions (or particles) lower the freezing point of water. If the concentration of salt particles is large, the freezing point may be depressed below the surrounding temperature, and the ice would melt.
7.55 0.5 M sucrose
7.57 0.5 M sucrose
7.59 24 atm at 25°C
7.61 polar, high boiling point, low vapor pressure, abundant, and easily purified
7.63 The ammonia converts to the extremely soluble and stable ammonium ion.
7.65

$$\begin{matrix} \text{H} & \text{H} \\ & \ddot{\text{O}}\!\!: \\ & \text{Na}^+ \\ \text{H}\!:\!\ddot{\text{O}}\!\!: & :\!\ddot{\text{O}}\!:\!\text{H} \\ \text{H} & \text{H} \end{matrix}$$

   Several water molecules "hydrate" each sodium ion.
7.67 A low sodium ion concentration in the dialysis solution favors transport of sodium ions from the blood.
7.69 confusion, stupor, or coma
7.71 diabetes, diarrhea, and certain high-protein diets

Chapter 8

8.1 a. Exothermic
   b. Exothermic
   c. Exothermic
8.3 13°C
8.5 2.7×10^3 J
8.7 $\dfrac{2.1 \times 10^2 \text{ nutritional Cal}}{\text{candy bar}}$
8.9 Heat energy produced by the friction of striking the match provides the activation energy for this combustion process.
8.11 If the enzyme catalyzed a process needed to sustain life, the substance interfering with that enzyme would be classified as a poison.
8.13 a. rate = $k[N_2]^n[O_2]^{n'}$ (n and n' are experimentally determined)
   b. rate = $k[C_4H_6]^n$ (n must be experimentally determined)
8.15 At rush hour, approximately the same number of passengers enter and exit the train at each stop. At any time the number of passengers may be essentially unchanged but the identity of the individual passengers is continually changing.
8.17 Measure the concentration of products and reactants at different times until no further concentration change is observed.
8.19 a. $K_{eq} = \dfrac{[N_2][O_2]^2}{[NO_2]^2}$
   b. $K_{eq} = [H_2]^2[O_2]$
8.21 A large K_{eq} favors products.

8.23 8.2×10^{-2}
8.25 a. decrease
   b. increase
   c. decrease
   d. remain the same
8.27 a. An exothermic reaction is one in which energy is released during chemical change.
   b. An endothermic reaction is one in which energy is absorbed during chemical change.
   c. A calorimeter is a device for measuring heat absorbed or released during chemical change.
8.29 Enthalpy is a measure of heat energy.
8.31 1.20×10^3 cal
8.33 5.02×10^3 J
8.35 a. Entropy increases. Conversion of a solid to a liquid results in an increase in disorder of the substance. Solids retain their shape whereas liquids will flow and their shape is determined by their container.
   b. Entropy increases. Conversion of a liquid to a gas results in an increase in disorder of the substance. Gas particles move randomly with very weak interactions between particles, much weaker than those interactions in the liquid state.
8.37 An increase in stability is equated with a decrease in energy (reaching a lower energy state). The energy of products is less than that of the reactants in an exothermic reaction; energy is given off in an exothermic reaction.
8.39 Isopropyl alcohol quickly evaporates after being applied to the skin. Conversion of a liquid to a gas requires heat energy. The heat energy is supplied by the skin. When this heat is lost, the skin temperature drops.
8.41 The activated complex is the arrangement of reactants in an unstable transition state as a chemical reaction proceeds. The activated complex must form to convert reactants to products.
8.43

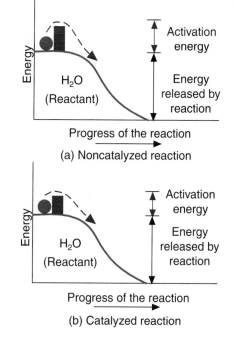

(a) Noncatalyzed reaction

(b) Catalyzed reaction

8.45 Enzymes are biological catalysts. The enzyme lysozyme catalyzes a process that results in the destruction of the cell

b.

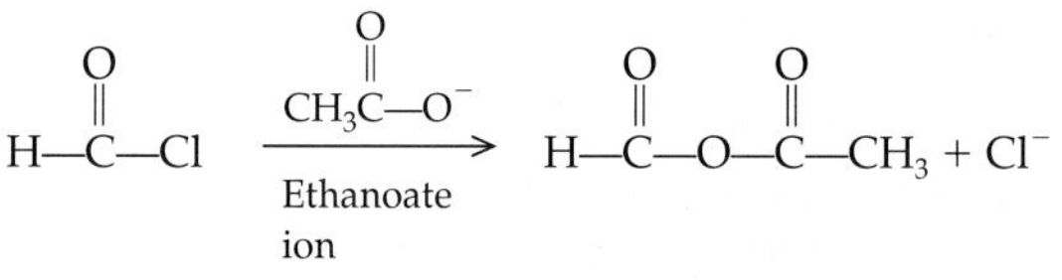

Methanoyl chloride → Ethanoate ion → Ethanoic methanoic anhydride

15.15 a.

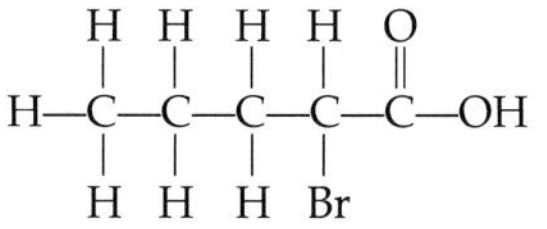

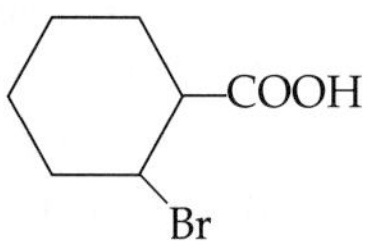

b.

c.

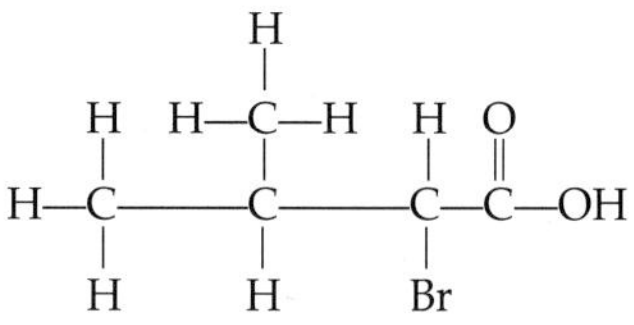

d.

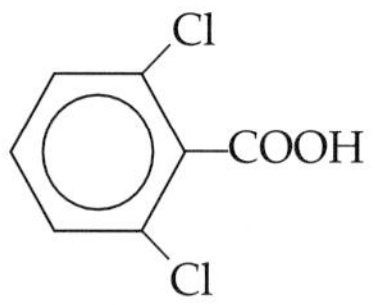

e.

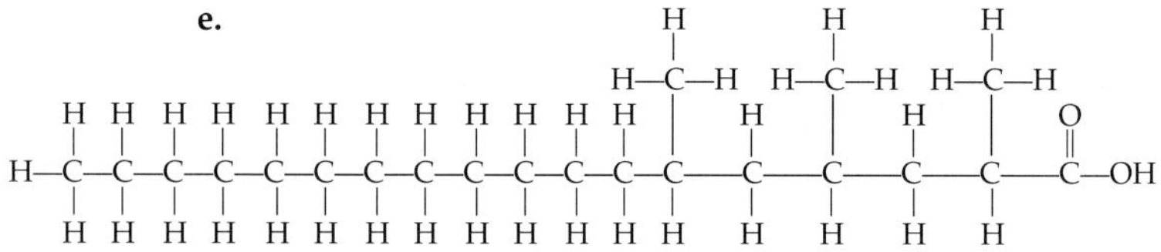

f.

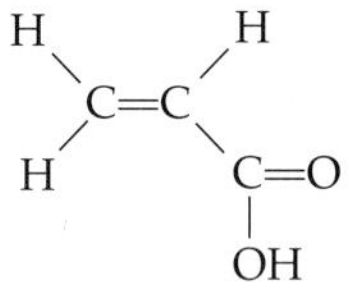

15.17

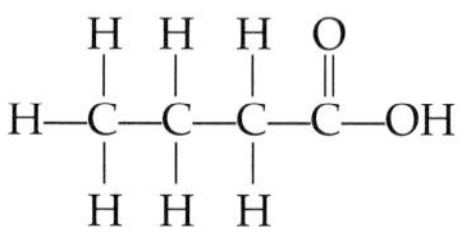

Butanoic acid 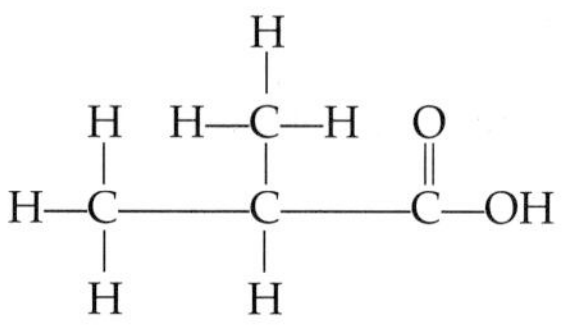   Methylpropanoic acid

15.19 a.

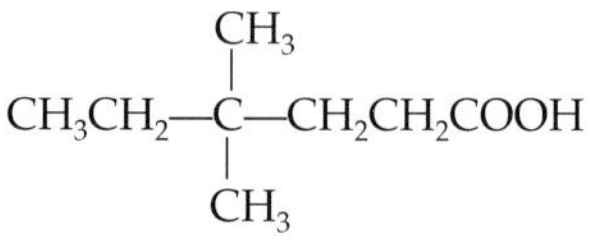

$$CH_3CH_2-\overset{CH_3}{\underset{CH_3}{C}}-CH_2CH_2COOH$$

b.
$$CH_3\underset{Br}{\overset{CH_3}{CHCHCH_2COOH}}$$

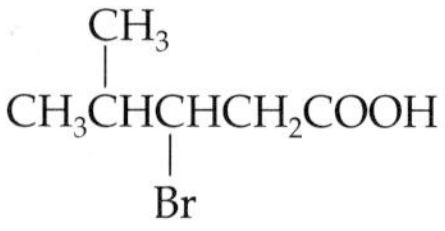

c.

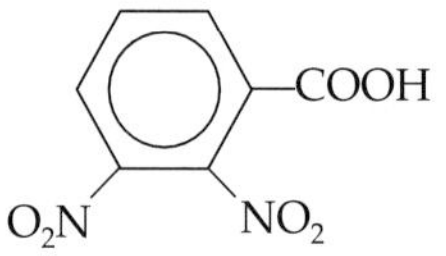

d.

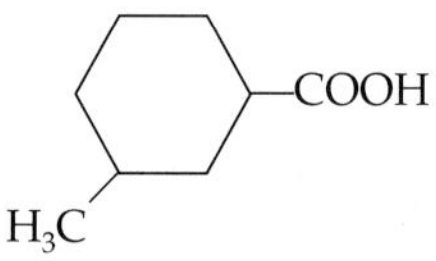

15.21 a. I.U.P.A.C. name: 2-Hydroxypropanoic acid
   Common name: α-Hydroxypropionic acid
b. I.U.P.A.C. name: 3-Hydroxybutanoic acid
   Common name: β-Hydroxybutyric acid

c. I.U.P.A.C. name: 4, 4-Dimethylpentanoic acid
   Common name: γ, γ-Dimethylvaleric acid
d. I.U.P.A.C. name: 3, 3-Dichloropentanoic acid
   Common name: β, β-Dichlorovaleric acid

15.23 a. Heptanoic acid
b. 1-Propanol
c. Pentanoic acid
d. Butanoic acid

15.25 The smaller carboxylic acids are water-soluble. They have sharp, sour tastes and unpleasant aromas.

15.27 Citric acid is added to foods to give them a tart flavor or to act as a food preservative and antioxidant. Adipic acid imparts a tart flavor to soft drinks and is a preservative.

15.29 Carboxylic acids are produced commercially by the oxidation of the corresponding alcohol or aldehyde.

15.31 Soaps are made from water, a strong base, and natural fats or oils.

15.33 a. CH_3COOH

b.
$$CH_3CH_2CH_2-\overset{O}{\overset{\|}{C}}-O-CH_3 + H_2O$$

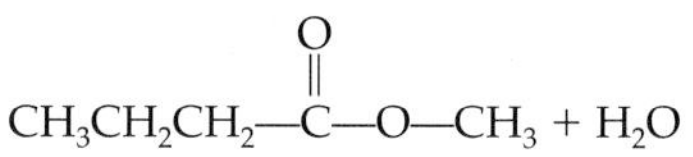

c. CH_3OH

15.35 a. The oxidation of 1-pentanol yields pentanal.
b. Continued oxidation of pentanal yields pentanoic acid.

15.37 a.

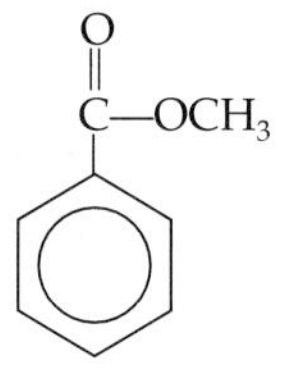

b.
$$CH_3CH_2CH_2CH_2CH_2CH_2CH_2CH_2CH_2-\overset{O}{\overset{\|}{C}}-O-CH_2CH_2CH_2CH_3$$

c.
$$CH_3CH_2-\overset{O}{\overset{\|}{C}}-O-CH_3$$

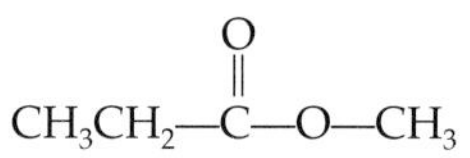

d.
$$CH_3CH_2-\overset{O}{\overset{\|}{C}}-O-CH_2CH_3$$

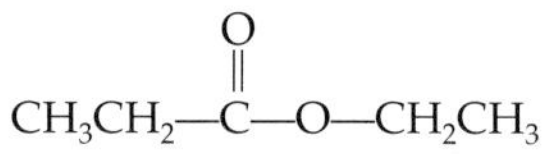

e.

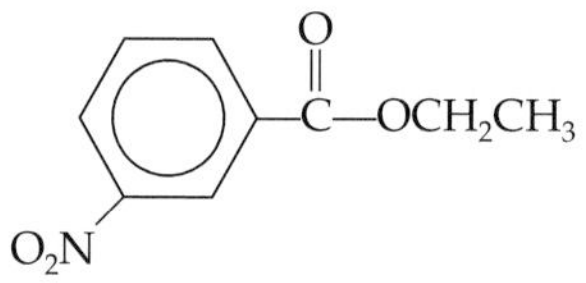

f.
$$CH_3-\overset{O}{\overset{\|}{C}}-O-\overset{CH_3}{\underset{}{CHCH_3}}$$

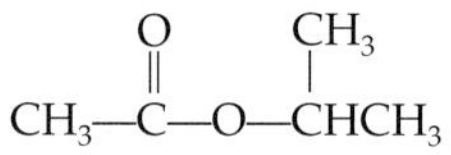

g.
$$CH_3CH_2CH_2-\overset{O}{\overset{\|}{C}}-O-CH_3$$

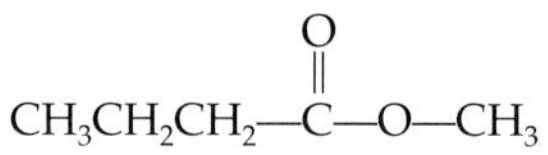

15.39 a.
$$CH_3CH_2CH_2-\overset{O}{\overset{\|}{C}}-O-CH_2CH_3$$

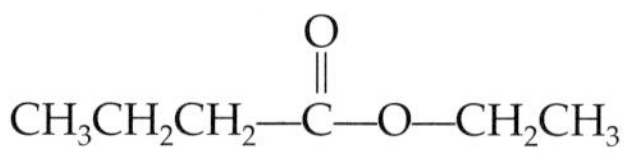

b.
$$CH_3CH_2-\overset{O}{\overset{\|}{C}}-OH + CH_3CH_2OH$$

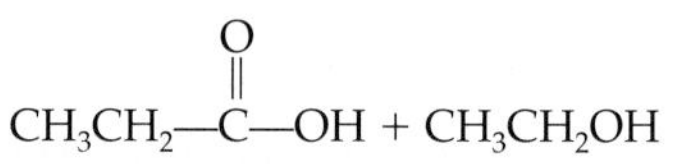

c. $CH_3CH_2CH_2OH$

d.
$$CH_3CH_2CHCH_2-\overset{O}{\overset{\|}{C}}-O^- + CH_3CH_2OH$$
with Br substituent

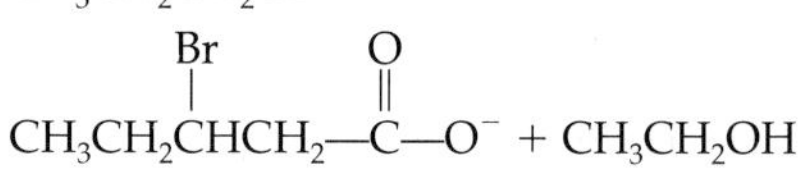

15.41 Saponification is a reaction in which a soap is produced. More generally, it is the hydrolysis of an ester in the presence of a base. The following reaction shows the base-catalyzed hydrolysis of an ester:

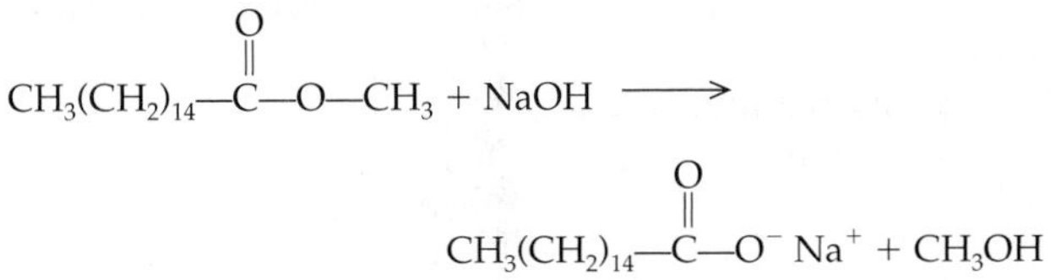

$$CH_3(CH_2)_{14}-\overset{\overset{\displaystyle O}{\|}}{C}-O-CH_3 + NaOH \longrightarrow$$

$$CH_3(CH_2)_{14}-\overset{\overset{\displaystyle O}{\|}}{C}-O^- Na^+ + CH_3OH$$

15.43

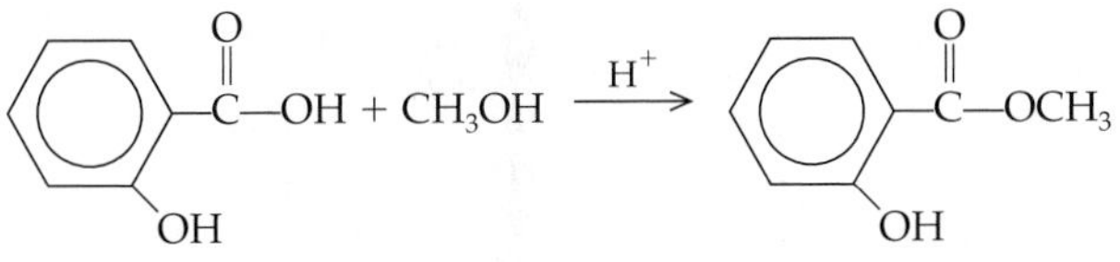

Salicylic acid   Methyl salicylate

15.45 Compound A is

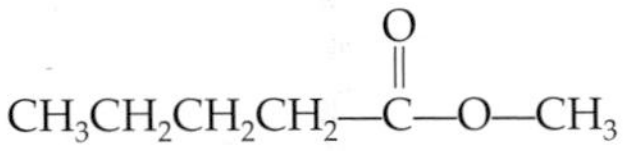

$$CH_3CH_2CH_2CH_2-\overset{\overset{\displaystyle O}{\|}}{C}-O-CH_3$$

Compound B is

$$CH_3CH_2CH_2CH_2-\overset{\overset{\displaystyle O}{\|}}{C}-OH$$

Compound C is CH_3OH

15.47 a. PCl_3, PCl_5, or $SOCl_2$

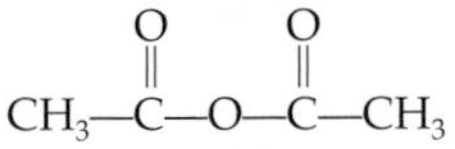

b. $CH_3-\overset{\overset{\displaystyle O}{\|}}{C}-O^-$

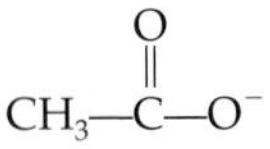

c.

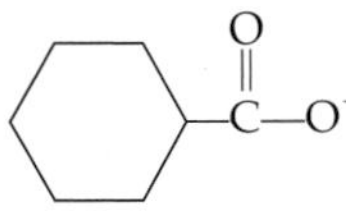

15.49 a.

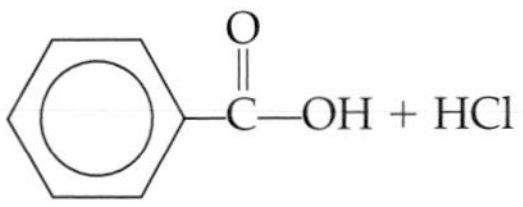

$+ HCl$

b. $2\ CH_3-\overset{\overset{\displaystyle O}{\|}}{C}-OH$

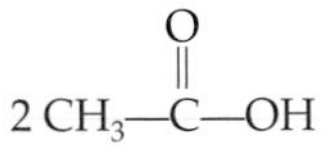

15.51 a.

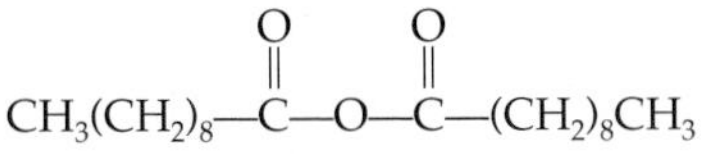

$$CH_3(CH_2)_8-\overset{\overset{\displaystyle O}{\|}}{C}-O-\overset{\overset{\displaystyle O}{\|}}{C}-(CH_2)_8CH_3$$

b.

$$CH_3-\overset{\overset{\displaystyle O}{\|}}{C}-O-\overset{\overset{\displaystyle O}{\|}}{C}-CH_3$$

c.

$$CH_3(CH_2)_3-\overset{\overset{\displaystyle O}{\|}}{C}-O-\overset{\overset{\displaystyle O}{\|}}{C}-(CH_2)_3CH_3$$

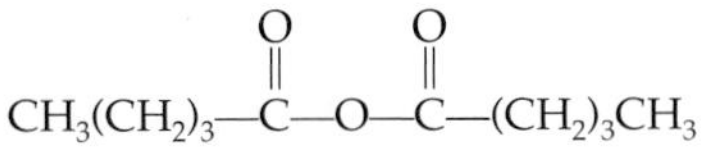

d.

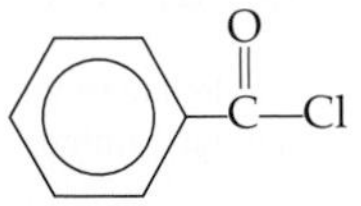

15.53 Acid chlorides are noxious, irritating chemicals. They are slightly polar and have boiling points similar to comparable aldehydes or ketones. They cannot be dissolved in water because they react violently with it.

15.55 a. Monoester:

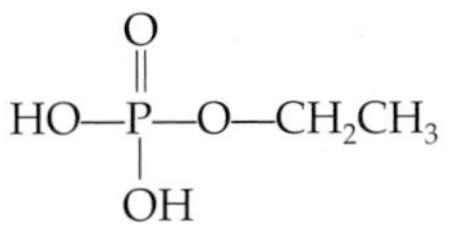

$$HO-\overset{\overset{\displaystyle O}{\|}}{\underset{\underset{\displaystyle OH}{|}}{P}}-O-CH_2CH_3$$

b. Diester:

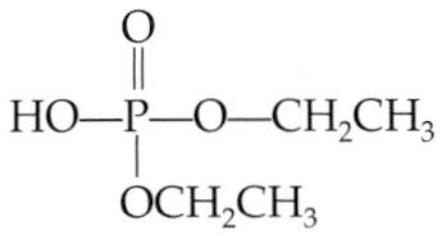

$$HO-\overset{\overset{\displaystyle O}{\|}}{\underset{\underset{\displaystyle OCH_2CH_3}{|}}{P}}-O-CH_2CH_3$$

c. Triester:

$$CH_3CH_2-O-\overset{\overset{\displaystyle O}{\|}}{\underset{\underset{\displaystyle OCH_2CH_3}{|}}{P}}-O-CH_2CH_3$$

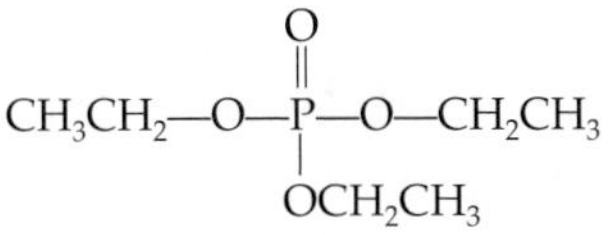

15.57 ATP is the molecule used to store the energy released in metabolic reactions. The energy is stored in the phosphoanhydride bonds between two phosphoryl groups. The energy is released when the bond is hydrolyzed. A portion of the energy can be transferred to another molecule if the phosphoryl group is transferred from ATP to the other molecule.

15.59

$$CH_3-\overset{\overset{\displaystyle O}{\|}}{C}-S-COENZYME\ A$$

15.61

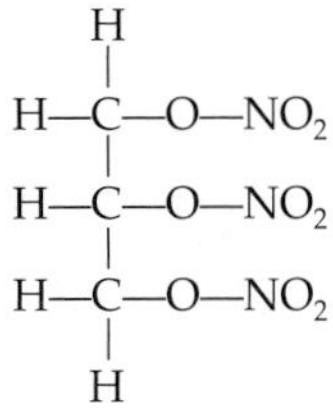

Chapter 16

16.1 a. Tertiary
  b. Primary
  c. Secondary

16.3

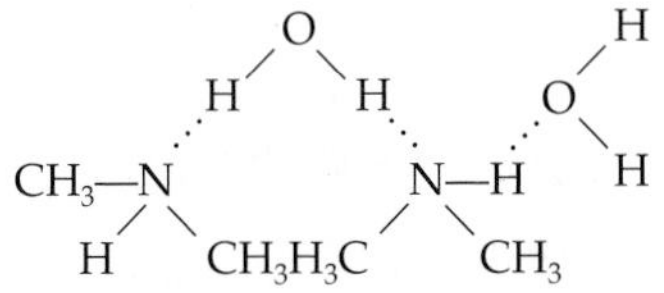

16.5 a. Methanol because the intermolecular hydrogen bonds between alcohol molecules will be stronger.
  b. Water because the intermolecular hydrogen bonds between water molecules will be stronger.
  c. Ethylamine because it has a higher molecular weight.
  d. Propylamine because propylamine molecules can form intermolecular hydrogen bonds while the nonpolar butane cannot do so.

16.7 a.

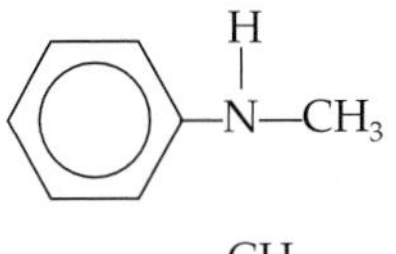

b.

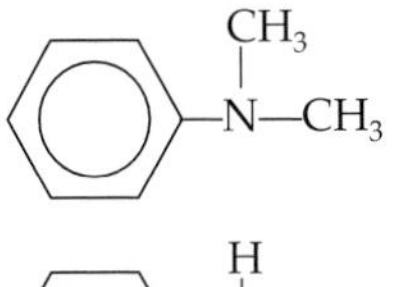

c.

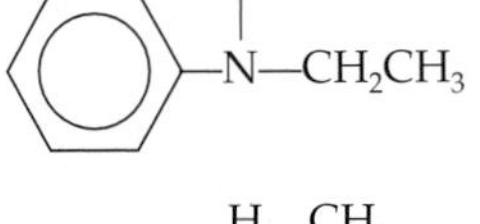

d.

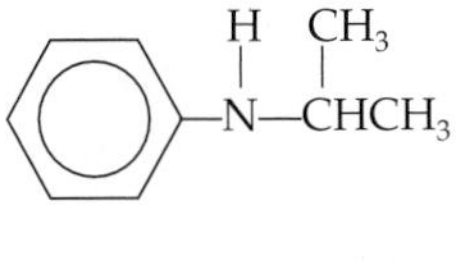

16.9 a.

```
      H   H   H
      |   |   |
  H—C—C—C—H
      |   |   |
      H   N   H
        /   \
       H     H
```

b.

```
            H       H
             \     /
  H   H   N   H   H   H   H   H
  |   |   |   |   |   |   |   |
H—C—C—C—C—C—C—C—C—H
  |   |   |   |   |   |   |   |
  H   H   H   H   H   H   H   H
```

c.

```
  H   H   H   H   H   H   H   H
  |   |   |   |   |   |   |   |
H—C—C—C—C—C—C—C—H
  |   |   |   |   |   |   |
  H       H   H   H   H   H
          |
          N—H
          |
      H—C—H
          |
      H—C—H
          |
          H
```

d.

```
      H       H
       \     /
  H   N   H   H   H
  |   |   |   |   |
H—C—C—C—C—C—H
  |   |   |   |   |
  H       H   H   H
      H—C—H
          |
          H
```

e.

```
  H       H
   \     /
    N   H   H   H   H   H   H   H
    |   |   |   |   |   |   |   |
H—C—C—C—C—C—C—C—C—H
    |   |   |   |   |   |   |   |
    H   H   H   Cl  I   H   H   H   H
```

f.

```
  H   H       H   H
  |   |       |   |
H—C—C—N—C—C—H
  |   |   |   |   |
  H   H   |   H   H
          H   H   H
          |   |   |
      H—C—C—C—C—H
          |   |   |   |
          H   H   H   H
```

16.11 a.

```
⬠—NH₃⁺ Br⁻
```

b.

```
              H
              |
CH₃CH₂—N⁺—CH₃ + OH⁻
              |
              H
```

c. CH₃—N⁺H₃ + OH⁻

16.13 a. CH₃—NH₂

b. CH₃—NH
```
      CH₃
      |
CH₃—NH
```

16.15 a. 1-Butanamine would be more soluble in water because it has a polar amine group that can form hydrogen bonds with water molecules.

b. 2-Pentanamine would be more soluble in water because it has a polar amine group that can form hydrogen bonds with water molecules.

16.17 Triethylamine molecules cannot form hydrogen bonds with one another, but 1-**hexanam**ine molecules are able to do so.

16.19 a. 2-Butanamine
b. 3-Hexanamine
c. Cyclopentanamine
d. 2-Methyl-2-propanamine

16.21 a. CH₃CH₂—NH—CH₂CH₃
b. CH₃CH₂CH₂CH₂NH₂

c. CH₃CH₂CHCH₂CH₂CH₂CH₂CH₂CH₂CH₃
```
         |
         NH₂
```

d.
```
         Br
         |
CH₃CHCHCH₂CH₃
         |
         NH₂
```

e.
(triphenylamine structure: three benzene rings attached to central N)

16.23 a. CH₃CHCH₂CH₂CH₃
```
         |
         NH₂
```

b.
```
              Br
              |
CH₃CH₂CHCH₂NH₂
```

c. CH₃CH₂—NH—CHCH₃
```
                   |
                   CH₃
```

d.
⬠—NH₂ (cyclopentane with NH₂)

16.25 CH₃CH₂CH₂CH₂NH₂   CH₃CH₂CHCH₃
```
                                        |
                                        NH₂
```

1-Butanamine   2-Butanamine
(Primary amine)   (Primary amine)

CH₃CHCH₂NH₂
```
   |
   CH₃
```
                              CH₃
                              |
                      CH₃—C—CH₃
                              |
                              NH₂

2-Methyl-1-propanamine   2-Methyl-2-propanamine
(Primary amine)   (Primary amine)

```
      CH₃
      |
CH₃CH₂—N—CH₃
```
                            CH₃CH₂—NH—CH₂CH₃

N,N-Dimethylethanamine   N-Ethylethanamine
(Tertiary amine)   (Secondary amine)

```
CH₃CHCH₃
   |
   NH—CH₃
```
                            CH₃CH₂CH₂—NH—CH₃

N-Methyl-2-propanamine   N-Methyl-1-propanamine
(Secondary amine)   (Secondary amine)

16.27 a. primary
b. secondary
c. primary
d. tertiary

16.29 a.

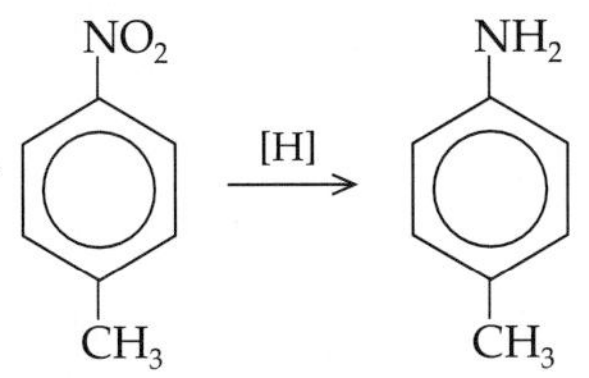

[structure: 4-methyl-nitrobenzene → [H] → 4-methyl-aniline (with NO_2 and CH_3 → NH_2 and CH_3)]

b.

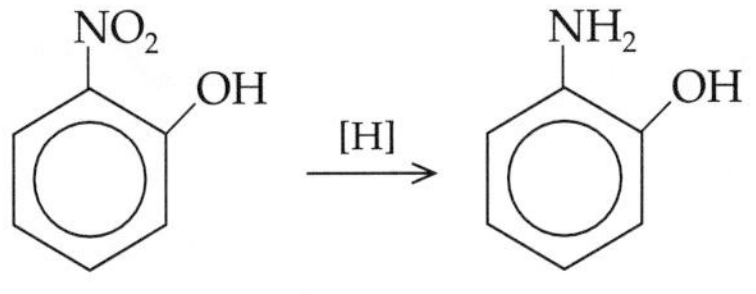

[structure: 2-nitrophenol with NO_2, OH → [H] → 2-aminophenol with NH_2, OH]

c.

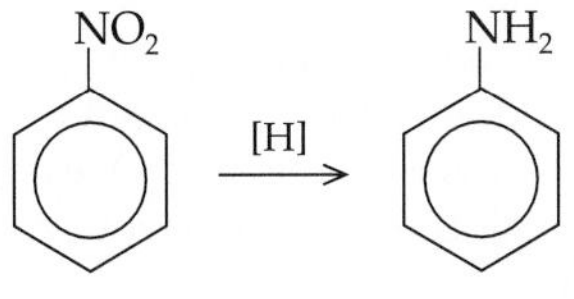

[structure: nitrobenzene NO_2 → [H] → aniline NH_2]

d.

[structure: nitro-methylbenzene NO_2, CH_2 → [H] → NH_2, CH_2]

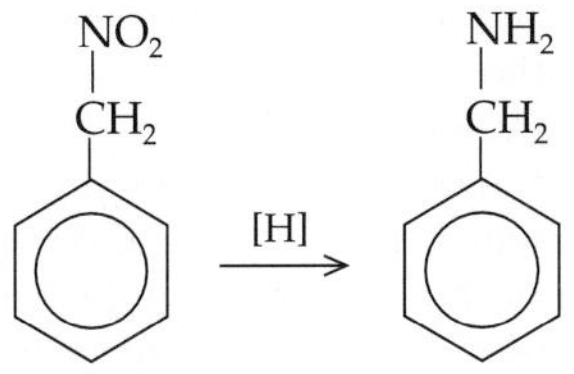

16.31 a. H_2O
b. HBr
c. $CH_3CH_2CH_2—N^+H_3$
d. $CH_3CH_2—N^+H_2Cl^-$
           |
           CH_2CH_3

16.33 Lower molecular weight amines are soluble in water because the N—H bond is polar and can form hydrogen bonds with water molecules.

16.35 Drugs containing amine groups are generally administered as ammonium salts because the salt is more soluble in water and, hence, in body fluids.

16.37 Putrescine (1,4-Diaminobutane):

$$\underset{NH_2}{CH_2}CH_2CH_2\underset{NH_2}{CH_2}$$

Cadaverine (1, 5-Diaminopentane):

$$\underset{NH_2}{CH_2}CH_2CH_2CH_2\underset{NH_2}{CH_2}$$

16.39 a.

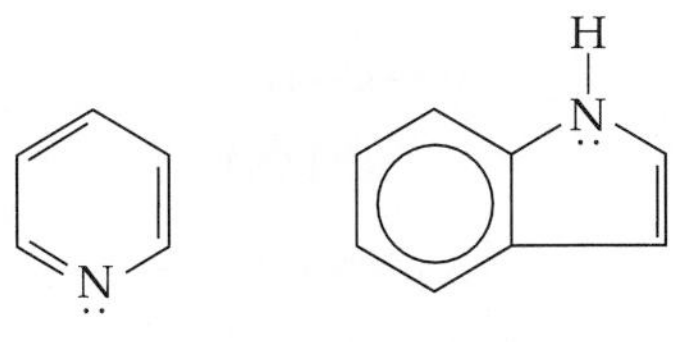

[structures: Pyridine and Indole]

Pyridine   Indole

b. The indole ring is found in lysergic acid diethylamide, which is a hallucinogenic drug. The pyridine ring is found in vitamin B_6, an essential water-soluble vitamin.

16.41 Morphine, codeine, quinine, and vitamin B_6

16.43 a. I.U.P.A.C. name: Propanamide
      Common name: Propionamide
b. I.U.P.A.C. name: Pentanamide
      Common name: Valeramide

c. I.U.P.A.C. name: *N,N*-Dimethylethanamide
     Common name: *N,N*-Dimethylacetamide

16.45 a. $CH_3—\overset{O}{\overset{\|}{C}}—NH_2$

b. $CH_3CH_2—\overset{O}{\overset{\|}{C}}—NH—CH_3$

c.

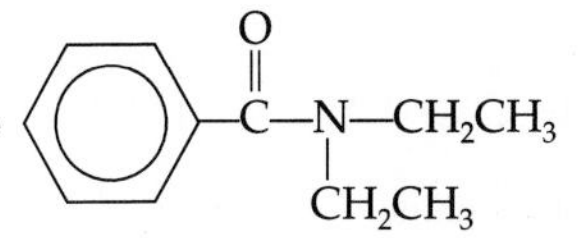

[structure: benzene ring—C(=O)—N(CH$_2$CH$_3$)—CH$_2$CH$_3$]

d. $CH_3CH_2\underset{Br}{CH}CHCH_2—\overset{O}{\overset{\|}{C}}—NH_2$ (with CH_3 on the CH)

e. $CH_3—\overset{O}{\overset{\|}{C}}—\underset{CH_3}{N}—CH_3$

16.47 *N, N*-Diethyl-*m*-toluamide:

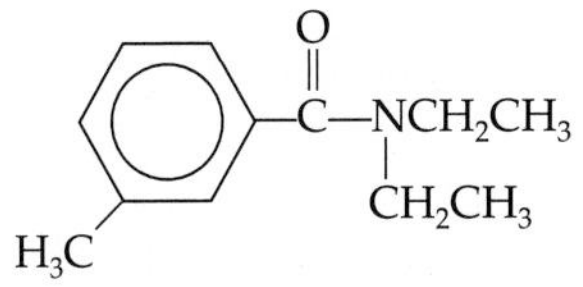

[structure: m-toluamide with ring, H_3C, —C(=O)—NCH$_2$CH$_3$ / CH$_2$CH$_3$]

Hydrolysis of this compound would release the carboxylic acid *m*-toluic acid and the amine *N*-ethylethanamine (diethylamine).

16.49 Amides are not proton acceptors (bases) because the highly electronegative carbonyl oxygen has a strong attraction for the nitrogen lone pair of electrons. As a result they cannot "hold" a proton.

16.51

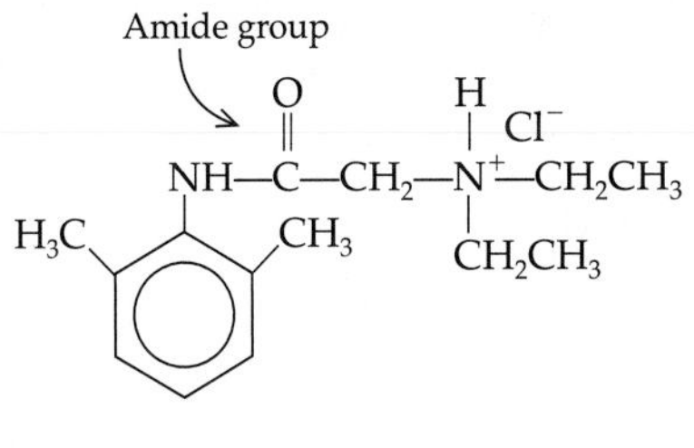

[structure labeled "Amide group" with ring, H_3C, CH_3; NH—C(=O)—CH$_2$—N$^+$(H)(CH$_2$CH$_3$)—CH$_2$CH$_3$, Cl^-]

Lidocaine hydrochloride

16.53

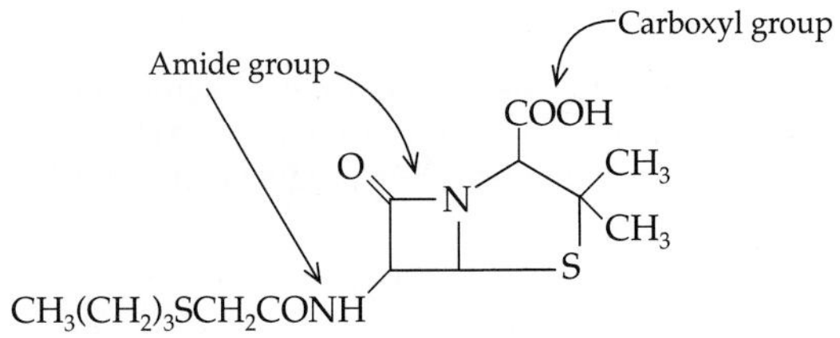

[structure of Penicillin BT with labels "Amide group", "Carboxyl group", COOH, CH_3, CH_3, S, N, O, $CH_3(CH_2)_3SCH_2CONH$]

Penicillin BT

16.55 a.

$CH_3—\overset{O}{\overset{\|}{C}}—NHCH_3 + H_3O^+ \longrightarrow$

N-Methylethanamide

$$CH_3COOH \;+\; CH_3NH_3^+$$

Ethanoic acid   Methanamine

b.

$$CH_3CH_2CH_2-\overset{\overset{\displaystyle O}{\|}}{C}-NH-CH_3 + H_3O^+ \longrightarrow$$

N-Methylbutanamide

$$CH_3CH_2CH_2-COOH + CH_3NH_3^+$$

Butanoic acid   Methanamine

c.

$$CH_3\overset{\overset{\displaystyle CH_3}{|}}{C}HCH_2-\overset{\overset{\displaystyle O}{\|}}{C}-NH-CH_2-CH_3 + H_3O^+ \longrightarrow$$

N-Ethyl-3-methylbutanamide   Hydronium ion
(Strong acid)

$$CH_3\overset{\overset{\displaystyle |}{C}}{H}CH_2-COOH + CH_3CH_2NH_3^+$$
$$\underset{CH_3}{}$$

3-Methylbutanoic acid   Ethanamine

16.57 a. $CH_3CH_2-\overset{\overset{\displaystyle O}{\|}}{C}-O-\overset{\overset{\displaystyle O}{\|}}{C}-CH_2CH_3$

b. $CH_3CH_2-\overset{\overset{\displaystyle O}{\|}}{C}-NH_2 + NH_4^+Cl^-$

c. $CH_3CH_2CH_2-\overset{\overset{\displaystyle O}{\|}}{C}-Cl + 2CH_3CH_2NH_2$

16.59

$$\overset{H}{\underset{H}{\nearrow}}N-\overset{\overset{\displaystyle H}{|}}{\underset{\underset{R}{|}}{C}}-\overset{\overset{\displaystyle O}{\|}}{C}-OH$$

16.61

Amide bond

$$\overset{H}{\underset{H}{\nearrow}}N-\overset{\overset{\displaystyle H}{|}}{\underset{\underset{H}{|}}{C}}-\overset{\overset{\displaystyle O}{\|}}{C}-N-\overset{\overset{\displaystyle H}{|}}{\underset{\underset{CH_3}{|}}{C}}-\overset{\overset{\displaystyle O}{\|}}{C}-OH$$

Glycyl alanine

16.63

$$\overset{H}{\underset{H}{\nearrow}}N-\overset{\overset{\displaystyle H}{\ast}}{\underset{\underset{CH_3}{|}}{C}}-\overset{\overset{\displaystyle O}{\|}}{C}-OH$$

16.65 In an acyl group transfer reaction, the acyl group of an acid chloride is transferred from the Cl of the acid chloride to the N of an amine or ammonia. The product is an amide.

Chapter 17

17.1 It is currently recommended that 58% of the calories in the diet should be carbohydrates. Of that amount, no more than 10% should be simple sugars.

17.3 An aldose is a sugar with an aldehyde functional group. A ketose is a sugar with a ketone functional group.

17.5 a. ketose   **d.** aldose
b. aldose   **e.** ketose
c. ketose   **f.** aldose

17.7

β-D-Galactose   α-D-Galactose

17.9 α-Amylase and β-amylase are digestive enzymes that break down the starch amylose. α-Amylase cleaves glycosidic bonds of the amylose chain at random, producing shorter polysaccharide chains. β-Amylase sequentially cleaves maltose (a disaccharide of glucose) from the reducing end of the polysaccharide chain.

17.11 A monosaccharide is the simplest sugar and consists of a single saccharide unit. A disaccharide is made up of two monosaccharides joined covalently by a glycosidic bond.

17.13 Mashed potato flakes, rice, and corn starch contain amylose and amylopectin, both of which are polysaccharides. A candy bar contains sucrose, a disaccharide. Orange juice contains fructose, a monosaccharide. It may also contain sucrose if the label indicates that sugar has been added.

17.15 Four

17.17

D-Galactose   D-Fructose
(An aldohexose)   (A ketohexose)

17.19 a. β-D-Glucose is a hemiacetal.
b. β-D-Fructose is a hemiketal.
c. α-D-Galactose is a hemiacetal.

17.21

D-Glyceraldehyde   L-Glyceraldehyde

17.23 Dextrose is a common name used for D-glucose.

17.25 D- and L-Glyceraldehyde are a pair of enantiomers, that is, they are nonsuperimposable mirror images of one another.

17.27 When the carbonyl group at C-1 of D-glucose reacts with the C-5 hydroxyl group, a new chiral carbon is created (C-1). In the α-isomer of the cyclic sugar the C-1 hydroxyl group is below the ring and in the β-isomer the C-1 hydroxyl group is above the ring.

17.29 β-Maltose and α-lactose would give positive Benedict's tests. Glycogen would give only a weak reaction because there are fewer reducing ends for a given mass of the carbohydrate.

17.31 Enantiomers are stereoisomers that are nonsuperimposable mirror images of one another. For instance:

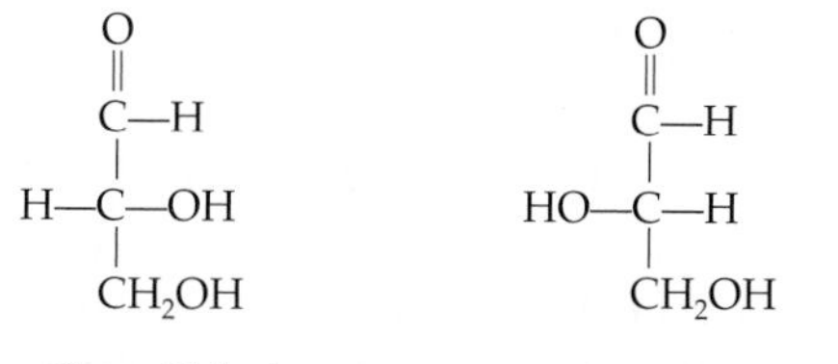

D-Glyceraldehyde L-Glyceraldehyde

17.33 An aldehyde sugar forms an intramolecular hemiacetal when the carbonyl group of the monosaccharide reacts with a hydroxyl group on one of the other carbon atoms.

17.35

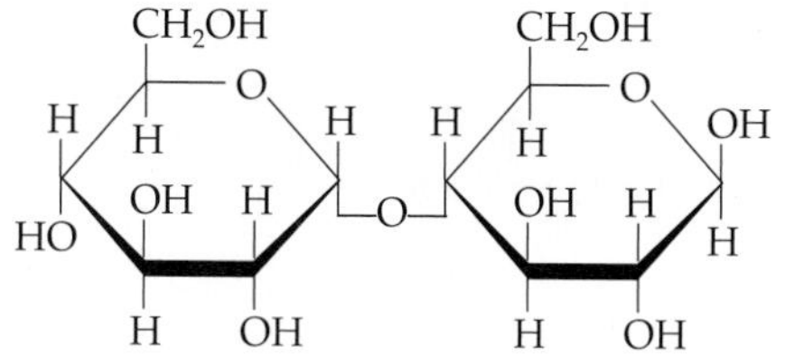

β-Maltose

17.37 Milk

17.39 eliminating milk and milk products from the diet

17.41 Lactose intolerance is the inability to produce the enzyme lactase that hydrolyzes the milk sugar lactose into its component monosaccharides, glucose and galactose.

17.43 The glucose units of amylose are joined by α (1 → 4) glycosidic bonds and those of cellulose are bonded together by β (1 → 4) glycosidic bonds.

17.45 Glycogen serves as a storage molecule for glucose.

17.47 the salivary glands and the pancreas

Chapter 18

18.1 **a.** $CH_3(CH_2)_7CH{=}CH(CH_2)_7COOH$
b. $CH_3(CH_2)_{10}COOH$
c. $CH_3(CH_2)_4CH{=}CH{-}CH_2{-}CH{=}CH(CH_2)_7COOH$
d. $CH_3(CH_2)_{16}COOH$

18.3 **a.** Esterification of lauric acid and ethanol

$$CH_3(CH_2)_{10}{-}\underset{\displaystyle O}{\overset{\|}{C}}{-}OH + CH_3CH_2OH \longrightarrow CH_3(CH_2)_{10}{-}\underset{\displaystyle O}{\overset{\|}{C}}{-}OCH_2CH_3 + H_2O$$

b. Reaction of oleic acid with NaOH

$$CH_3(CH_2)_7CH{=}CH(CH_2)_7{-}\underset{\displaystyle O}{\overset{\|}{C}}{-}OH + NaOH \longrightarrow CH_3(CH_2)_7CH{=}CH(CH_2)_7{-}\underset{\displaystyle O}{\overset{\|}{C}}{-}O^-\,Na^+ + H_2O$$

c. Hydrogenation of arachidonic acid

$$CH_3(CH_2)_4CH{=}CHCH_2CH{=}CHCH_2CH{=}CHCH_2CH{=}CH(CH_2)_3{-}\underset{\displaystyle O}{\overset{\|}{C}}{-}OH + 4H_2 \xrightarrow{Ni} CH_3(CH_2)_{18}{-}\underset{\displaystyle O}{\overset{\|}{C}}{-}OH$$

18.5 **a.**

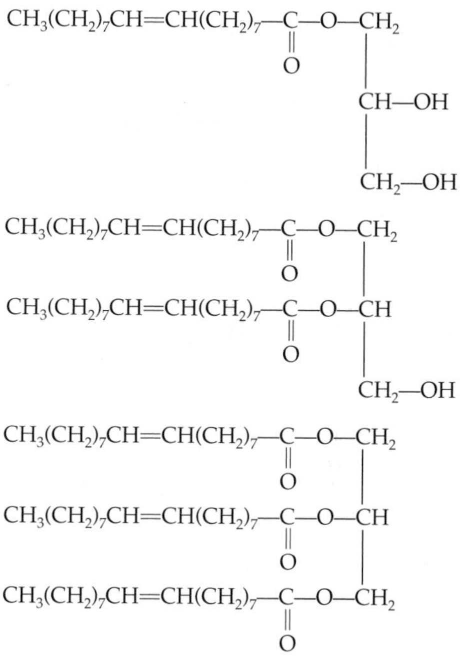

b.

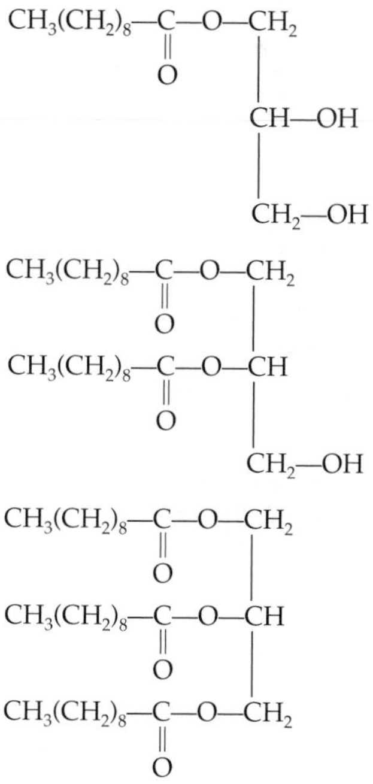

c. 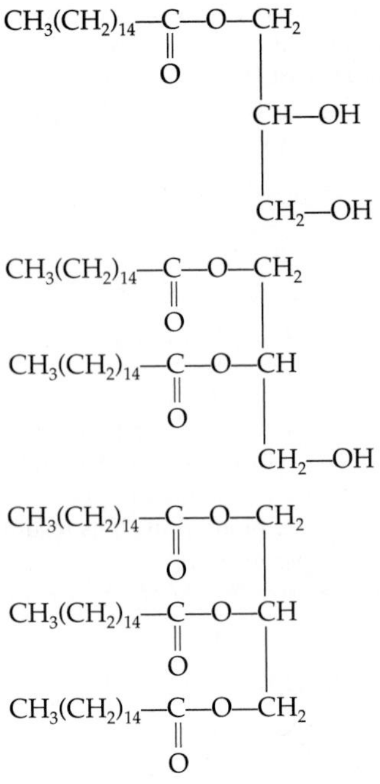 $CH_3(CH_2)_{14}$—C—O—CH_2

||

O

    CH—OH

    CH_2—OH

$CH_3(CH_2)_{14}$—C—O—CH_2

||

O

$CH_3(CH_2)_{14}$—C—O—CH

||

O

    CH_2—OH

$CH_3(CH_2)_{14}$—C—O—CH_2

||

O

$CH_3(CH_2)_{14}$—C—O—CH

||

O

$CH_3(CH_2)_{14}$—C—O—CH_2

||

O

d. 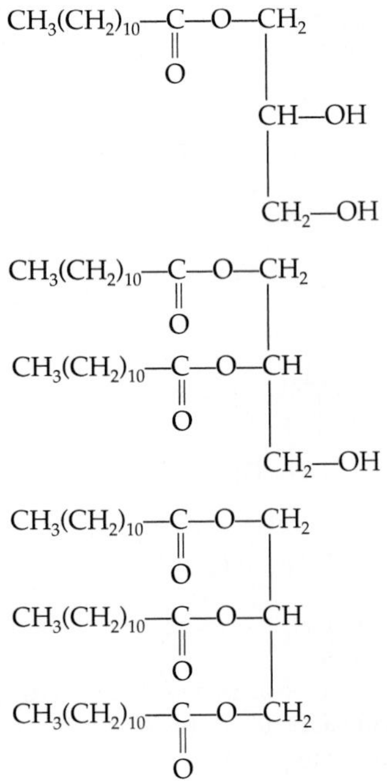 $CH_3(CH_2)_{10}$—C—O—CH_2

||

O

    CH—OH

    CH_2—OH

$CH_3(CH_2)_{10}$—C—O—CH_2

||

O

$CH_3(CH_2)_{10}$—C—O—CH

||

O

    CH_2—OH

$CH_3(CH_2)_{10}$—C—O—CH_2

||

O

$CH_3(CH_2)_{10}$—C—O—CH

||

O

$CH_3(CH_2)_{10}$—C—O—CH_2

||

O

18.7

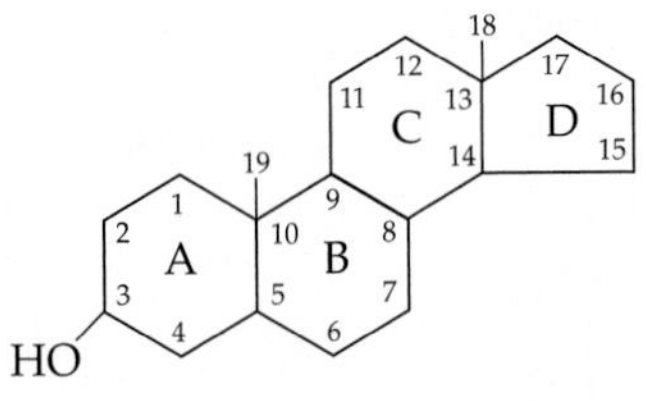

Steroid nucleus

18.9 receptor-mediated endocytosis

18.11 Membrane transport resembles enzyme catalysis because both processes exhibit a high degree of specificity.

18.13 fatty acids, glycerides, nonglyceride lipids, and complex lipids

18.15 A saturated fatty acid is one in which the hydrocarbon tail has only carbon-to-carbon single bonds. An unsaturated fatty acid has at least one carbon-to-carbon double bond.

18.17 The melting points increase.

18.19 **a.** Decanoic acid

$CH_3(CH_2)_8COOH$

b. Stearic acid

$CH_3(CH_2)_{16}COOH$

c. *trans*-5-Decenoic acid

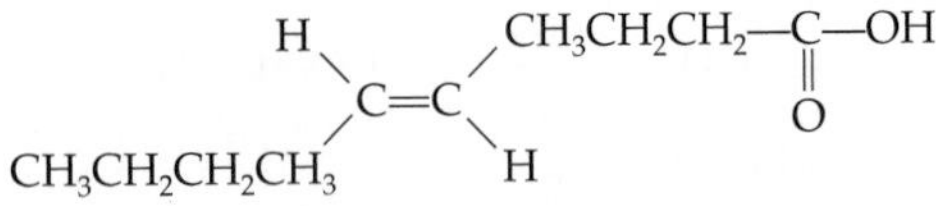

d. *cis*-5-Decenoic acid

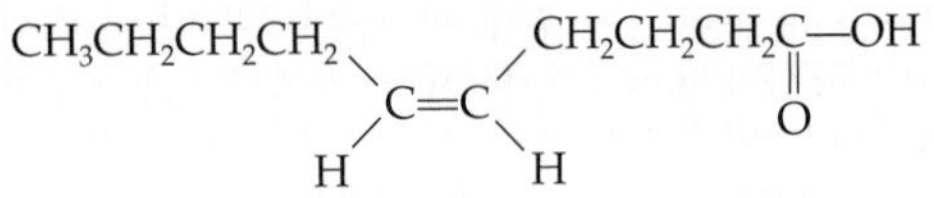

18.21 **a.** 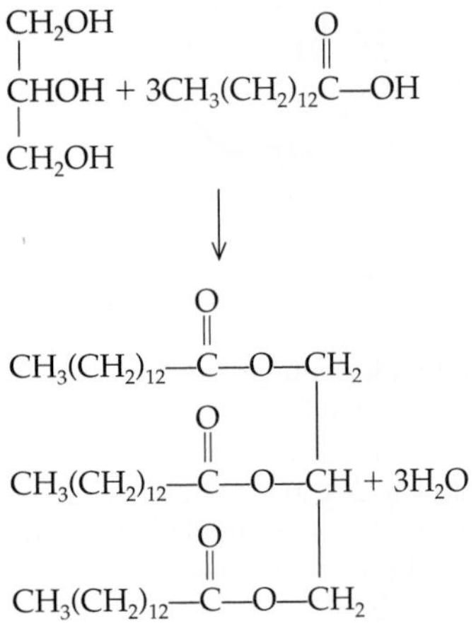 CH_2OH

CHOH + $3CH_3(CH_2)_{12}$C—OH

    ||

    O

CH_2OH

↓

$CH_3(CH_2)_{12}$—C—O—CH_2

||

O

$CH_3(CH_2)_{12}$—C—O—CH + $3H_2O$

||

O

$CH_3(CH_2)_{12}$—C—O—CH_2

||

O

b. $CH_3(CH_2)_{16}$—C—O—CH_2

||

O

$CH_3(CH_2)_{16}$—C—O—CH + $3H_2O$

||

O

$CH_3(CH_2)_{16}$—C—O—CH_2

||

O

↓

$3CH_3(CH_2)_{16}$—C—OH + CHOH

|| CH_2OH

O

            CH_2OH

c. $CH_3CH_2CH_2CH_2CH_2CH_2CH_2CH_2CH_2$—C—OH

||

O

↓ KOH

$CH_3CH_2CH_2CH_2CH_2CH_2CH_2CH_2CH_2$—C—O$^-$ K$^+$ + H_2O

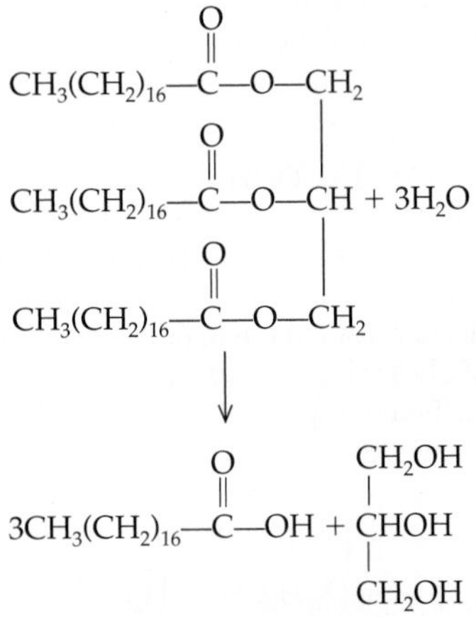

||

O

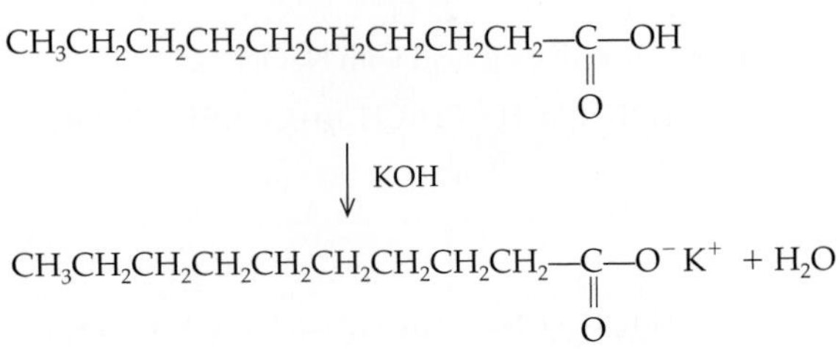

d. $CH_3(CH_2)_4CH=CHCH_2CH=CH(CH_2)_7-C-OH + 2H_2$
  $\quad\quad\quad\quad\quad\quad\quad\quad\quad\quad\quad\quad\quad\quad\quad\;\; \overset{\|}{O}$

$\downarrow$ Ni

$CH_3(CH_2)_{16}-C-OH$
  $\quad\quad\quad\quad\;\; \overset{\|}{O}$

18.23 The essential fatty acid linoleic acid is required for the synthesis of arachidonic acid, a precursor for the synthesis of the prostaglandins, a group of hormonelike molecules.

18.25 Aspirin effectively decreases the inflammatory response by inhibiting the synthesis of all prostaglandins. Aspirin works by inhibiting cyclooxygenase, the first enzyme in prostaglandin biosynthesis. This inhibition results from the transfer of an acetyl group from aspirin to the enzyme. Because cyclooxygenase is found in all cells, synthesis of all prostaglandins is inhibited.

18.27 smooth muscle contraction, enhancement of fever and swelling associated with the inflammatory response, bronchial dilation, inhibition of secretion of acid into the stomach

18.29

18.31

18.33 Triglycerides consist of three fatty acids esterified to the three hydroxyl groups of glycerol. In phospholipids there are only two fatty acids esterified to glycerol. A phosphoryl group is esterified (phosphoester linkage) to the third hydroxyl group.

18.35 Sphingolipids are phospholipids that are derived from sphingosine rather than glycerol. Sphingosine is a nitrogen-containing (amino) alcohol.

18.37 Cholesterol is readily soluble in the hydrophobic region of biological membranes. It is involved in regulating the fluidity of the membrane.

18.39 Progesterone is the most important hormone associated with pregnancy. Testosterone is needed for development of male secondary sexual characteristics. Estrone is required for proper development of female secondary sexual characteristics.

18.41 Cortisone is used to treat rheumatoid arthritis, asthma, gastrointestinal disorders, and many skin conditions.

18.43 Myricyl palmitate (beeswax) is made up of the fatty acid palmitic acid and the alcohol myricyl alcohol—$CH_3(CH_2)_{28}CH_2OH$.

18.45 Isoprenoids are a large, diverse collection of lipids that are synthesized from the isoprene unit:

$$\begin{array}{c} CH_3 \\ | \\ CH_2=C-CH=CH_2 \end{array}$$

18.47 steroids and bile salts, lipid-soluble vitamins, certain plant hormones, and chlorophyll

18.49 chylomicrons, high-density lipoproteins, low-density lipoproteins, and very low density lipoproteins

18.51 Atherosclerosis results when cholesterol and other substances coat the arteries causing a narrowing of the passageways. As the passageways become narrower, greater pressure is required to provide adequate blood flow. This results in higher blood pressure (hypertension).

18.53 If the LDL receptor is defective, it cannot function to remove cholesterol-bearing LDL particles from the blood. The excess cholesterol, along with other substances, will accumulate along the walls of the arteries, causing atherosclerosis.

18.55 If the fatty acyl tails of membrane phospholipids are converted from saturated to unsaturated, the fluidity of the membrane will increase.

18.57 The basic structure of a biological membrane is a bilayer of phospholipid molecules arranged so that the hydrophobic hydrocarbon tails are packed in the center and the hydrophilic head groups are exposed on the inner and outer surfaces.

18.59 A peripheral membrane protein is bound to only one surface of the membrane, either inside or outside the cell.

18.61 Cholesterol is freely soluble in the hydrophobic layer of a biological membrane. It moderates the fluidity of the membrane by disrupting the stacking of the fatty acid tails of membrane phospholipids.

18.63 L. Frye and M. Edidin carried out studies in which specific membrane proteins on human and mouse cells were labeled with red and green fluorescent dyes, respectively. The human and mouse cells were fused into single-celled hybrids and were observed using a microscope with an ultraviolet light source. The ultraviolet light caused the dyes to fluoresce. Initially the dyes were localized in regions of the membrane representing the original human or mouse cell. Within an hour, the proteins were evenly distributed throughout the membrane of the fused cell.

18.65 In simple diffusion the molecule moves directly across the membrane, whereas in facilitated diffusion a protein channel through the membrane is required.

18.67 Active transport requires an energy input to transport molecules or ions against the gradient (from an area of lower concentration to an area of higher concentration). Facilitated diffusion is a means of passive transport in which molecules or ions pass from regions of higher concentration to regions of lower concentration through a permease protein. No energy is expended by the cell in facilitated diffusion.

18.69 An antiport transport mechanism is one in which one molecule or ion is transported into the cell while a different molecule or ion is transported out of the cell.

18.71 Each permease or channel protein has a binding site that has a shape and charge distribution that is complementary to the molecule or ion that it can bind and transport across the cell membrane.

18.73 One ATP molecule is hydrolyzed to transport 3 Na^+ out of the cell and 2 K^+ into the cell.

18.75 Active transport is the movement of molecules or ions across a membrane against a concentration gradient (from a region of lower concentration to a region of higher concentration).

Chapter 19

19.1 a. Glycine (gly):

$$\begin{array}{c} COO^- \\ | \\ H_3^+N—C—H \\ | \\ H \end{array}$$

b. Proline (pro):

$$\begin{array}{c} COO^- \\ | \\ H_2^+N———CH \\ | \quad\quad | \\ H_2C\diagdown{}_{\underset{H_2}{C}}\diagup CH_2 \end{array}$$

c. Threonine (thr):

$$\begin{array}{c} COO^- \\ | \\ H_3^+N—C—H \\ | \\ H—C—OH \\ | \\ CH_3 \end{array}$$

d. Aspartate (asp):

$$\begin{array}{c} COO^- \\ | \\ H_3^+N—C—H \\ | \\ H—C—H \\ | \\ COO^- \end{array}$$

e. Lysine (lys):

$$\begin{array}{c} COO^- \\ | \\ H_3^+N—C—H \\ | \\ H—C—H \\ | \\ H—C—H \\ | \\ H—C—H \\ | \\ H—C—H \\ | \\ N^+H_3 \end{array}$$

19.3 a. Alanyl-phenylalanine:

$$\begin{array}{c} \overset{H}{|}\quad\overset{O}{\|}\quad\overset{H}{|}\quad\overset{H}{|} \\ H_3^+N—C—C—N—C—COO^- \\ | \quad\quad\quad\quad | \\ CH_3 \quad\quad\quad CH_2 \\ \quad\quad\quad\quad\quad\bigcirc \end{array}$$

b. Lysyl-alanine:

$$\begin{array}{c} \overset{H}{|}\quad\overset{O}{\|}\quad\overset{H}{|}\quad\overset{H}{|} \\ H_3^+N—C—C—N—C—COO^- \\ | \quad\quad\quad\quad | \\ CH_2 \quad\quad\quad CH_3 \\ | \\ CH_2 \\ | \\ CH_2 \\ | \\ CH_2 \\ | \\ N^+H_3 \end{array}$$

c. Phenylalanyl-tyrosyl-leucine:

$$\begin{array}{c} \overset{H}{|}\;\overset{O}{\|}\;\overset{H}{|}\;\overset{H}{|}\;\overset{O}{\|}\;\overset{H}{|}\;\overset{H}{|} \\ H_3^+N—C—C—N—C—C—N—C—COO^- \\ | \quad\quad\quad | \quad\quad\quad | \\ CH_2 \quad\quad CH_2 \quad\quad CH_2 \\ \bigcirc \quad\quad \bigcirc \quad\quad CHCH_3 \\ \quad\quad\quad | \quad\quad\quad | \\ \quad\quad\quad OH \quad\quad CH_3 \end{array}$$

19.5 The primary structure of a protein is the amino acid sequence of the protein chain. Regular, repeating folding of the peptide chain caused by hydrogen bonding between the amide nitrogens and carbonyl oxygens of the peptide bond is the secondary structure of a protein. The two most common types of secondary structure are the α-helix and the β-pleated sheet. Tertiary structure is the further folding of the regions of α-helix and β-pleated sheet into a compact, spherical structure. Formation and maintenance of the tertiary structure results from weak attractions between amino acid R groups. The binding of two or more peptides to produce a functional protein defines the quaternary structure.

19.7 Oxygen is efficiently transferred from hemoglobin to myoglobin in the muscle because myoglobin has a greater affinity for oxygen.

19.9 High temperature disrupts the hydrogen bonds and other weak interactions that maintain protein structure.

19.11 Vegetables vary in amino acid composition. No single vegetable can provide all of the amino acid requirements of the body. By eating a variety of different vegetables, all the amino acid requirements of the human body can be met.

19.13 Five of the biological functions carried out by proteins include serving as enzymes to speed up biochemical reactions, acting as antibodies to protect the body against disease, transport of materials throughout the body and into and out of cells, regulation of cellular function, and serving as structural support for animals.

19.15

$$\begin{array}{c} COO^- \\ | \\ H_3^+N—C—H \\ | \\ R \end{array}$$

19.17 Interactions between the R groups of the amino acids in a polypeptide chain are important for the formation and maintenance of the tertiary and quaternary structures of proteins.

19.19

Glycine
$$H_3{}^+N-C-COO^-$$
(with H above and H below)

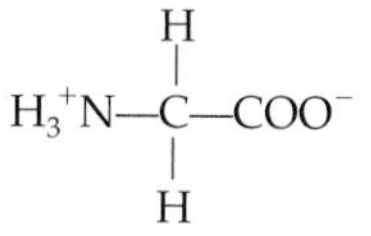

Alanine
$$H_3{}^+N-C-COO^-$$
(with H above and CH_3 below)

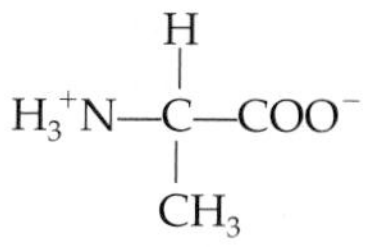

Valine
$$H_3{}^+N-C-COO^-$$
(with H above and CH below, CH bonded to H_3C and CH_3)

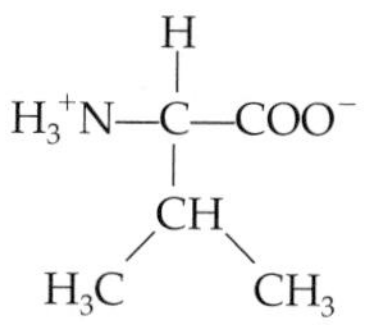

Leucine
$$H_3{}^+N-C-COO^-$$
(with H above; below CH_2, then CH bonded to H_3C and CH_3)

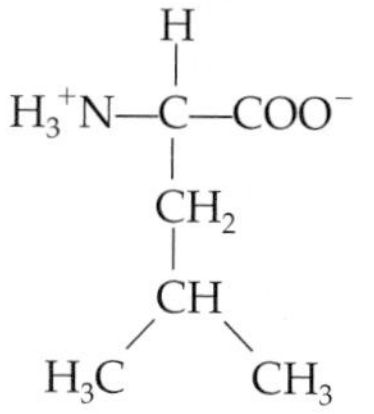

Isoleucine
$$H_3{}^+N-C-COO^-$$
(with H above; below $H-C-CH_3$, then CH_2, then CH_3)

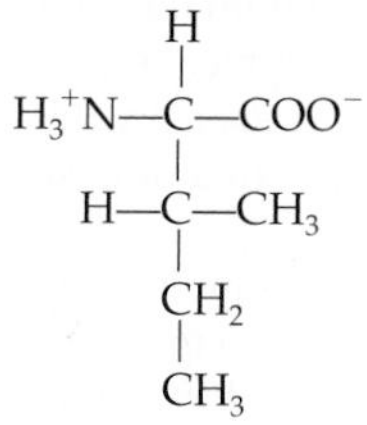

Phenylalanine
$$H_3{}^+N-C-COO^-$$
(with H above; below CH_2, then benzene ring)

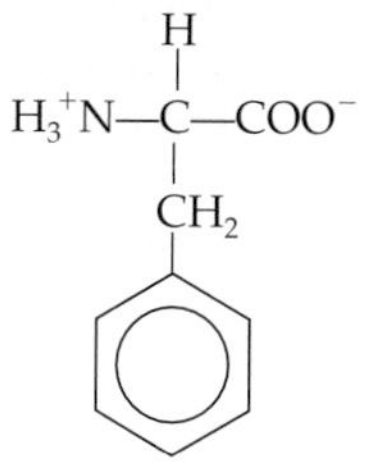

Proline
$$H_2N^+ \!-\!-\!C-COO^-$$
(with H above; below CH_2 and CH_2 joined by CH_2, forming ring to N)

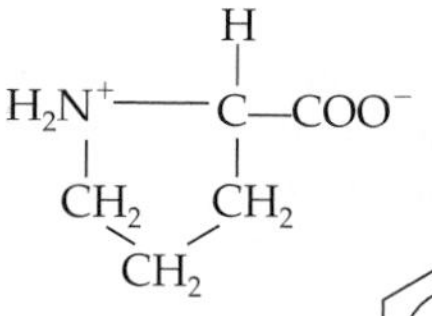

Tryptophan
$$H_3{}^+N-C-COO^-$$
(with H above; below CH_2, then C=CH, fused indole ring with NH)

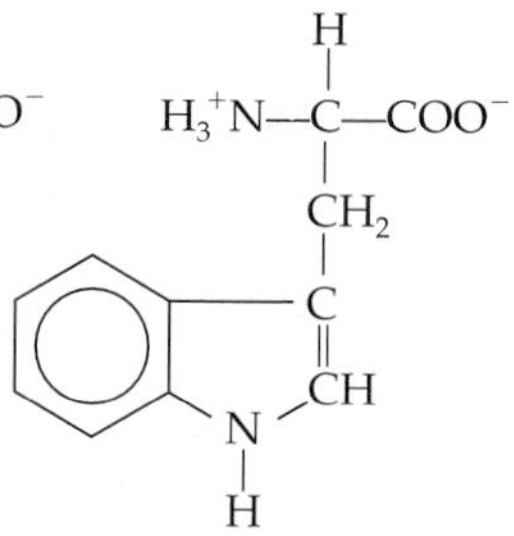

Methionine
$$H_3{}^+N-C-COO^-$$
(with H above; below CH_2, CH_2, S, CH_3)

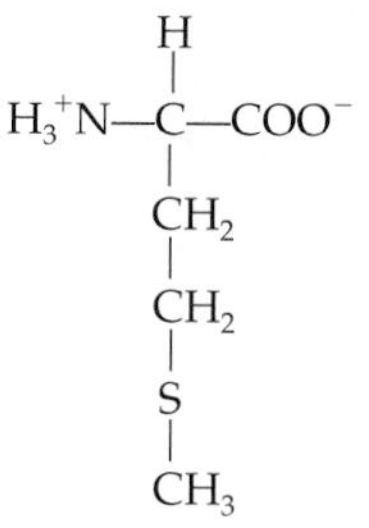

19.21 a. His-trp-cys:

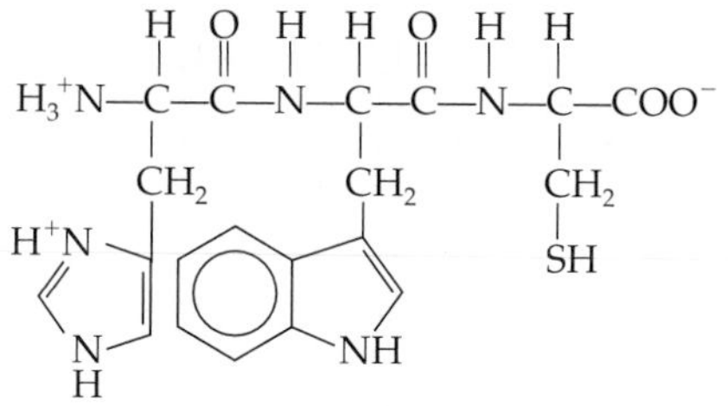

$$H_3{}^+N-C-C-N-C-C-N-C-COO^-$$
(backbone with H, O, H, H, O, H, H substituents; side chains CH_2 to imidazole (H^+N ring with N, H), CH_2 to indole (NH), CH_2 to SH)

b. Gly-leu-ser:

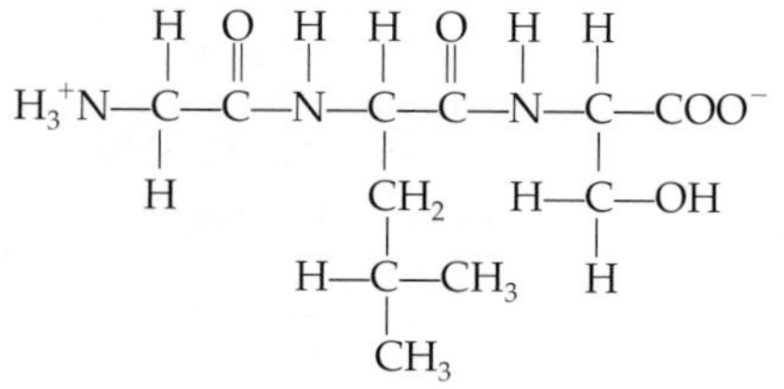

$$H_3{}^+N-C-C-N-C-C-N-C-COO^-$$
(backbone with H, O, H, H, O, H, H; side chains: H; CH_2, $H-C-CH_3$, CH_3; $H-C-OH$, H)

c. Arg-ile-val:

$$H_3{}^+N-C-C-N-C-C-N-C-COO^-$$
(backbone with H, O, H, H, O, H, H; side chains: CH_2, CH_2, CH_2, NH, $C=N^+H_2$, NH_2; $CHCH_3$, CH_2, CH_3; $CHCH_3$, CH_3)

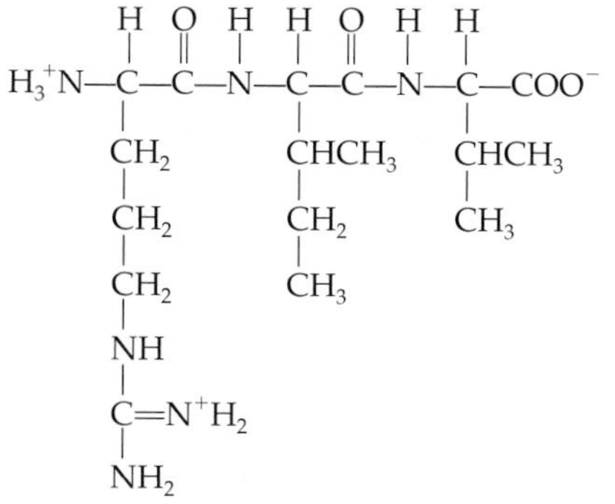

19.23 The peptide bond consists of an amide group. There is no free rotation around the peptide bond because the lone pair of electrons of the nitrogen atom interacts with the carbon and oxygen of the carbonyl group. This results in a resonance structure with a partially double bonded character.

19.25 The genetic information in the DNA dictates the order in which amino acids will be added to the protein chain. The order of the amino acids is the primary structure of the protein.

19.27 The primary structure of a protein is the linear arrangement of amino acids joined to one another by peptide bonds.

19.29 The secondary structure of a protein is the folding of the primary structure into an α-helix or β-pleated sheet.

19.31 a. α-Helix
  b. β-Pleated sheet

19.33 A fibrous protein is one that is composed of peptides arranged in long sheets or fibers.

19.35 A parallel β-pleated sheet is one in which the hydrogen bonded peptide chains have their amino-termini aligned head-to-head.

19.37 The tertiary structure of a protein is the globular, three-dimensional structure of a protein that results from folding the regions of secondary structure.

19.39

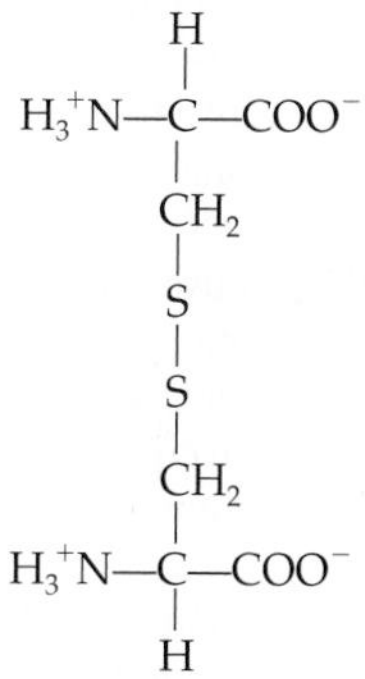

$$H_3{}^+N-C-COO^-$$
(top residue: H above; below CH_2, S, S, CH_2; bottom residue)
$$H_3{}^+N-C-COO^-$$
(with H below)

19.41 The tertiary structure is a level of folding of a protein chain that has already undergone secondary folding. The regions of α-helix and β-pleated sheet are folded into a globular structure.

19.43 Quaternary protein structure is the aggregation of two or more folded peptide chains to produce a functional protein.

19.45 A conjugated protein is a protein that requires an additional nonprotein group to be functional.

19.47 Hydrogen bonding maintains the secondary structure of a protein and contributes to the stability of the tertiary and quaternary levels of structure.

19.49 The peptide bond exhibits resonance, which results in a partially double bonded character. This causes the rigidity of the peptide bond.

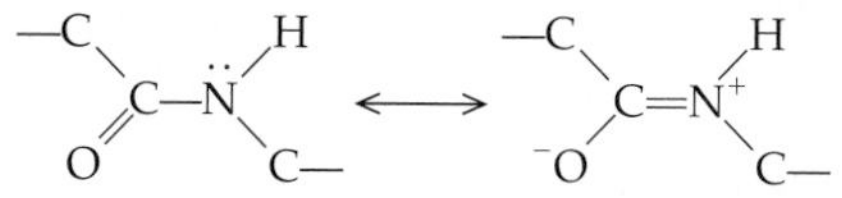

(resonance structures: amide form with $C=O$ and $C-N{:}-H$ ⟷ iminol form with $C-O^-$ and $C=N^+$)

19.51 The code for the primary structure of a protein is carried in the genetic information (DNA).

19.53 The function of hemoglobin is to carry oxygen from the lungs to oxygen-demanding tissues throughout the body. Hemoglobin is found in red blood cells.

19.55 Hemoglobin is a protein composed of four subunits—two α-globin and two β-globin subunits. Each subunit holds a heme group, which in turn carries an Fe^{2+} ion.

19.57 The function of the heme group in hemoglobin and myoglobin is to bind to molecular oxygen.

19.59 Because carbon monoxide binds tightly to the heme groups of hemoglobin, it is not easily removed or replaced by oxygen. As a result, the effects of oxygen deprivation (suffocation) occur.

19.61 When sickle cell hemoglobin (HbS) is deoxygenated, the amino acid valine fits into a hydrophobic pocket on the surface of another HbS molecule. Many such sickle cell hemoglobin molecules polymerize into long rods that cause the red blood cell to sickle. In normal hemoglobin, glutamic acid is found in the place of the valine. This negatively charged amino acid will not "fit" into the hydrophobic pocket.

19.63 When individuals have one copy of the sickle cell gene and one copy of the normal gene, they are said to carry the *sickle cell trait*. These individuals will not suffer serious side-effects, but may pass the trait to their offspring. Individuals with two copies of the sickle cell globin gene exhibit all the symptoms of the disease and are said to have *sickle cell anemia.*

19.65 *Denaturation* is the process by which the organized structure of a protein is disrupted, resulting in a completely disorganized, nonfunctional form of the protein.

19.67 Heat is an effective means of sterilization because it destroys the proteins of microbial life-forms, including fungi, bacteria, and viruses.

19.69 Even relatively small fluctuations in blood pH can be life-threatening. It is likely that these small changes would alter the normal charges on the proteins and modify their interactions. These changes can render a protein incapable of carrying out its functions.

19.71 Proteins become polycations at low pH because the additional protons will protonate the carboxylate groups. As these negative charges are neutralized, the charge on the proteins will be contributed only by the protonated amino groups ($-N^+H_3$).

19.73 The low pH of the yogurt denatures the proteins of microbial contaminants, inhibiting their growth.

19.75 In a vegetarian diet, vegetables are the only source of dietary protein. Because individual vegetable sources do not provide all the needed amino acids, vegetables must be mixed to provide all the essential and nonessential amino acids in the amounts required for biosynthesis.

19.77 Nonessential amino acids can be synthesized by the body and are, therefore, not required in the diet. Essential amino acids cannot be synthesized by the body and must be provided by the diet.

19.79 Synthesis of digestive enzymes must be carefully controlled because the active enzyme would digest, and thus destroy, the cell that produces it.

Chapter 20

20.1 **a.** sucrose
    b. pyruvate
    c. succinate

20.3 **a.** transferase
    b. ligase
    c. isomerase
    d. oxidoreductase
    e. isomerase

20.5 The induced fit model assumes that the enzyme is flexible. Both the enzyme and the substrate are able to change shape to form the enzyme–substrate complex. The lock-and-key model assumes that the enzyme is inflexible (the lock) and the substrate (the key) fits into a specific rigid site (the active site) on the enzyme to form the enzyme–substrate complex.

20.7 An enzyme might put pressure on a bond, thereby catalyzing bond breakage. An enzyme could bring two reactants into close proximity and in the proper orientation for the reaction to occur. Finally, an enzyme could alter the pH of the microenvironment of the active site, thereby serving as a transient donor or acceptor of H^+.

20.9 Water-soluble vitamins are required by the body for the synthesis of coenzymes that are required for the function of a variety of enzymes.

20.11 A decrease in pH will change the degree of ionization of the R groups within a peptide chain. This disturbs the weak interactions that maintain the structure of an enzyme, which may denature the enzyme. Less drastic alterations in the charge of R groups in the active site of the enzyme can inhibit enzyme–substrate binding or destroy the catalytic ability of the active site.

20.13 Irreversible inhibitors bind very tightly, sometimes even covalently, to an R group in enzyme active sites. They generally inhibit many different enzymes. The loss of enzyme activity impairs normal cellular metabolism, resulting in death of the cell or the individual.

20.15 A structural analog is a molecule that has a structure and charge distribution very similar to that of the natural substrate of an enzyme. Generally they are able to bind to the enzyme active site. This inhibits enzyme activity because the normal substrate must compete with the structural analog to form an enzyme–substrate complex.

20.17 **a.**

ala-phe-ala

b.

tyr-ala-tyr

c.

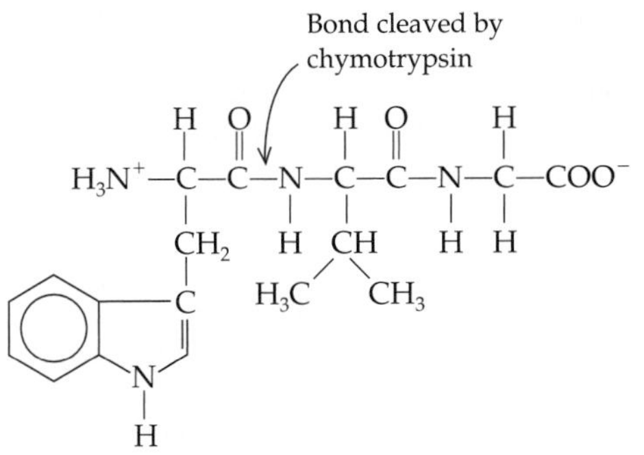

Bond cleaved by
chymotrypsin

$$H_3N^+-C-C-N-C-C-N-C-COO^-$$

trp-val-gly

d.

Bond cleaved by
chymotrypsin

$$H_3N^+-C-C-N-C-C-N-C-COO^-$$

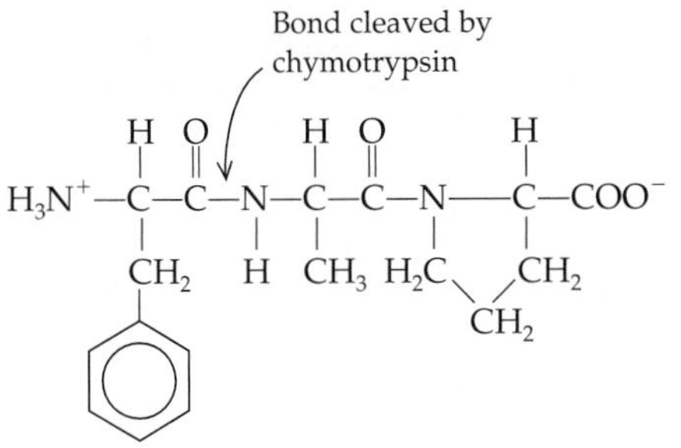

phe-ala-pro

20.19 1. urease
2. peroxidase
3. lipase
4. aspartase
5. glucose-6-phosphatase
6. sucrase

20.21 **a.** Citrate decarboxylase catalyzes the cleavage of a carboxyl group from citrate.
b. Adenosine diphosphate phosphorylase catalyzes the addition of a phosphate group to ADP.
c. Oxalate reductase catalyzes the reduction of oxalate.
d. Nitrite oxidase catalyzes the oxidation of nitrite.
e. *cis-trans* Isomerase catalyzes interconversion of *cis* and *trans* isomers.

20.23 The activation energy of a reaction is the energy required for the reaction to occur.

20.25 The equilibrium constant for a chemical reaction is a reflection of the difference in energy of the reactants and products. Consider the following reaction:

$$aA + bB \rightarrow cC + dD$$

The equilibrium constant for this reaction is:

$$K_{eq} = [D]^d[C]^c/[A]^a[B]^b = [products]/[reactants]$$

Because the difference in energy between reactants and products is the same regardless of what path the reaction takes, an enzyme does not alter the equilibrium constant of a reaction.

20.27 The rate of an uncatalyzed chemical reaction typically doubles every time the substrate concentration is doubled.

20.29

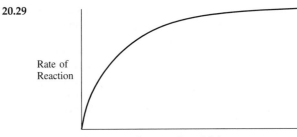

Rate of Reaction

Concentration of Substrate

20.31 Enzyme active sites are pockets in the surface of an enzyme that include R groups involved in binding and R groups involved in catalysis. The shape of the active site is complementary to the shape of the substrate. Thus, the conformation of the active site determines the specificity of the enzyme. Enzyme–substrate binding involves weak, noncovalent interactions.

20.33 The lock-and-key model of enzyme–substrate binding was proposed by Emil Fischer in 1894. He thought that the active site was a rigid region of the enzyme into which the substrate fit perfectly. Thus, the model purports that the substrate simply snaps into place within the active site, like two pieces of a jigsaw puzzle fitting together.

20.35 *Absolute specificity*—an enzyme catalyzes the reaction of only one substrate.
Group specificity—an enzyme catalyzes processes involving similar molecules having the same functional group.
Linkage specificity—an enzyme catalyzes the formation or breakage of only one type of bond.
Stereochemical specificity—an enzyme distinguishes one enantiomer from another.

20.37 The first step of an enzyme-catalyzed reaction is the formation of the enzyme–substrate complex. In the second step, the transition state is formed. This is the state in which the substrate assumes a form intermediate between the original substrate and the product. In step 3 the substrate is converted to product and the enzyme–product complex is formed. Step 4 involves the release of the product and regeneration of the enzyme in its original form.

20.39 In a reaction involving bond breaking, the enzyme might put pressure on a bond, producing a transition state in which the bond is stressed. An enzyme could bring two reactants into close proximity and in the proper orientation for the reaction to occur, producing a transition state in which the proximity of the reactants facilitates bond formation. Finally, an enzyme could alter the pH of the microenvironment of the active site, thereby serving as a transient donor or acceptor of H^+.

20.41 A cofactor helps maintain the shape of the active site of an enzyme.

20.43 NAD^+ serves as a donor or acceptor of hydride anions in biochemical reactions. NAD^+ serves as a coenzyme for oxidoreductases.

20.45 Changes in pH or temperature affect the activity of enzymes, as can changes in the concentration of substrate and the concentrations of certain ions.

20.47 A drastic change in pH above or below the pH optimum for an enzyme will denature the protein. Because a change in the conformation of the protein will drastically alter its active site, it will no longer be able to bind the substrate and catalyze the reaction.

20.49 High temperature denatures bacterial enzymes and structural proteins. Because the life of the cell is dependent on the function of these proteins, the cell dies.

20.51 A lysosome is a membrane-bound vesicle in the cytoplasm of cells which contains approximately fifty hydrolytic enzymes.

20.53 Enzymes used for clinical assays in hospitals are typically stored at refrigerator temperatures to ensure that they are not denatured by heat. In this way they retain their activity for long periods.

20.55 **a.** Cells regulate the level of enzyme activity to conserve energy. It is a waste of cellular energy to produce an enzyme if its substrate is not present or if its product is in excess.

   b. Production of proteolytic digestive enzymes must be carefully controlled because the active enzyme could destroy the cell that produces it. Thus, they are produced in an inactive form in the cell and are only activated at the site where they carry out digestion.

20.57 In positive allosterism, binding of the effector molecule turns the enzyme on. In negative allosterism, binding of the effector molecule turns the enzyme off.

20.59 A zymogen is the inactive form of an enzyme that is converted to the active form of the enzyme at the site of its activity.

20.61 *Competitive enzyme inhibition* occurs when a structural analog of the normal substrate occupies the enzyme active site so that the reaction cannot occur. The structural analog and the normal substrate compete for the active site. Thus, the rate of the reaction will depend on the relative concentrations of the two molecules.

20.63 A structural analog has a shape and charge distribution that are very similar to those of the normal substrate for an enzyme.

20.65 Irreversible inhibitors bind tightly to and block the active site of an enzyme and eliminate catalysis at the site.

20.67 The compound would be a competitive inhibitor of the enzyme.

20.69 The structural similarities among chymotrypsin, trypsin, and elastase suggest that these enzymes evolved from a single ancestral gene that was duplicated. Each copy then evolved independently.

20.71

tyr-lys-ala-phe

20.73 Elastase will cleave the peptide bonds on the carbonyl side of alanine and glycine. Trypsin will cleave the peptide bonds on the carbonyl side of lysine and arginine. Chymotrypsin will cleave the peptide bonds on the carbonyl side of tryptophan and phenylalanine.

20.75 Creatine phosphokinase (CPK), lactate dehydrogenase (LDH), and aspartate aminotransferase (AST/SGOT)

Chapter 21

21.1 ATP is called the universal energy currency because it is the major molecule used by all organisms to store energy.

21.3 The first stage of catabolism is the digestion (hydrolysis) of dietary macromolecules in the stomach and intestine.

   In the second stage of catabolism, monosaccharides, amino acids, fatty acids, and glycerol are converted by metabolic reactions into molecules that can be completely oxidized.

   In the third stage of catabolism, the two-carbon acetyl group of acetyl CoA is completely oxidized by the reactions of the citric acid cycle. The energy of the electrons harvested in these oxidation reactions is used to make ATP.

21.5 Substrate level phosphorylation is one way the cell can make ATP. In this reaction, a high-energy phosphoryl group of a substrate in the reaction is transferred to ADP to produce ATP.

21.7 Glycolysis is a pathway involving ten reactions. In reactions 1–3, energy is invested in the beginning substrate, glucose. This is done by transferring high-energy phosphoryl groups from ATP to the intermediates in the pathway. The product is fructose-1,6-bisphosphate. In the energy-harvesting reactions of glycolysis, fructose-1,6-bisphosphate is split into two three-carbon molecules that begin a series of rearrangement, oxidation–reduction, and substrate level phosphorylation reactions that produce four ATP, two NADH, and two pyruvate molecules. Because of the investment of two ATP in the early steps of glycolysis, the net yield of ATP is two.

21.9 Both the alcohol and lactate fermentations are anaerobic reactions that use the pyruvate and re-oxidize the NADH produced in glycolysis.

21.11 Gluconeogenesis (synthesis of glucose from noncarbohydrate sources) appears to be the reverse of glycolysis (the first stage of carbohydrate degradation) because the intermediates in the two pathways are the same. However, reactions 1, 3, and 10 of glycolysis are not reversible reactions. Thus, the reverse reactions must be carried out by different enzymes.

21.13 The enzyme glycogen phosphorylase catalyzes the phosphorolysis of a glucose unit at one end of a glycogen molecule. The reaction involves the displacement of the glucose by a phosphate group. The products are glucose-1-phosphate and a glycogen molecule that is one glucose unit shorter.

21.15 Glucokinase traps glucose within the liver cell by phosphorylating it. Because the product, glucose-6-phosphate, is charged, it cannot be exported from the cell.

21.17 Glucagon indirectly stimulates glycogen phosphorylase, the first enzyme of glycogenolysis. This speeds up glycogen degradation. Glucagon also inhibits glycogen synthase, the first enzyme in glycogenesis. This inhibits glycogen synthesis.

21.19 ATP

21.21

Adenosine triphosphate

Adenosine diphosphate

Inorganic phosphate group

21.23 Glycolysis requires NAD^+ for reaction 6 in which glyceraldehyde-3-phosphate dehydrogenase catalyzes the oxidation of glyceraldehyde-3-phosphate. NAD^+ is reduced.

21.25 Two ATP per glucose

21.27 Although muscle cells only have enough ATP stored for a few seconds of activity, glycolysis speeds up dramatically when there is a demand for more energy. If the cells have a sufficient supply of oxygen, aerobic respiration (the citric acid cycle and oxidative phosphorylation) will contribute large amounts of ATP. If oxygen is limited, the lactate fermentation will speed up. This will use up the pyruvate and re-oxidize the NADH produced by glycolysis and allow continued synthesis of ATP for muscle contraction.

21.29 $C_6H_{12}O_6 + 2ADP + 2P_i + 2NAD^+ \rightarrow$

Glucose

$$2C_3H_3O_3 + 2ATP + 2NADH + 2H_2O$$

Pyruvate

21.31 **a.** Hexokinase catalyzes the phosphorylation of glucose.
  b. Pyruvate kinase catalyzes the transfer of a phosphoryl group from phosphoenolpyruvate to ADP.
  c. Phosphoglyceromutase catalyzes the isomerization reaction that converts 3-phosphoglycerate to 2-phosphoglycerate.
  d. Glyceraldehyde-3-phosphate dehydrogenase catalyzes the oxidation and phosphorylation of glyceraldehyde-3-phosphate and the reduction of NAD^+ to NADH.

21.33 To optimize efficiency and minimize waste, it is important that energy-harvesting pathways, such as glycolysis, respond to the energy demands of the cell. If energy in the form of ATP is abundant, there is no need for the pathway to continue at a rapid rate. When this is the case, allosteric enzymes that catalyze the reactions of the pathway are inhibited by binding to their negative effectors. Similarly, when there is a great demand for ATP, the pathway speeds up as a result of the action of allosteric enzymes binding to positive effectors.

21.35 ATP and citrate are allosteric inhibitors of phosphofructo-kinase, whereas AMP and ADP are allosteric activators.

21.37 Citrate, which is the first intermediate in the citric acid cycle, is an allosteric inhibitor of phosphofructokinase. The citric acid cycle is a pathway that results in the complete oxidation of the pyruvate produced by glycolysis. A high concentration of citrate signals that sufficient substrate is entering the citric acid cycle. The inhibition of phosphofructokinase by citrate is an example of *feedback inhibition:* the product, citrate, allosterically inhibits the activity of an enzyme early in the pathway.

21.39

Acetaldehyde   Ethanol

21.41 The lactate fermentation

21.43 yogurt and some cheeses

21.45 Lactate dehydrogenase

21.47 This child must have the enzymes to carry out the alcohol fermentation. When the child exercised hard, there was not enough oxygen in the cells to maintain aerobic respiration. As a result, glycolysis and the alcohol fermentation were responsible for the majority of the ATP production by the child. The accumulation of alcohol (ethanol) in the child caused the symptoms of drunkenness.

21.49 The first stage of the pentose phosphate pathway is an oxidative stage in which glucose-6-phosphate is converted to ribulose-5-phosphate. Two NADPH molecules and one CO_2 molecule are also produced in these reactions. The second stage of the pentose phosphate pathway involves isomerization reactions that convert ribulose-5-phosphate into other five-carbon sugars, ribose-5-phosphate and xylulose-5-phosphate. The third stage of the pathway involves a complex series of rearrangement reactions that result in the production of two fructose-6-phosphate and one glyceraldehyde-3-phosphate molecules from three molecules of pentose phosphate.

21.51 The ribose-5-phosphate is used for the biosynthesis of nucleotides. The erythrose-4-phosphate is used for the biosynthesis of aromatic amino acids.

21.53 The liver

21.55 Lactate is first converted to pyruvate.

21.57 Because steps 1, 3, and 10 of glycolysis are irreversible, gluconeogenesis is not simply the reverse of glycolysis. The reverse reactions must be carried out by different enzymes.

21.59 Steps 1, 3, and 10 of glycolysis are irreversible. Step 1 is the transfer of a phosphoryl group from ATP to carbon-6 of glucose and is catalyzed by hexokinase. Step 3 is the transfer of a phosphoryl group from ATP to carbon-1 of fructose-6-phosphate and is catalyzed by phosphofructokinase. Step 10 is the substrate level phosphorylation in which a phosphoryl group is transferred from phosphoenolpyruvate to ADP and is catalyzed by pyruvate kinase.

21.61 The liver and pancreas

21.63 *Hypoglycemia* is the condition in which blood glucose levels are too low.

21.65 **a.** Insulin stimulates glycogen synthase, the first enzyme in glycogen synthesis. It also stimulates uptake of glucose from the bloodstream into cells and phosphorylation of glucose by the enzyme glucokinase.
  b. This traps glucose within liver cells and increases the storage of glucose in the form of glycogen.
  c. These processes decrease blood glucose levels.

21.67 Any defect in the enzymes required to degrade glycogen or export glucose from liver cells will result in a reduced ability of the liver to provide glucose at times when blood glucose levels are low. This will cause hypoglycemia.

Chapter 22

22.1 Mitochondria are the organelles responsible for aerobic respiration.

22.3

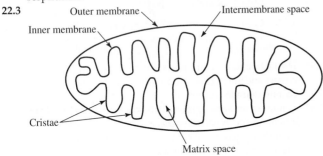

22.5 Pyruvate is converted to acetyl CoA by the pyruvate dehydrogenase complex. This huge enzyme complex requires four coenzymes, each of which is made from a different vitamin. The four coenzymes are thiamine pyrophosphate (made from thiamine), FAD (made from riboflavin), NAD^+ (made from niacin), and coenzyme A (made from the vitamin pantothenic acid). The coenzyme lipoamide is also involved in this reaction.

22.7 *Oxidative phosphorylation* is the process by which the energy of electrons harvested from oxidation of a fuel molecule is used to phosphorylate ADP to produce ATP.

22.9 $NAD^+ + H:^- \rightarrow NADH$

22.11 During transamination reactions, the α-amino group is transferred to the coenzyme pyridoxal phosphate. In the last part of the reaction, the α-amino group is transferred from pyridoxal phosphate to an α-keto acid.

22.13 The purpose of the urea cycle is to convert toxic ammonium ions to urea, which is excreted in the urine of land animals.

22.15 An amphibolic pathway is a metabolic pathway that functions both in anabolism and catabolism. The citric acid cycle is amphibolic because it has a catabolic function—it completely oxidizes the acetyl group carried by acetyl CoA to provide electrons for ATP synthesis. Because citric acid cycle intermediates are precursors for the biosynthesis of many other molecules, it also serves a function in anabolism.

22.17 The intermembrane compartment is the location of the high-energy proton (H^+) reservoir produced by the electron transport system. The energy of this H^+ reservoir is used to make ATP.

22.19 The outer mitochondrial membrane is freely permeable to substances of molecular weight less than 10,000. The inner mitochondrial membrane is highly impermeable. Embedded within the inner mitochondrial membrane are the electron carriers of the electron transport system, and ATP synthase, the multisubunit enzyme that makes ATP.

22.21 Under aerobic conditions pyruvate is converted to acetyl CoA.

22.23 The coenzymes NAD^+, FAD, thiamine pyrophosphate, and coenzyme A are required by the pyruvate dehydrogenase complex for the conversion of pyruvate to acetyl CoA. These coenzymes are synthesized from the vitamins niacin, riboflavin, thiamine, and pantothenic acid, respectively. If the vitamins are not available, the coenzymes will not be available and pyruvate cannot be converted to acetyl CoA. Because the complete oxidation of the acetyl group of acetyl

CoA produces the vast majority of the ATP for the body, ATP production would be severely inhibited by a deficiency of any of these vitamins.

22.25 a. False
   b. False
   c. True
   d. True

22.27 a. The acetyl group of acetyl CoA is transferred to oxaloacetate.
   b. The product is citrate.

22.29 Three

22.31 Two ATP per glucose

22.33 The function of acetyl CoA in the citric acid cycle is to bring the two-carbon remnant (acetyl group) of pyruvate from glycolysis and transfer it to oxaloacetate. In this way the acetyl group enters the citric acid cycle for the final stages of oxidation.

22.35 The high-energy phosphoryl group of the GTP is transferred to ADP to produce ATP. This reaction is catalyzed by the enzyme dinucleotide diphosphokinase.

22.37 Energy-harvesting pathways, such as the citric acid cycle, must be responsive to the energy needs of the cell. If the energy requirements are high, as during exercise, the reactions must speed up. If energy demands are low and ATP is in excess, the reactions of the pathway slow down.

22.39 ADP

22.41 Three ATP

22.43 The oxidation of a variety of fuel molecules, including carbohydrates, the carbon skeletons of amino acids, and fatty acids provides the electrons. The energy of these electrons is used to produce an H^+ reservoir. The energy of this proton reservoir is used for ATP synthesis.

22.45 The electron transport system passes electrons harvested during oxidation of fuel molecules to molecular oxygen. At three sites protons are pumped from the mitochondrial matrix into the intermembrane compartment. Thus, the electron transport system builds the high-energy H^+ reservoir that provides energy for ATP synthesis.

22.47 a. Two ATP per glucose (net yield) are produced in glycolysis, whereas the complete oxidation of glucose in aerobic respiration (glycolysis, the citric acid cycle, and oxidative phosphorylation) results in the production of thirty-six ATP per glucose.
   b. Thus, aerobic respiration harvests nearly 40% of the potential energy of glucose, and anaerobic glycolysis harvests only about 2% of the potential energy of glucose.

22.49 Aminotransferases (transaminases) transfer amino groups from amino acids to ketoacids.

22.51 The glutamate family of transaminases is very important because the ketoacid corresponding to glutamate is α-ketoglutarate, one of the citric acid cycle intermediates. This provides a link between the citric acid cycle and amino acid metabolism. These transaminases provide amino groups for amino acid synthesis and collect amino groups during catabolism of amino acids.

22.53 Fifteen ATP

22.55

$$
\begin{array}{ccc}
\underset{\text{α-Ketoglutarate}}{\begin{array}{c} \text{O} \\ \| \\ \text{C--COO}^- \\ | \\ \text{CH}_2 \\ | \\ \text{CH}_2 \\ | \\ \text{COO}^- \end{array}}
& \underset{\text{Ammonia}}{+\ \text{NADPH} + \text{N}^+\text{H}_4\ \longrightarrow}
& \underset{\text{Glutamate}}{\begin{array}{c} \text{N}^+\text{H}_3 \\ | \\ \text{H--C--COO}^- \\ | \\ \text{CH}_2 \\ | \\ \text{CH}_2 \\ | \\ \text{COO}^- \end{array}}
\end{array}
+\ \text{NADP}^+ + \text{H}_2\text{O}
$$

22.57 Hyperammonemia

22.59 a. The source of one amino group of urea is the ammonium ion and the source of the other is the α-amino group of the amino acid aspartate.

    b. The carbonyl group of urea is derived from CO_2.

22.61 α-ketoglutarate

22.63 Citric acid cycle intermediates are the starting materials for the biosynthesis of many biological molecules.

22.65 An essential amino acid is one that cannot be synthesized by the body and must be provided in the diet.

22.67

$$\underset{\text{Pyruvate}}{\overset{\displaystyle O}{\underset{\displaystyle CH_3}{\overset{\displaystyle \|}{C}}\!-\!COO^-}} + CO_2 + ATP \longrightarrow \underset{\text{Oxaloacetate}}{\overset{\displaystyle O}{\underset{\displaystyle \underset{\displaystyle COO^-}{CH_2}}{\overset{\displaystyle \|}{C}}\!-\!COO^-}} + ADP + P_i$$

Chapter 23

23.1 Because dietary lipids are hydrophobic, they arrive in the small intestine as large fat globules. The bile salts emulsify these fat globules into tiny fat droplets. This greatly increases the surface area of the lipids, allowing them to be more accessible to pancreatic lipases and thus more easily digested.

23.3 a. Four acetyl CoA, one benzoate, four NADH, and four $FADH_2$

    b. Three acetyl CoA, one phenyl acetate, three NADH, and three $FADH_2$

    c. Three acetyl CoA, one benzoate, three NADH, and three $FADH_2$

    d. Five acetyl CoA, one phenyl acetate, five NADH, and five $FADH_2$

23.5

$$CH_3CH_2CH_2{-}\overset{\displaystyle O}{\overset{\displaystyle \|}{C}}{-}S{-}CoA + FAD \longrightarrow$$

$$CH_3CH{=}CH{-}\overset{\displaystyle O}{\overset{\displaystyle \|}{C}}{-}S{-}CoA + FADH_2$$

$$\Big\downarrow\!\!\diagdown H_2O$$

$$\longleftarrow CH_3{-}\underset{\displaystyle H}{\overset{\displaystyle OH}{\overset{\displaystyle |}{C}}}{-}CH_2{-}\overset{\displaystyle O}{\overset{\displaystyle \|}{C}}{-}S{-}CoA + NAD^+$$

$$CH_3{-}\overset{\displaystyle O}{\overset{\displaystyle \|}{C}}{-}CH_2{-}\overset{\displaystyle O}{\overset{\displaystyle \|}{C}}{-}S{-}CoA + NADH$$

$$\Big\downarrow\!\!\diagdown\text{Coenzyme A}$$

$$2CH_3{-}\overset{\displaystyle O}{\overset{\displaystyle \|}{C}}{-}S{-}CoA$$

23.7 Starvation, a diet low in carbohydrates, and diabetes mellitus are conditions that lead to the production of ketone bodies.

23.9 (1) Fatty acid biosynthesis occurs in the cytoplasm whereas β-oxidation occurs in the mitochondria.

    (2) The acyl group carrier in fatty acid biosynthesis is acyl carrier protein while the acyl group carrier in β-oxidation is coenzyme A.

    (3) The seven enzymes of fatty acid biosynthesis are associated as a multienzyme complex called *fatty acid synthase.* The enzymes involved in β-oxidation are not physically associated with one another.

    (4) NADPH is the reducing agent used in fatty acid biosynthesis. NADH and $FADH_2$ are produced by β-oxidation.

23.11 The liver regulates blood glucose levels under the control of the hormones insulin and glucagon. When blood glucose levels are too high, insulin stimulates the uptake of glucose by liver cells and the storage of the glucose in glycogen polymers. When blood glucose levels are too low, the hormone glucagon stimulates the breakdown of glycogen and release of glucose into the bloodstream. Glucagon also stimulates the liver to produce glucose for export into the bloodstream by the process of gluconeogenesis.

23.13 Insulin stimulates uptake of glucose and amino acids by cells, glycogen and protein synthesis, and storage of lipids. It inhibits glycogenolysis, gluconeogenesis, breakdown of stored triglycerides, and ketogenesis.

23.15 Triglycerides

23.17 the large fat globule that takes up nearly the entire cytoplasm

23.19 Lipases catalyze the hydrolysis of the ester bonds of triglycerides.

23.21 Acetyl CoA is the precursor for fatty acids, several amino acids, cholesterol, and other steroids.

23.23 Chylomicrons are plasma lipoproteins (aggregates of protein and triglycerides) that carry dietary triglycerides from the intestine to all tissues via the bloodstream.

23.25 Bile salts serve as detergents. Fat globules stimulate their release from the gallbladder. The bile salts then emulsify the lipids, increasing their surface area and making them more accessible to digestive enzymes (pancreatic lipases).

23.27 When dietary lipids in the form of fat globules reach the duodenum, they are emulsified by bile salts. The triglycerides in the resulting tiny fat droplets are hydrolyzed into monoglycerides and fatty acids by the action of pancreatic lipases, assisted by colipase. The monoglycerides and fatty acids are absorbed by cells lining the intestine.

23.29 Six acetyl CoA, one phenyl acetate, six NADH, and six $FADH_2$.

23.31 112 ATP

23.33 Two ATP

23.35 The acetyl CoA produced by β-oxidation will enter the citric acid cycle.

23.37

$$\underset{\text{Acetoacetate}}{CH_3{-}\overset{\displaystyle O}{\overset{\displaystyle \|}{C}}{-}CH_2{-}\overset{\displaystyle O}{\overset{\displaystyle \|}{C}}{-}O^-} \qquad\qquad \underset{\text{β-Hydroxybutyrate}}{CH_3{-}\underset{\displaystyle OH}{\overset{\displaystyle |}{C}H}CH_2{-}\overset{\displaystyle O}{\overset{\displaystyle \|}{C}}{-}O^-}$$

23.39 In those suffering from uncontrolled diabetes, the glucose in the blood cannot get into the cells of the body. The excess glucose is excreted in the urine. Body cells degrade fatty acids because glucose is not available. β-Oxidation of fatty acids yields enormous quantities of acetyl CoA, so much acetyl CoA, in fact, that it cannot all enter the citric acid cycle because there is not enough oxaloacetate available. Excess acetyl CoA is used for ketogenesis.

23.41 Ketone bodies are the preferred energy source of the heart.

23.43 The phosphopantetheine group allows formation of a high-energy thioester bond with a fatty acid. It is derived from the vitamin pantothenic acid and β-mercaptoethylamine.

23.45 Fatty acid synthase is a huge multienzyme complex consisting of the seven enzymes involved in fatty acid synthesis. It is found in the cell cytoplasm. The enzymes involved in β-oxidation are not physically associated with one another. They are free in the mitochondrial matrix space.

23.47 The major metabolic function of the liver is to regulate blood glucose levels.

23.49 Ketone bodies are the major fuel for the heart. Glucose is the major energy source of the brain, and the liver obtains most of its energy from the oxidation of amino acid carbon skeletons.

23.51 Fatty acids are absorbed from the bloodstream by adipocytes. Using glycerol-3-phosphate, produced as a by-product of glycolysis, triglycerides are synthesized. Triglycerides are constantly being hydrolyzed and resynthesized in adipocytes. The rates of hydrolysis and synthesis are determined by lipases that are under hormonal control.

23.53 Insulin is produced in the β-cells of the islets of Langerhans in the pancreas.

23.55 Insulin stimulates the uptake of glucose from the blood into cells. It enhances glucose storage by stimulating glycogenesis and inhibiting glycogen degradation and gluconeogenesis.

23.57 Insulin stimulates synthesis and storage of triglycerides.

23.59 Untreated diabetes mellitus is starvation in the midst of plenty because blood glucose levels are very high. However, in the absence of insulin, blood glucose can't be taken up into cells. The excess glucose is excreted into the urine while the cells of the body are starved for energy.

Chapter 24

24.1 **a.** Adenosine diphosphate:

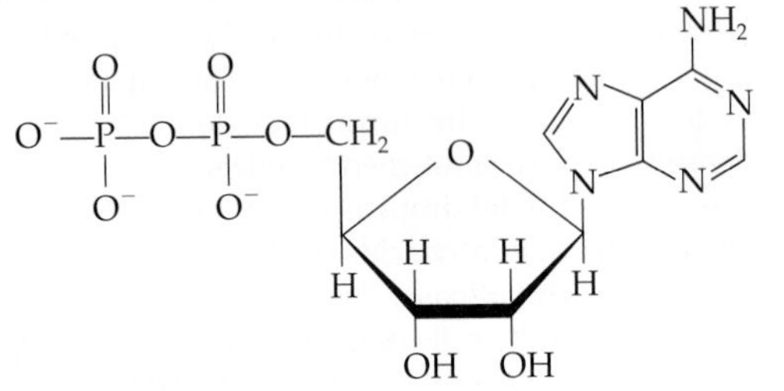

b. Deoxyguanosine triphosphate:

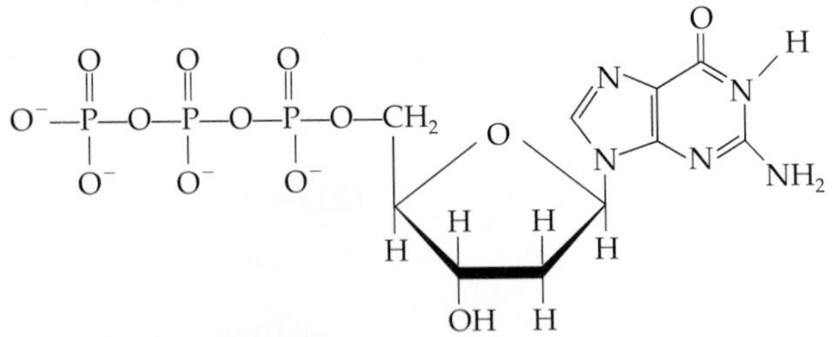

24.3 The deoxyribonucleotides of guanine are:
Deoxyguanosine monophosphate (dGMP)
Deoxyguanosine diphosphate (dGDP)
Deoxyguanosine triphosphate (dGTP)

The ribonucleotides of guanine are:
Guanosine monophosphate (GMP)
Guanosine diphosphate (GDP)
Guanosine triphosphate (GTP)

24.5 The RNA polymerase recognizes the promoter site for a gene, separates the strands of DNA, and catalyzes the polymerization of an RNA strand complementary to the DNA strand that carries the genetic code for a protein. It recognizes a termination site at the end of the gene and releases the RNA molecule.

24.7 The genetic code is said to be degenerate because several different triplet codons may serve as code words for a single amino acid.

24.9 The nitrogenous bases of the codons are complementary to those of the anticodons. As a result they are able to hydrogen bond to one another according to the base pairing rules.

24.11 The ribosomal P-site holds the peptidyl tRNA during protein synthesis. The peptidyl tRNA is the tRNA carrying the growing peptide chain. The only exception to this is during initiation of translation when the P-site holds the initiator tRNA.

24.13 The normal mRNA sequence, AUG-CCC-GAC-UUU, would encode the peptide sequence methionine-proline-aspartate-phenylalanine. The mutant mRNA sequence, AUG-CGC-GAC-UUU, would encode the mutant peptide sequence methionine-arginine-aspartate-phenylalanine. This would not be a silent mutation because a hydrophobic amino acid (proline) has been replaced by a positively charged amino acid (arginine).

24.15 It is the N-9 of the purine that forms the N-glycosidic bond with C-1 of the five-carbon sugar. The general structure of the purine ring is shown below:

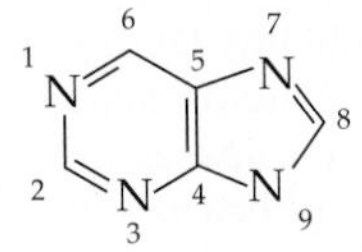

24.17 The ATP nucleotide is composed of the five-carbon sugar ribose, the purine adenine, and a triphosphate group.

24.19 The two strands of DNA in the double helix are said to be *antiparallel* because they run in opposite directions. One strand progresses in the $5' \rightarrow 3'$ direction, and the opposite strand progresses in the $3' \rightarrow 5'$ direction.

24.21 Two

24.23

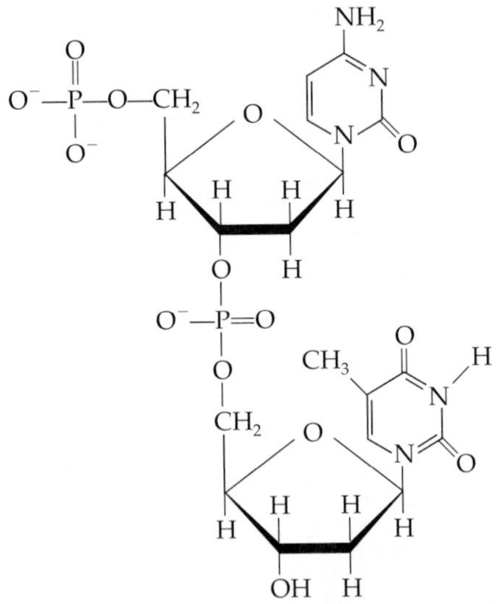

24.25 The term *semiconservative DNA replication* refers to the fact that each parental DNA strand serves as the template for the synthesis of a daughter strand. As a result, each of the daughter DNA molecules is made up of one strand of the original parental DNA and one strand of newly synthesized DNA.

24.27 The two primary functions of DNA polymerase are to read a template DNA strand and catalyze the polymerization of a new daughter strand, and to proofread the newly synthesized strand and correct any errors by removing the incorrectly inserted nucleotide and adding the proper one.

24.29 3'-TACGCCGATCTTATAAGGT-5'.

24.31 The *replication origin* of a DNA molecule is the unique sequence on the DNA molecule where DNA replication begins.

24.33 DNA → RNA → Protein

24.35 Anticodons are found on transfer RNA molecules.

24.37 3'-AUGGAUCGAGACCAGUAAUUCCGUCAU-5'.

24.39 *RNA splicing* is the process by which the noncoding sequences (introns) of the primary transcript of a eukaryotic mRNA are removed and the protein coding sequences (exons) are spliced together.

24.41 messenger RNA, transfer RNA, and ribosomal RNA

24.43 Spliceosomes are small ribonucleoprotein complexes that carry out RNA splicing.

24.45 The *poly(A) tail* is a stretch of 100–200 adenosine nucleotides polymerized onto the 3' end of a mRNA by the enzyme poly(A) polymerase.

24.47 The *cap structure* is made up of the nucleotide 7-methylguanosine attached to the 5' end of a mRNA by a 5'-5' triphosphate bridge. Generally the first two nucleotides of the mRNA are also methylated.

24.49 Sixty-four

24.51 The reading frame of a gene is the sequential set of triplet codons that carries the genetic code for the primary structure of a protein.

24.53 Methionine and tryptophan

24.55 The codon 5'-UUU-3' encodes the amino acid phenylalanine. The mutant codon 5'-UUA-3' encodes the amino acid leucine. Both leucine and phenylalanine are hydrophobic amino acids, however, leucine has a smaller R group. It is possible that the smaller R group would disrupt the structure of the protein.

24.57 The ribosomes serve as a platform on which protein synthesis can occur. They also carry the enzymatic activity that forms peptide bonds.

24.59 In the initiation of translation, initiation factors, methionyl tRNA (the initiator tRNA), the mRNA, and the small and large ribosomal subunits form the initiation complex. During the elongation stage of translation an aminoacyl tRNA binds to the A-site of the ribosome. Peptidyl transferase catalyzes the formation of a peptide bond and the peptide chain is transferred to the tRNA in the A-site. Translocation shifts the peptidyl tRNA from the A-site into the P-site, leaving the A-site available for the next aminoacyl tRNA. In the termination stage of translation, a termination codon is encountered. A release factor binds to the empty A-site and peptidyl transferase catalyzes the hydrolysis of the bond between the peptidyl tRNA and the completed peptide chain.

24.61 an ester bond

24.63 UV light causes the formation of pyrimidine dimers, the covalent bonding of two adjacent pyrimidine bases. Mutations occur when the UV damage repair system makes an error during the repair process. This causes a change in the nucleotide sequence of the DNA.

24.65 **a.** A *carcinogen* is a compound that causes cancer. Cancers are caused by mutations in the genes responsible for controlling cell division.

   b. Carcinogens cause DNA damage that results in changes in the nucleotide sequence of the gene. Thus, carcinogens are also mutagens.

24.67 A *restriction enzyme* is a bacterial enzyme that "cuts" the sugar–phosphate backbone of DNA molecules at a specific nucleotide sequence.

24.69 Nucleic acid hybridization is based on the fact that complementary DNA or RNA sequences will hydrogen bond to one another according to the base pairing rules.

24.71 Human insulin, interferon, human growth hormone, and human blood clotting factor VIII

24.73 2048 copies

Credits

Photos

Chapter 1
Opener: © Paul Barton/Stock Market; **1.2a:** © Geoff Tompkinson/Science Photo Library/Photo Researchers, Inc.; **1.2b:** © T. J. Florian/Rainbow; **1.2c:** © David Parker/Seagate Microelectronics Ltd./Photo Researchers, Inc.; **1.2d:** © APHIS, PPQ, Otis Methods Development Center, Otis, MA; **1.4a–c, 1.6a–d, 1.8:** © Louis Rosenstock.

Chapter 2
Opener: © Yoav Levy/Phototake; **2.1a:** © Louis Rosenstock; **2.1b:** © Jeff Topping; **2.1c:** © Louis Rosenstock; **2.2:** © Ken Karp; **2.5:** © P. Plaily/Science Photo Library/Photo Researchers, Inc.; **2.7a,b:** © Louis Rosenstock; **2.12:** © Yoav Levy/Phototake; p. 51: © Eric Kamp/Phototake; p. 52: © Earth Satellite Corp./Science Photo Library/Photo Researchers, Inc.; p. 53 (*bottom*): © Louis Rosenstock; p. 53 (*top*): © Dan McCoy/Rainbow.

Chapter 3
Opener: © Louis Rosenstock.

Chapter 4
Opener: © Photri/Stock Market; **4.13:** © Charles D. Winters/Photo Researchers, Inc.

Chapter 5
Opener: © McGraw-Hill Higher Education/Stephen Frisch, photographer; **5.1,5.3:** © Louis Rosenstock; p. 144: © George W. Disario/Stock Market.

Chapter 6
Opener: Courtesy of Robert Shoemaker; p. 150: © Corbis; p. 156: © SIU/Visuals Unlimited; **6.4:** © Peter/Stef Lamberti/Tony Stone Images; **6.5a,b:** © Louis Rosenstock.

Chapter 7
Opener: © Richard Megna/Fundamental Photographs; **7.1:** © Kip and Pat Peticolas/Fundamental Photographs; p. 185: © J. W. Mowbray/Photo Researchers, Inc.; **7.2,7.5a,b,** p. 198: © Louis Rosenstock; p. 200: © David Joel/Tony Stone Images.

Chapter 8
Opener: © Richard Megna/Fundamental Photographs; **8.9:** © Ken Karp; p. 218(both), **8.12:** © Louis Rosenstock; **8.16,8.17:** © Ken Karp.

Chapter 9
Opener: © Richard Megna/Fundamental Photographs; **9.1:** © Dr. E. R. Degginger/Color-Pic Inc.; **9.3a,b,9.4a,b,9.6a,b,** p. 253, **9.7,9.8** (*left*): © Louis Rosenstock; **9.8** (*right*): © Phil Degginger/Color-Pic Inc.; p. 261: © AAA Photo/Phototake; **9.9:** © McGraw-Hill Higher Education/Stephen Frisch, photographer.

Chapter 10
Opener: © David Job/Tony Stone Images; **10.4:** © U.S. Department of Energy/SPL/Photo Researchers, Inc.; **10.5:** © Gianni Tortoli/Photo Researchers, Inc.; p. 281: © NASA/The Image Works, Inc.; **10.6:** © Blair Seitz/Photo Researchers, Inc.; **10.7b:** © Dupont Pharmaceuticals; p. 286 (*left*): © SIU Biomed/Custom Medical Stock Photo; p. 286 (*right*): © Louis Rosenstock; **10.8:** © US Department of Energy/Mark Marten/Photo Researchers, Inc.; **10.9:** © Louis Rosenstock; **10.10:** © Scott Camazine/Photo Researchers, Inc.

Chapter 11
Opener: © Tom McHugh/Photo Researchers, Inc.

Chapter 12
Opener: © Bob Thomason/Tony Stone Images; **12.4:** © Ken Karp.

Chapter 13
Opener: © Bruce Forster/Tony Stone Images.

Chapter 14
Opener: © David S. Addison/Visuals Unlimited.

Chapter 15
Opener: © Christel Rosenfeld/Tony Stone Images; p. 416: © Louis Rosenstock.

Chapter 16
Opener: © Gail Shumway/FPG International.

Chapter 17
Opener: © Louis Rosenstock; p. 489: © R. Feldman/Dan McCoy/Rainbow.

Chapter 18
Opener: © Peter Poulides/Tony Stone Images; p. 515: © Hans Pfletschinger/Peter Arnold, Inc.; **18.10a:** © James Dennis/Phototake; **18.20a–c:** © David Phillips/Visuals Unlimited.

Chapter 19
Opener: © Francoise Sauze/SPL/Photo Researchers, Inc.; **19.17a:** © SIU/Peter Arnold, Inc.; **19.17b:** © David Scharf/Peter Arnold, Inc.; **19.17c:** © Jackie Lewin/SPL/Photo Researchers, Inc.

Chapter 20
Opener: © David Eisenberg; p. 577 (both): © Hank Morgan/Photo Researchers, Inc.

Chapter 21
Opener: © Louis Rosenstock; p. 626: © Louis Rosenstock.

Chapter 22
Opener: © Bob Thomas/Tony Stone Images; **22.1a:** © CNRI/Phototake.

Chapter 23
Opener: © Louis Rosenstock.

Chapter 24
Opener: © Douglas Struthers/Tony Stone Images; **24.19:** © Yoav-Simon/Phototake; p. 720: © Mark Stolorow/Cellmark Diagnostics.

Text and Line Art

Chapter 1
1.5: From Raymond Chang, *Chemistry*, 6th ed. Copyright © 1998. The McGraw-Hill Companies, Inc., Dubuque, Iowa. All Rights Reserved. Reprinted by permission, p. 21; **Caloric expenditure table:** This table has been reproduced from *The Book of Health*, E.L. Wynder, Editor. American Health Foundation. New York, Franklin Watts, 1981, with permission of the American Health Foundation, p. 25.

Chapter 2
2.6: From Raymond Chang, *Chemistry*, 6th ed. Copyright © 1998. The McGraw-Hill Companies, Inc., Dubuque, Iowa. All Rights Reserved. Reprinted by permission, p. 45; **2.11:** From Raymond Chang, *Chemistry*, 6th ed. Copyright © 1998. The McGraw-Hill Companies, Inc., Dubuque, Iowa. All Rights Reserved. Reprinted by permission, p. 47; **2.13:** From Martin Silberberg, *Chemistry: The Molecular Nature of Matter and Change*, 2nd ed. Copyright © 2000. The McGraw-Hill Companies, Inc., Dubuque, Iowa. All Rights Reserved. Reprinted by permission, p. 48;

Index

Note: Page numbers followed by B indicate boxed material; those followed by F indicate figures; those followed by T indicate tables.

A

ethylamine (ethanamine), 443, 443T, 446, 449

3-ethylaniline, 344

meta-ethylaniline, 345

ethylbenzene, 343

ethyl butanoate (ethyl butyrate), 418–420, 431B

2-ethyl-1-butanol, 362

ethyl butyrate (ethyl butanoate), 418–420, 431B

ethyl chloride (chloroethane), 315B, 316

ethyl dodecanoate, 502

ethylene (ethene), 326, 327F, 336–337, 339, 341, 366

ethylene glycol (1,2-ethanediol), 193, 339, 359–361, 430B

ethyl formate (ethyl methanoate), 431B

ethyl groups, 302, 303T

5-ethylheptanal, 387

3-ethyl-3-hexene, 331

ethyl isopropyl ether, 374

ethyl methanoate (ethyl formate), 431B

ethylmethylamine, 444, 446

ethyl methyl ether (methoxyethane), 374, 385, 409

4-ethyloctane, 305

6-ethyl-2-octanone, 389

ethyl pentyl ketone, 389

ethyl propanoate, 420–421

ethyne (acetylene), 326, 327F

eukaryotes, 702, 705–706

evaporation, 166, 167F

evolution, 544
  divergent, 591
  mitochondria, 636B, 637

excitation, 49–50, 49F, 51B

excited state, 49

excited state atom, 49

exercise
  Calories used in, 25B
  energy metabolism and, 638B–639B
  for weight loss, 673B

exergonic reactions, 601–602

exons, 706–707, 707F

EXOSURF Neonatal, 496B

exothermic reactions, 207–208, 207F, 208F

expanded octet, 104–105

experiment, 2B, 4, 4B

experimental control, 4, 5B

exponent(s)
  equilibrium-constant expression, 227–228
  significant figures, 18–19

exponential notation. *See* scientific notation

extensive properties, 36–37

F

Fabry's disease, 510B

facilitated diffusion, 525–526, 526F

factor VIII, 723

factor IX, 723

factor-label method, 9

FAD (flavin adenine dinucleotide), 581, 581T, 582F, 640
  in citric acid cycle, 642, 643F, 645
  FADH$_2$, electrons for oxidative phosphorylation, 646–648, 647F
  in β-oxidation, 671, 671F, 675, 683

Fahrenheit scale, 22–23, 154

fast twitch muscle fibers, 639B

fat(s), 422, 423F
  Calories per gram, 25B
  catabolism, 604
  conversion of oil to, 334, 334F
  dietary, 496
  digestion, 605–606, 606F, 607F

fat cells. *See* adipocytes

fat globules, 667

fatty acid(s), 409, 414, 497, 497F. *See also* glycerides
  essential, 503
  melting points, 499–500, 500F
  membrane, 520–521
  monounsaturated, 325, 325F
  naming, 499T
  omega-labeled, 670
  polyunsaturated, 325
  reactions, 501–503
  saturated, 325, 498–500, 499T, 500F, 519, 521
  structure and properties, 498–500
  unsaturated, 498–500, 499T, 500F, 520–521

fatty acid metabolism
  animals, 667–670
  degradation, 666, 670–677
  insulin, 687
  lipid storage, 668–670
  β-oxidation, 432, 670–677, 671F, 682–683
  reactions, 674–677, 676F
  regulation
    adipose tissue, 685, 685F
    brain, 686
    liver, 684–685, 684F
    muscle tissue, 685, 685F
  synthesis, 619, 666, 682–683, 682F, 683F

fatty acid salts, 422

fatty acid synthase, 683

fatty acyl CoA, 674–675

feedback inhibition, 586

fermentation, 357, 360–361, 616–618, 617F
  alcohol, 600B, 617–618, 617F, 626B–627B
  butyric acid, butanol, acetone, 627B
  lactate, 414, 559, 616–618, 617F

Fermi, Enrico, 6

fetal alcohol effects, 356B

fetal alcohol syndrome, 356B, 371B

fetal hemoglobin, 408B, 556

fetus, immune system, 554B

F$_0$F$_1$ complex. *See* ATP synthase

fiber
  dietary, 468, 489
  synthetic, 340

fibrils, cellulose, 488

fibrous proteins, 545–546

"fight or flight" response, 455B

film badges, 289

fingerprinting, DNA, 720B

fireworks, 51B

fission, nuclear, 278, 279F, 280F

flavin adenine dinucleotide. *See* FAD

flavin mononucleotide (FMN), 581T

flavorings, 346B, 373, 391F, 431B

Fleming, Alexander, 5B, 524B

Flotte, Terry, 724B

fluidity, membrane, 519–520, 523F

fluid mosaic model, 519–522, 520F, 521F, 522F

fluoride ions, 72, 90T

fluorine
  diatomic, 85
  electron configuration, 64, 65T, 69
  Lewis structure, 83

FMN (flavin mononucleotide), 581T

folic acid, 581T, 588–590

food(s)
  calories, 25B
  copper content, 62B
  fuel value, 213–215
  nitrites, 451B
  sodium and potassium content, 96B

food additives, 390

food chain, 324B

food pyramid, 469, 469F

f orbital, 67–68

forensic science, 720B, 724

formaldehyde (methanal), 315, 368, 384B, 386, 387T, 390, 391F, 392B

formalin, 390, 391F, 392B

formamide (methanamide), 457T

formic acid (methanoic acid), 409, 412T

formula
  condensed, 301–302
  covalent compounds, 93–94
  ionic compounds, 87–92
  molecular, 300
  structural, 300

formula unit, 123, 125F

formula weight, 124–126

fossil fuels
  combustion, 180, 206B, 253B, 260–261, 314–315
  origin, 296B

foxglove, 515B

fragrances, 346B, 373, 390, 418, 431B

Franklin, Rosalind, 697

free energy, 210

freeze-fracture, 521

freezing point, water, 22, 22F

freezing point depression, 192–193

Friedman, Jeffrey, 666B